Sustainable Minerals Operations in the Developing World

Geological Society Special Publications
Society Book Editors

R. J. PANKHURST (CHIEF EDITOR)
P. DOYLE
F. J. GREGORY
J. S. GRIFFITHS
A. J. HARTLEY
R. E. HOLDSWORTH
J. A. HOWE
P. T. LEAT
A. C. MORTON
N. S. ROBINS
J. P. TURNER

Special Publication reviewing procedures

The Society makes every effort to ensure that the scientific and production quality of its books matches that of its journals. Since 1997, all book proposals have been refereed by specialist reviewers as well as by the Society's Books Editorial Committee. If the referees identify weaknesses in the proposal, these must be addressed before the proposal is accepted.

Once the book is accepted, the Society has a team of Book Editors (listed above) who ensure that the volume editors follow strict guidelines on refereeing and quality control. We insist that individual papers can only be accepted after satisfactory review by two independent referees. The questions on the review forms are similar to those for *Journal of the Geological Society*. The referees' forms and comments must be available to the Society's Book Editors on request.

Although many of the books result from meetings, the editors are expected to commission papers that were not presented at the meeting to ensure that the book provides a balanced coverage of the subject. Being accepted for presentation at the meeting does not guarantee inclusion in the book.

Geological Society Special Publications are included in the ISI Index of Scientific Book Contents, but they do not have an impact factor, the latter being applicable only to journals.

More information about submitting a proposal and producing a Special Publication can be found on the Society's web site: www.geolsoc.org.uk.

It is recommended that reference to all or part of this book should be made in one of the following ways:

MARKER, B. R., PETTERSON, M. G. & MCEVOY, F. (eds) 2005. *Sustainable Minerals Operations in the Developing World*. Geological Society, London, Special Publications, **250**.

STEPHENSON, M. & PENN, I. 2005. Capacity building of developing country public sector institutions in the natural resource sector. *In*: MARKER, B. R., PETTERSON, M. G. & MCEVOY, F. (eds) 2005. *Sustainable Minerals Operations in the Developing World*. Geological Society, London, Special Publications, **250**, 183–192.

GEOLOGICAL SOCIETY SPECIAL PUBLICATION NO. 250

Sustainable Minerals Operations in the Developing World

EDITED BY

B. R. MARKER
Office of the Deputy Prime Minister, UK

M. G. PETTERSON, F. MCEVOY and M. H. STEPHENSON
British Geological Survey, UK

2005
Published by
The Geological Society
London

THE GEOLOGICAL SOCIETY

The Geological Society of London (GSL) was founded in 1807. It is the oldest national geological society in the world and the largest in Europe. It was incorporated under Royal Charter in 1825 and is Registered Charity 210161.

The Society is the UK national learned and professional society for geology with a worldwide Fellowship (FGS) of 9000. The Society has the power to confer Chartered status on suitably qualified Fellows, and about 2000 of the Fellowship carry the title (CGeol). Chartered Geologists may also obtain the equivalent European title, European Geologist (EurGeol). One fifth of the Society's fellowship resides outside the UK. To find out more about the Society, log on to www.geolsoc.org.uk.

The Geological Society Publishing House (Bath, UK) produces the Society's international journals and books, and acts as European distributor for selected publications of the American Association of Petroleum Geologists (AAPG), the American Geological Institute (AGI), the Indonesian Petroleum Association (IPA), the Geological Society of America (GSA), the Society for Sedimentary Geology (SEPM) and the Geologists' Association (GA). Joint marketing agreements ensure that GSL Fellows may purchase these societies' publications at a discount. The Society's online bookshop (accessible from www.geolsoc.org.uk) offers secure book purchasing with your credit or debit card.

To find out about joining the Society and benefiting from substantial discounts on publications of GSL and other societies worldwide, consult www.geolsoc.org.uk, or contact the Fellowship Department at: The Geological Society, Burlington House, Piccadilly, London W1J 0BG: Tel. +44 (0)20 7434 9944; Fax +44 (0)20 7439 8975; E-mail: enquiries@geolsoc.org.uk.

For information about the Society's meetings, consult *Events* on www.geolsoc.org.uk. To find out more about the Society's Corporate Affiliates Scheme, write to enquiries@geolsoc.org.uk.

Published by The Geological Society from:
The Geological Society Publishing House
Unit 7, Brassmill Enterprise Centre
Brassmill Lane
Bath BA1 3JN, UK

Orders: Tel. +44 (0)1225 445046
Fax +44 (0)1225 442836
Online bookshop: www.geolsoc.org.uk/bookshop

The publishers make no representation, express or implied, with regard to the accuracy of the information contained in this book and cannot accept any legal responsibility for any errors or omissions that may be made.

© The Geological Society of London 2005. All rights reserved. No reproduction, copy or transmission of this publication may be made without written permission. No paragraph of this publication may be reproduced, copied or transmitted save with the provisions of the Copyright Licensing Agency, 90 Tottenham Court Road, London W1P 9HE. Users registered with the Copyright Clearance Center, 27 Congress Street, Salem, MA 01970, USA: the item-fee code for this publication is 0305-8719/05/$15.00.

British Library Cataloguing in Publication Data

A catalogue record for this book is available from the British Library.

ISBN 1-86239-188-2

Typeset by Techset Composition, Salisbury, UK
Printed by MPG Books Ltd, Bodmin, UK

Distributors

USA
AAPG Bookstore
PO Box 979
Tulsa
OK 74101-0979
USA
Orders: Tel. +1 918 584-2555
Fax +1 918 560-2652
E-mail bookstore@aapg.org

India
Affiliated East-West Press Private Ltd
Marketing Division
G-1/16 Ansari Road, Darya Ganj
New Delhi 110 002
India
Orders: Tel. +91 11 2327-9113/2326-4180
Fax +91 11 2326-0538
E-mail affiliat@vsnl.com

Japan
Kanda Book Trading Company
Cityhouse Tama 204
Tsurumaki 1-3-10
Tama-shi, Tokyo 206-0034
Japan
Orders: Tel. +81 (0)423 57-7650
Fax +81 (0)423 57-7651
E-mail geokanda@ma.kcom.ne.jp

Contents

MARKER, B. R., PETTERSON, M. G., MCEVOY, F. & STEPHENSON, M. H. Sustainable Minerals Operations in the Developing World: introduction — 1

PETTERSON, M. G., MARKER, B. R., MCEVOY, F., STEPHENSON, M. & FALVEY, D. A. The need and context for sustainable mineral development — 5

HOBBS, J. C. A. Enhancing the contribution of mining to sustainable development — 9

RICHARDS, J. P. The role of minerals in sustainable human development — 25

HARRISON, D. J., FIDGETT, S., SCOTT, P. W., MACFARLANE, M., MITCHELL, P., EYRE, J. M. & WEEKS, J. M. Sustainable river mining of aggregates in developing countries — 35

SCOTT, P. W., EYRE, J. M., HARRISON, D. J. & BLOODWORTH, A. J. Markets for industrial mineral products from mining waste — 47

ESHUN, P. A. Sustainable small-scale gold mining in Ghana: setting and strategies for sustainability — 61

NYAMBE, I. A. & KAWAMYA, V. M. Approaches to sustainable minerals development in Zambia — 73

DAVIES, T. C. & OSANO, O. Sustainable mineral development: case study from Kenya — 87

D'SOUZA, K. P. C. J. Artisanal and small-scale mining in Africa: the poor relation — 95

MITCHELL, C. J. FarmLime: low-cost lime for small-scale farming — 121

AKHTAR, A. Mineral resources and their economic significance in national development: Bangladesh perspective — 127

HUSAIN, V. Obstacles in the sustainable development of artisanal and small-scale mines in Pakistan and remedial measures — 135

MISHRA, P. P. Mining and environmental problems in the Ib valley coalfield of Orissa, India — 141

TOLIA, D. H. & PETTERSON, M. G. The Gold Ridge Mine, Guadalcanal, Solomon Islands' first gold mine: a case study in stakeholder consultation — 149

CARVALHO, J. F. & LISBOA, J. V. Construction raw materials in Timor Leste and sustainable development — 161

STEPHENSON, M. H. & PENN, I. E. Capacity building of developing country public sector institutions in the natural resource sector — 185

SHIELDS, D. J. & ŠOLAR, S. V. Sustainable development and minerals: measuring mining's contribution to society   195

O'REGAN, B. & MOLES, R. System dynamics modelling: a more effective tool for assessing the impact of sustainable development policies on the mining industry   213

VAN DER MEULEN, M. J. Sustainable mineral development: possibilities and pitfalls illustrated by the rise and fall of Dutch mineral planning guidance   225

WHITMORE, A. The emperor's new clothes: sustainable mining?   233

Index   243

Sustainable Minerals Operations in the Developing World: introduction

B. R. MARKER[1], M. G. PETTERSON[2], F. MCEVOY[2] & M. H. STEPHENSON[2]

[1]*Office of the Deputy Prime Minister, Eland House, Bressenden Place, London, SW1E 5DU, UK*
[2]*British Geological Survey, Nicker Hill, Keyworth, Nottingham, NG2 SGG, UK*

Sustainable development requires an appropriate balance between social, economic and environmental well-being, now and for the future. Since most minerals are non-renewable resources, sustainability of supply can only be addressed by extracting, processing and distributing raw materials in the least environmentally damaging ways, using minerals wisely, and recycling as much as possible. However, there also is significant scope for improved sustainability in terms of economic and social aspects.

Minerals are essential raw materials but high-quality deposits have become depleted in many developed countries. These countries have increasingly turned to developing countries for supplies and it is in these that most high-quality untapped future prospects remain. For countries with limited export opportunities, minerals are often a mainstay of the domestic economy. However, low selling prices may reflect limited environmental regulation and low wages. This can lead to charges that the rich countries are exporting their environmental damage to, and exploiting, poorer countries. As more countries develop, the global demand for supplies of essential raw materials increases, and resources will be depleted more quickly. Therefore, sustainable minerals supply from the developing countries is an important global issue.

In this Special Report, general aspects of sustainable minerals operations in the developing world are reviewed by Petterson *et al.*, Hobbs, and Richards while the remaining papers consider specific issues in more detail. Hobbs, in particular, emphasizes the need to give proper weight each to human capital, financial capital, manufactured capital, and environmental capital in any full analysis as a context for sustainable development and effective aid.

Geographical coverage

The included papers highlight the diversity of sustainability issues in parts of Africa (**Davies & Osano, D'Souza, Eshun, Mitchell, Nyambe & Kawamya**), Asia (**Akhtar, Hussain, Mishra**), the Pacific (**Carvalho & Lisboa, Tolia & Petterson, Whitmore**), and Jamaica (**Harrison *et al.***). Some comparative material from parts of Europe is also included (**Scott *et al.*, Shields & Šolar, van der Meulen**).

Scale of extractive operations

Minerals operations in the developing world are varied. Papers in this report consider all scales including:

(a) major underground and open-pit mines run by international corporations, which are particularly vulnerable to changes in world commodity prices and investment opportunities (**Richards, Tolia & Petterson, Whitmore**);
(b) substantial mines and quarries operated by major national firms (**Mishra**); and
(c) smaller scale mines and quarries run by local firms (especially **Harrison *et al.*, Mitchell**), and informal or 'artisanal' mining, often undertaken without regulatory permission by local people (especially **D' Souza, Eshun, Hussain**).

Carvalho & Lisboa, and Stephenson & Penn concentrate on medium-sized to artisanal operations, while Akhtar, Nyambe & Kawamya, and Davies & Osano consider all three levels.

It is a fact of life that significant numbers of people have to work in artisanal mining because of a lack of any other means to earn a livelihood (**Hussain, Eshun**). D'Souza makes it clear that artisanal mining can support very large numbers of people, directly or indirectly, and fully or seasonally, in areas that are otherwise at poverty level. However, he makes it clear that artisanal mining is not a single phenomenon and may range from ephemeral to permanent operations. Because of this wide range in scale, investment, plant and machinery, techniques, education, and training, there can be

From: MARKER, B. R., PETTERSON, M. G., MCEVOY, F. & STEPHENSON, M. H. (eds) 2005. *Sustainable Minerals Operations in the Developing World*. Geological Society, London, Special Publications, **250**, 1–4.
0305-8719/05/$15.00 © The Geological Society of London 2005.

no 'one size fits all' solution to problems of sustainability. A particular problem can be competition for land between legal major operations and informal artisanal mining (**Davies & Osano, Eshun**).

Economic aspects

There is often an emphasis in developing countries on attracting international investment in large-scale minerals operations, rather than developing smaller scale enterprises to meet local needs, because of lack of local investment capital (**Akhtar**), the need for foreign exchange (**Nyambe & Kawamya**), and lack of internal infrastructure or of local markets, which inhibit smaller operations in remote locations (**Hussain, Davies & Osano, Carvalho & Lisboa**). Therefore, the contribution to GDP from minerals operations, while valuable, is often lower than it could be (**Hussain**). In particular, there are limited facilities for 'value added' manufacturing, so much of the 'downstream' wealth is realized in other economies (**D'Souza, Hussain**).

In respect of large-scale mining enterprises, an important factor is the attractiveness or otherwise of countries to international mining companies. The ambient regulatory and fiscal regime may deter or encourage exploration and development. There may be an economic need for governments to take steps to liberalize administrative processes, but that may work counter to sustainable development (**Nyambe & Kawamya**) and the interests of indigenous peoples in particular (**Whitmore**).

Particular problems for small-scale and artisanal mining include:

- inability to borrow money because of lack of collateral (**D'Souza, Hussain**);
- lack of training in marketing skills such that middle men buy cheaply and make the profit (**Hussain**);
- lack of clear business plans and weak 'project ownership' (**Stephenson & Penn**); and
- lack of technical expertise, especially on mineral resource evaluation and processing, and availability only of outdated equipment (**Hussain, Davies & Osano**)

In these circumstances, a crucial step towards improvement is the development of appropriate funding and support models (**Nyambe & Kawamya**) to secure proper investment in plant and equipment, alongside capacity building. However, circumstances are sometimes extreme, for instance when trying to rebuild small local mining industries, while attempting to diversify and extend operations to contribute to rebuilding a post-conflict economy (**Carvalho & Lisboa**).

Despite such limitations, there may be opportunities for diversification, for instance in the gemstone and industrial minerals sectors in Zambia (**Nyambe & Kawamya**), development of niche products such as agricultural lime (**Mitchell**), or greater use of mineral wastes where processing costs permit. While mining wastes containing rare and valuable minerals are attractive already, at least to artisanal miners, there is scope for more use of lower priced commodities in local economies (**Scott et al.**).

Social aspects

There are strong social issues surrounding sustainable minerals operations in the developing world. Medium- to large-scale mining was largely undertaken by companies from the developed countries but often operating to lower standards in terms of matters such as health and safety than they would in their own countries, and paying low wages, leading to criticisms of exploitation. In recent years, many major companies have taken steps to address these issues (see below). Even so, major mining proposals can significantly affect people through displacement of communities (**Mishra, Whitmore**).

However, small-scale and artisanal mining was, and remains, a dangerous activity linked to poverty and ill health (**Eshun**), the exploitation of women, and use of child labour (**D'Souza**). Tensions may exist between major minerals operations and small-scale miners who compete for the same land (**Richards**).

There is a clear need for support, training, education and, overall, holistic capacity building (**D'Souza, Eshun, Hussain, Nyambe & Kawamya, Stephenson & Penn**) including engagement with government, industry, non-governmental organizations (NGOs), landowners and, crucially, local communities and their traditional authorities. However, that requires, in turn, willingness and commitment on the part of miners and their dependents (**Eshun**). A particular problem is the lack of materials in local languages (**Hussain**). However, overall, views of stakeholders should influence the ways in which aid and assistance are provided (**Hobbs, Stephenson & Penn**) and whether, or not, mining should proceed (**Whitmore**).

Holistic capacity building is important at all levels – government, industry, landowners communities. But the necessary information, education experience and training is impaired by insufficient budgets (**Stephenson & Penn**). Stakeholder views should influence the way that aid is provided.

Communication is also of importance in relation to proposed major mining. For example, Tolia & Petterson review issues involved in community consultation on a major mining proposal in a tribal area of the Solomon Islands with no previous experience of the industry and different land ownership concepts. Negotiations were undertaken with all stakeholders and agreement was secured on the basis of a compensation 'package' including money at the outset and during the life of the operation, provision of relocation and commercial land, social and educational and infrastructure facilities, and, wisely, the basis for investment in skills and occupations for the post-mining phase. In contrast, Whitmore gives examples where corporate policy may have been poorly communicated or lacking in flexibility.

Environmental aspects

The environmental impacts of mining and quarrying, including land degradation and emissions, are well known and have been particularly severe where environmental management systems and restoration procedures are not in place (**Mishra**).

In recent years many major mining corporations have re-examined their operations in terms of corporate and social responsibility (**Richards**), for instance through the Global Mining Initiative of the 1990s and World Bank Extractive Industry Review of 2004. This has led to the recognition of the importance of good governance and observance of human rights in this sector, as well as introducing social and environmental safeguards (**Hobbs**). However some groups representing indigenous peoples feel that the views of those people have often been overlooked in this process (**Whitmore**). There are also encouraging examples within developing countries of national operations making strong efforts to improve environmental performance (**Mishra**).

To take a specific example, Harrison et al. review practices in river mining of sand and gravel for construction in Jamaica. These operations give rise to bank erosion, increased flooding and damage to ecosystems. There is a clear need for environmental impacts assessment, a sound code of practice, and technical improvements. However, it will not be easy to implement these without a significant capacity-building initiative.

Richards emphasizes the need for all mining operations to internalize environmental costs, but, while environmental management schemes are now commonplace in large companies, there is a long way to go before this can be carried through to smaller formal operations (**Richards**). Meanwhile, in the artisanal sector, it is probably not possible to make significant environmental progress until the social and economic aspects have been addressed (**Nyambe & Kawamya**).

Regulation and governance

The regulatory situation is very variable. This ranges from:

(a) a lack of public policy in post-conflict situations (**Carvalho & Lisboa**);
(b) old legislation that was fashioned for colonial mining, essentially favouring foreign companies, rather than modern circumstances (**Eshun**), and which may have little regard for environmental issues (**Davies & Osano**); or
(c) new national laws that are not transposed into regional or local law, are difficult to enforce in remote and difficult terrain, and which may be cumbersome, 'non-transparent' and poorly applied due sometimes to the in experience of officials (**Hussain**) and which may favour corporate interests over those of local people (**Whitmore**).

While there have been some attempts to make artisanal mining illegal, these regulations, in particular, are seldom enforceable and stand in the way of dialogue and improvement (**Davies & Osano**). Therefore, an important aspect is the need to regularize illegal artisanal mining operations so that the lives of miners and their dependents can be improved and concepts of sustainability can begin to be introduced (**Eshun, Nyambe & Kawamya, Davies & Osano**).

A paper from van der Meulen sets out recent changes to Dutch national mineral planning practices to illustrate that well-intentioned approaches to the supply of construction aggregates are not necessarily successful when the public opposes extraction. That is a common feature in developed countries, but it is less widely appreciated that it can also be a factor in developing countries (**Davies & Osano**). Caution is needed in attempting to transfer experiences in minerals planning from one place to another.

Monitoring and assessment

It is difficult to measure sustainability (**Richards**). However, key steps are to assess the contribution that sustainable development can make to an economy, to set out a plan of action towards improvement, to monitor progress, and to reassess the initiative as time passes. Shields & Šolar point out that approaches often over- or under-emphasize the influence of the minerals industry on economic growth when considering the balance between income growth and alleviation of poverty. It is important to weigh social equity, environmental health, and economic growth as a basis for managing change. To that end, they present a sustainable resource management model for minerals development, and aggregates in particular.

O'Regan & Moles focus on computer modelling of the effects of economic, fiscal and corporate policies on the flow of investment funds and the development of mineral resources as a basis for assessing perceived investment risks with particular reference to base metal mining.

Conclusion

Some key points that emerge from the Report are that:

- the emphasis in many developing countries on earning revenue from mining operations of international companies may work against the development of a minerals supply to local markets and 'value added' manufacturing;
- major mining companies have done much in recent years to review their operations and to introduce better codes of practice, but much remains to be done at all levels, especially in the small-scale and artisanal sectors;
- improvement requires good dialogue between all stakeholders at all stages of minerals operations, from exploration to rehabilitation and, ideally, with investment from revenue gained from mining in post-mining employment opportunities;
- regulations need to be fitted to the circumstances, but are only effective if they are implemented by suitably experienced administrators;
- key stakeholders are largely trying to do the right things, but within a dauntingly complex situation;
- circumstances vary greatly from area to area, so there is no 'one size fits all' approach to improved sustainability;
- it is important to properly analyse and to monitor approaches towards more sustainable development; however, the first priority in the artisanal sector is to regularize operations so that improvements can be sought, rather than marginalizing these as illegal activities; and
- good dialogue and coordination is needed between all interested parties, especially local communities, so that issues can be addressed positively, while avoiding unintended consequences.

History tells us that artisanal mining tends to a build to a medium scale and then, later, to large-scale mining enterprises accompanied by growing awareness of good practices. The question is how to accelerate that process in the developing countries to the well-being of all concerned and their environments.

The conference that gave rise to the papers contained in this Special Report was convened by the Joint Association of Geoscientists for International Development in collaboration with the Environment Group of the Geological Society and the British Geological Survey. Thanks are due to the following organizations that supported the conference, particularly those who helped to pay for speakers from developing countries to attend: Department for International Development (UK), Anglo American plc, Rio Tinto plc, Wardell Armstrong plc.

The need and context for sustainable mineral development

M. G. PETTERSON[1], B. R. MARKER[2], F. MCEVOY[1],
M. STEPHENSON[1] & D. A. FALVEY[1]

[1]*British Geological Survey, Nicker Hill, Keyworth, Nottingham NG12 5GG, UK*
[2]*Office of the Deputy Prime Minister, Eland House, Bressenden Place, London, SW1E 5DU, UK*

Abstract: A special thematic conference was organized at the Geological Society of London in November 2003, aimed at bringing together experts in minerals development in the Developing Countries. Representatives of many aspects of mineral development attended, including mining companies, governments, aid agencies, non-governmental organizations (NGOs), academics and consultants. The opening address to the conference is given in this paper. Mining is an ancient human activity developed through essential societal demand. As society and technology have developed, they have inevitably become ever-more materials hungry. This demand will remain for the foreseeable future. Many areas of the Developed World have depleted high-grade mineral deposits, and remaining resources are subject to strong environmental constraints. This increases pressure on the Developing World to generate the mineral commodities upon which society depends. Mineral resources are also a potential source of capital over which Developing Countries can have their own decision-making powers (in contrast to aid money for example). Sustainable mineral development is all about balance. Achieving the dynamic balance between supply and demand, equitable capital distribution, good financial and environmental management and governance, economics, and social stability is the challenge the world faces in the twenty-first century and beyond.

Historical context

As long as there has been human society there has been mining. Stone Age peoples quarried stone resources to manufacture knives, arrowheads, axes, and other essential implements of their day. With the development of society through the Bronze and Iron Ages and, more recently, through the Industrial Revolution and the current Information and Computer Age, human global demand for mineral resources has inexorably increased in terms of required tonnages and the range of mineral commodities. Modern jet airlines, cars, and mobile telephones require a large cocktail of numerous individual mineral commodities, all used for specific physical, chemical, or electrical conductivity purposes.

Demand for minerals

If we examine a simple graph of global mine production of copper since 1830 (Fig. 1), we see the remarkable correlation between economic activity and the consumption of a basic metallic commodity. This graph reflects the electrification of the Developed World in the nineteenth century, the two World Wars and the Depression periods of the first half of the twentieth century, post-war reconstruction and the industrialization of Asia, South America, and other parts of the world, the impact of rising oil prices in the 1970s and, most recently, wiring the world for the Internet. The second graph (Fig. 2) leaves us in no doubt of the very close correlation between industrial production and base metal consumption for the world's richest countries. Both illustrations demonstrate the intrinsic value and importance of mineral commodities to human society and economic activity. The current economic revolution occurring in China, with year on year economic growth currently at 8–12%, clearly underlines the close association between raw materials and development. The predicted 50% increase in global population during this century and the ever-growing demands of an increasingly affluent world will inevitably require escalating levels of mineral exploration and extraction. Mining and minerals are as much a part of the world of today and tomorrow as they are of yesterday.

From: MARKER, B. R., PETTERSON, M. G., MCEVOY, F. & STEPHENSON, M. H. (eds) 2005. *Sustainable Minerals Operations in the Developing World.* Geological Society, London, Special Publications, **250**, 5–8.
0305-8719/05/$15.00   © The Geological Society of London 2005.

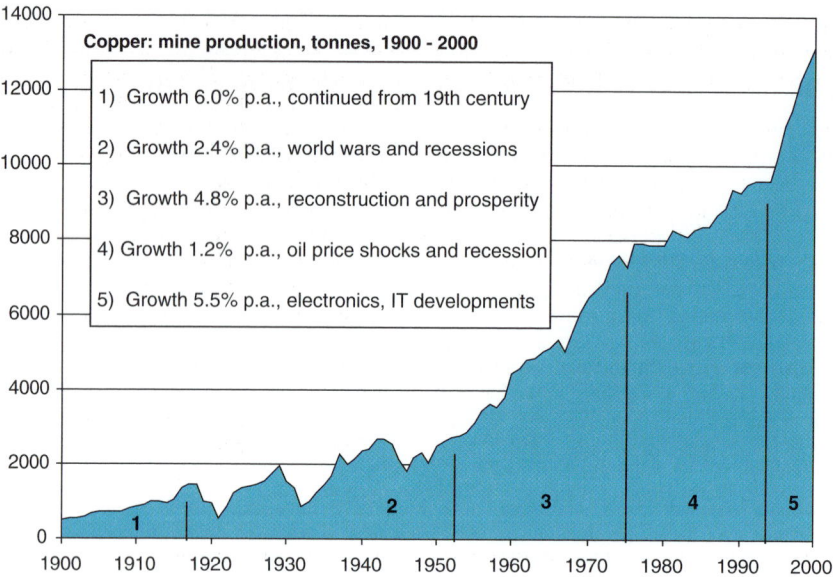

Fig. 1. Global copper production 1830–2000 (*source*: British Geological Survey).

The sustainability paradigm

Sustainability theory and practice is about getting the balance right. The sustainable minerals paradigm is about securing a viable dynamic equilibrium between industry, government, environment, community, and other stakeholders. The early history of mining was dominated by mining benefits staying for too long in the hands of too few people with too little regard for local government, community, and environment. In the last few decades there has been an inexorable shift towards the recognition that extractive industries must work with the consent of local communities and governments, who need to be convinced of the benefits of mining for themselves. The mining industry commissioned the *Global Mining Initiative* project, which published a seminal volume, *Breaking New Ground: Mining, Minerals, and Sustainable Development*, published in 2002 (IIED 2002). This report examined, in a detailed and highly informed way, a wide range of mining-related issues, including economics and society, health and environment, and small-scale mining across the globe. The project involved a major consultation process with a

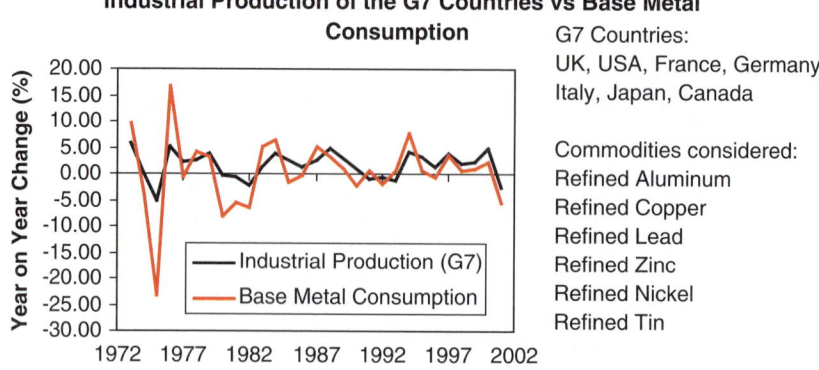

Fig. 2. Correlation between base metal consumption and industrial production (*source*: British Geological Survey).

wide range of mining stakeholders from a truly global community. These results are certainly a key milestone in the evolution of sustainable minerals.

The challenge of today and tomorrow is to move the concept of sustainable minerals forwards still further, partly through the sharing of experiences, examples and philosophies, from across the globe. The idealistic end-goal of *Sustainable Minerals Development* is the production of mineral commodities based on human need producing a minimal negative environmental footprint, using mined commodities in the most energy-efficient ways, recycling raw materials wherever possible, and rewarding appropriately the relevant range of stakeholders linked to mineral-generated wealth. Consultation and stakeholder sharing in decision making will always be a key process. This is a time-consuming affair, but ultimately leads to international best practice in business. Communities who feel involved in local mining ventures and empowered by local government and industry are much more likely to live peaceably alongside a mine than those communities who feel alienated and excluded. There needs to be a clearer demonstration of the links between direct benefits accrued from mining at a range of community and government levels. Mineral generated wealth is by its nature ephemeral, and must be transferred to more lasting areas of economic endeavour: benefits that will survive well beyond mine closure. In small Developing World economies, the impact of even one mine can be very significant – it is imperative that the window of economic opportunity offered by a mine in such a situation is turned into real, lasting and tangible benefits for the country concerned.

Inequitable consumption

As for everything material, the rich world uses far more minerals per head of population than the poor world. For example, in a 77-year lifespan, the average North American will consume around 600 tonnes of primary aggregate and 550 tonnes of fuel, while the average 77-year-old Ethiopian, Bangladeshi or Nepali will have consumed less than 5 tonnes of these commodities (BGS 2004). In spite of the unequal distribution in consumption, many of the richer countries, particularly the more densely populated areas, are drafting increasingly prescriptive and constraining planning policies and regulations. Environmentally designated areas, for example, may preclude mining. Tight regulations increase costs. This makes some prospects economically unviable, at least at present. This makes mineral extraction within their own country difficult, yet their year-on-year hunger for the raw materials that support their high quality of life increases.

In addition, a number of Developing Countries are proceeding towards developed status and have added significantly to the overall demand for materials. This will continue as more countries become developed.

Many of the remaining high-quality mineral deposits occur in the Developing World. With increasing areas of the more Developed World becoming effectively closed to mining, pressure will inevitably increase to develop those deposits. Some deposits are located within states that have relatively undeveloped, small and undiversified economies. These states largely welcome natural resource development, as this may be one of the few avenues available to generate wealth over which they have decision-making powers. In theory, the impact of mining on such countries can be of great benefit: it can alleviate poverty, stimulate economic and infrastructure development, allow governments to develop education and health care systems, and provide livelihoods for employment-starved nations, particularly in rural areas. However, in the absence of a strongly enforced regulatory framework or well-trained workforce, significant environmental damage may result. We are all aware of examples of good and not-so-good mining practices in the Developing World: examples where mining has demonstrably benefited countries in the longer term and examples where it has not. In addition, there are extensive informal or 'artisanal' mining operations in Developing Countries that are mainly unregulated, with resulting risks to both the miners and the surrounding environment. It can be a difficult task bringing these into a formal system to reduce associated problems.

A sharing of global expertise

This conference explores a range of diverse experiences and ways of thinking from across the globe that attempt to ensure that all stakeholders achieve lasting benefits from mining in the Developing World. These stakeholders include financial risk takers, investors, industry, governments at all levels, local communities affected by mining, and society at large. Everyone attending the conference has a stake in ensuring that mining progresses in the twenty-first century in a balanced and enlightened manner.

This will involve culture change and appropriate communication at a wide range of levels: within individual companies, in the consumer mindset, within government, and within society at large. Many positive starts and developments have been made in this area and this conference will allow further cross-fertilization of ideas and discussion and help us all become increasingly wise stewards of the world's natural resources.

References

BRITISH GEOLOGICAL SURVEY. 2004. *World Mineral Statistics 1999–2003*. Minerals Publication No 16. British Geological Survey, Keyworth, Nottingham.

IIED (INTERNATIONAL INSTITUTE FOR ENVIRONMENT AND DEVELOPMENT). 2002. *Breaking New Ground. Mining Minerals and Sustainable Development*. Earthscan Publications, London, UK, and Sterling, VA, USA.

Enhancing the contribution of mining to sustainable development

JONATHAN C. A. HOBBS

Sustainable Development Group, Policy Division, Department for International Development, 1 Palace Street, London SW1E 5HE, UK (e-mail: j-hobbs@dfid.gov.uk)

Abstract: This paper reviews recent developments aimed at improving the mining sector's contribution to sustainable development. Mineral endowments are regarded by many development and environmental non-governmental organizations (NGOs), as a 'curse' and counterproductive to long-term growth and poverty reduction goals, even antithetical to sustainable development in developing countries. This paper argues that, in spite of some empirical evidence in some countries, this is not an inevitable general rule and that the mining sector offers numerous possibilities for catalysing sustainable development and attainment of the millennium development goals. This is, however, conditional upon adequate governance and social and environmental safeguards being in place. The heterogeneity of the mining sector is considered and concern is expressed for the under-management of the growing, albeit not new, phenomenon of artisanal and small-scale mining in developing countries. Without better management of this sector any attempts to improve the contribution of mining to sustainable development will be severely limited.

How can the extractives industries (normally taken to refer to mining, oil and gas, but in this paper limited to the former) be sustainable? The very nature of the term *extractive* implies that this sector exists by exploiting and depleting non-renewable, finite, natural resources. The sustainability of extractive industries is, therefore, a difficult message to market and relatively new concepts such as 'sustainable mining' or 'sustainable minerals', which are attempts to do so, are considered by many as no more than illusory contradictions in terms.

The prevailing perception in many quarters is that there is little that is 'sustainable' about extractive industries, even if we accept the long time horizons that geologists adopt to convince sceptics that minerals are renewable resources, that is, resources that, on human time-scales, are essentially inexhaustible when managed correctly and exploited with sustainable yields.

Non-renewable resources exist in a fixed quantity in the Earth's crust and thus, theoretically, can be completely depleted. On a time-scale of millions to billions of years, geological processes can renew such resources. However, on human time-scales, they can be depleted quicker than they are formed, although some non-renewable resources can nonetheless be recycled and reused (Miller 2000). (Figs 1–3).

The only way for extractive industries to have a seat at the sustainable development 'table' is through broadening our understanding of the nature of not only the concept of sustainability, but also the dimensions of *capital*. We need to look beyond traditional definitions of capital that are limited to the financial and tangible factors of production – investment and equipment – to four different types of capital: human, financial, manufactured, and natural capital. The latter are usually described as comprising resources, living systems and ecosystem services (Hawken *et al.* 1996). We can, however, usefully qualify the definition of natural capital by subdividing the concept into two to illustrate two dimensions of resources, renewable and non-renewable, and thereby accommodate mining in sustainability discussions.

The legitimacy for the extractive industries to be a party to the sustainable development agenda is achieved when the sector is seen to be using the first three forms of capital to transform non-renewable, natural capital, not only into the material benefits that fulfil society's need for goods and services, but also into more sustainable opportunities and livelihoods for society (Fig. 5).

In some countries, the longevity of mining operations has spanned many centuries. For example, Sri Lanka's gemstone mining is reputed to have lasted two-and-half thousand years thus far. The key to the success of the role of minerals in sustainable development, however, is in the utilization of this resource as a platform for achieving economic

Fig. 1. Fulfilling the goods and services demanded by society: off-shore diamond dredging, west coast of South Africa (photo: Hobbs).

diversification, growth and broad development goals, while maintaining social stability and protecting environmental integrity.

While a number of developed countries (USA, Canada, Sweden, Finland and Australia, among others) and a few developing countries (Botswana and Chile; Hope 2003) have achieved considerable economic success through reliance on minerals in their development processes, we also find regions with extensive histories of mineral exploitation where the direct benefits to those regions have been less obvious (Figs 6–8).

The mineral wealth of Cornwall in the United Kingdom, for example, has been exploited since Phoenician times, reaching its peak at the turn of the 19th–20th centuries. One would have expected this mineral wealth to make the county an enduring region of prosperity, ahead of those without such mineral resource endowments. Instead, it was recorded during the death throws

Fig. 2. Fulfilling the goods and services demanded by society: Mozal, aluminium smelter, Mozambique (photo: Hobbs).

Fig. 3. Fulfilling the goods and services demanded by society: coal mining, Witbank, South Africa (photo: Hobbs).

of the tin and copper mining era to be the county with the most extensive land dereliction and social deprivation in the UK, its only merit being, in some places at least, its industrial archaeology value. (At its worst period, UK Government figures for 1966 showed that Cornwall was the county with the highest degree of dereliction; Barr 1970.)

In spite of centuries of prosperous mining activity, Cornwall is now officially classified as one of the poorest regions in Europe and is a recipient of European Union development assistance, much of which has been used to clean up the physical legacy of past tin, copper, and kaolin mining. No wonder then that it has been selected as the home for a new Post Mining Alliance Initiative, working in partnership with the Eden Project, which has created an educational and tourist resource in a disused china clay quarry.

Momentum for change

Up to 20 years or so ago, the only reference material available in Schools of Mines or university engineering departments that could be considered as, in any obvious way, linking mining with sustainable development were those addressing the technicalities of tailings design and management and rehabilitation of disturbed and derelict post-mining lands.

Fig. 4. Fulfilling the goods and services demanded by society: phosphate mining, Morocco (photo: Hobbs).

The mining industry has lagged behind other 'primary' industries, notably chemicals and oil and gas, in their understanding of, and commitment to, the broader concept of sustainable development. Sustainable development was taken as being synonymous with environmental management and was characterized by resignation to acceptance that a mining operation was a 'necessary evil', restricted in locational options and only amenable to mitigation and amelioration of its impacts, not actively seeking out the positive development opportunities it created.

This was noticeable at the Earth Summit (UN World Conference on Environment and Development) held in Rio de Janeiro in 1992, where the chemical industry took 'centre stage' with their *Responsible Care*© initiative. The (then) Business Council for Sustainable

Fig. 5. Diamond mining, Orapa, Botswana (photo: D'Souza).

Fig. 6. Cornwall's mining legacy. The Red River draining into to St. Ives Bay, Cornwall; 'Red' because it conveys tailings from the Tolvadon valley, the scene of centuries of tin and copper mining and milling (photo: Hobbs).

Fig. 7. Cornwall's mining legacy. Derelict tin mine workings near Redruth, Cornwall (photo: Hobbs).

Development (now World Business Council for Sustainable Development; WBCSD), which led business leaders' participation in the Earth Summit, had only one chairman of a company with mining as its core business among its 48 members, Rio Doce International of Brazil.

The seminal publication presented by WBCSD's business leaders to the Earth Summit, *Changing Course – A global business perspective on development and the environment* (Schmidheiny 1992), included only one, three-page case study on the mining sector's contribution to sustainable development, ALCOA's 'Sustainable mining in the Jarrah Forest, Australia', which predictably focused on land rehabilitation issues. The supportive regional reports similarly only included one mining case study from Africa, rehabilitation of limestone quarries at Bamburi Portland Cement, Kenya (Haller & Baer 1994).

Fig. 8. Cornwall's mining legacy. Wheal Coates engine house, now a National Trust property, near St. Agnes, Cornwall (photo: Hobbs).

Fig. 9. East African limestone quarrying, a rehabilitation challenge, Twiga Portland Cement, near Dar es Salaam, Tanzania (photo: Hobbs).

Other sectors had been galvanized into action beyond minimum regulatory compliance by a series of environmental, social, safety, financial and public relations disasters and adversarial campaigns by an increasingly activist civil society. These included:

- the Bhopal (Union Carbide) disaster that killed 5100 and seriously injured 200 000 in India;
- the incipient impact of pesticides on raptors through food chains (notably dichlorodiphenyltrichloroethane; DDT);
- the realization of the contribution of chlorofluorocarbons (CFCs) to stratospheric ozone depletion;
- the 1989 Exxon Valdez oil spill in Prince William Sound;
- Shell's experiences over the Brent Spar oil rig disposal; and
- community conflicts in the oil-rich Niger delta.

All serve as prominent examples along with growing concerns about climate change.

This was not to say that the mining sector was inactive nor that there was any shortage of conflicts, antagonistic campaigns, environmental and safety disasters or communication breakdowns that confronted the mining industry:

- protracted difficulties in gaining access to new mineral resources were common, such as those faced in the late 1980s and early 1990s by Richards Bay Minerals when trying to extend mining of heavy mineral bearing sands (ilmenite, rutile and zircon) further into the coastal dunes of St. Lucia in South Africa;
- the Romanian Baia Mare cyanide spill in 2000 entered the Danube and adversely impacted four countries;
- community conflicts and cessationist civil war were attributed to Bougainville copper resources exploitation in Papua New Guinea.

These and others contributed to mounting pressures and incentives for change.

Regular media headlines exposed the need for a greater responsiveness to society's concerns by the mining sector. Accepting physical environmental damage as an inevitable consequence of mining, it was also inevitable that the interpretation of sustainability in the sector was centred on legacy issues and rehabilitation. This was an important but far from sufficient response (Fig. 9).

A considerable momentum for change built up, resulting in the sector having the highest profile of all industrial sectors at the 2002 World Summit on Sustainable Development (WSSD) in Johannesburg (Figs 10 & 11).

Broadening the concept of sustainable development

The response in the 1990s had been to improve planning and operational environmental management through better environmental assessment and auditing, pollution prevention and control, and the development of integrated environmental management systems. International agencies developed environmental guidelines for the mining operations (the Berlin Guidelines 1991, revised 1999). The Australian Environmental Protection Agency's 'Best Practice Environmental Management in Mining' (1995, updated 2002) advisory and training modules and the various technical report publications of the United Nations Environment Programme (UNEP) (see www.mineralsresourcesforum.org) serve as good examples of the initiatives that guided a greater responsiveness in the industry in the post-Rio and pre-Johannesburg climate of change.

These responses demonstrated the truism that environmental management supported, rather than acted as a constraint to good business management. They were driven by the practical realization that good environmental management

Fig. 10. Broadening the concept of sustainable development, mining often requires sensitive relocation: coal mining for power generation, Majuba coal mine, South Africa (photo: Eskom).

Fig. 11. Broadening concerns of mining to development opportunities, small-scale gold mining, Ghana (photo: Hobbs).

reduces liabilities, cuts costs, improves efficiency and quality, and, for the more insightful companies, was increasingly interpreted as identifying new business opportunities.

Clearly, however, the gains were still preoccupied with the environmental dimensions of sustainable development and largely limited to core business issues within physical operational confines. The social dimension was one where the benefits were less obvious to the mining companies, other than serving some nebulous concept of 'social licence to operate' and easing permitting and regulatory processes.

Slowly there was a recognition that sustainable development is as much about the external socio-economic environment of the company (social justice, community relations and good governance and the interplay between these) than, generally, site-specific, environmental issues.

Public relations professionals started to populate embryonic sustainable development units within mining companies, frequently separate from the more technically oriented environmental divisions, which maintained an engineering profile. Their programmes geared to the sociocratic rather than technocratic goals. Glossy corporate sustainability reports emerged that were as important in changing internal organizational mindsets in the process of their formulation as in communicating to external stakeholders a company's commitment to sustainable development.

However, important pioneering initiatives emerged that reflected a more comprehensive approach to sustainable development within the sector. This was illustrated in the agenda of the International Council on Mining and the Environment (ICME), World Bank, UNEP and United Nations Commission on Trade and Development (UNCTAD) hosted Conference on 'Development, Environment and Mining' in Washington, DC (1994), a conference that addressed the contribution that the sector could make to international sustainable development.

Another initiative launched in 1998 was the Business Partners for Development (BPD) (Natural Resource Clusters) programme hosted by the World Bank. This was styled as 'a new way to manage social issues in the extractive industries' and gave particular attention to developing models of 'tri sector partnering' between private, public and civil society sectors based on the practical experiences of specific natural resource operations. The BPD pioneered a trend to move the sector towards participation in 'partnerships' and 'multistakeholder processes' and encouraged the industry to move from an enclave mentality to one in which it could meaningfully address sustainable development issues in partnerships through more effective social investment, engagement in community affairs, establishing new communications links and networks, and so on.

The BPD started to address the difficult grey boundaries where the moral and legal responsibilities of a mining company's role in community development end and the role of government takes over. Through case studies such as the Las Cristinas gold mine project in Venezuela and the Sarshatali coal mine in India, the BPD attempted to explore ways of enhancing the sustainability of development through focussing on poverty mitigation, growing human capital, community participation, environmental benefits, increased social cohesion, and improving access to basic services (www.bpdweb.org).

Business takes the lead

Further significant progress resulted, when the WBCSD added sector-specific action to the previously generic agenda of business advocacy that it had pursued (as the BCSD) at the Earth Summit in 1992. Following success in focussing attention on sustainable development issues in the paper and pulp sector, it moved into other realms of business where there was a need for a significant change, including the mining sector. Consequently, some mining companies joined the growing organization.

The WBCSD's resultant Global Mining Initiative (1999) was endorsed by 25 mining company CEOs and later transformed into the, two year 'Mining, Minerals and Sustainable Development' (MMSD) initiative. This was the most comprehensive analysis the sector had ever been subject to regarding sustainable development performance. A framework was developed to guide its role in the sustainable development agenda (IIED 2002).

The MMSD report led to adoption of a Toronto declaration (2002) – a commitment from multinational companies to improve performance towards sustainability – and created a reorganized mining industry association, the International Council on Mining and Metals (ICMM), mandated to develop a work programme to implement the Toronto Declaration and MMSD recommendations. The ICMM Sustainable Development Framework Principles (2003) (Table 1) followed, adding the mining sector's interpretation to the many voluntary principles developed by other business associations during the previous decade.

The MMSD report, as input to the 2002 Johannesburg WSSD, ensured that the sector now had a high profile on the sustainability agenda.

Table 1. *The 10 point ICMM principles*

- Implement and maintain ethical business practices and sound systems of corporate governance
- Integrate sustainable development considerations within the corporate decision – making process
- Uphold fundamental human rights and respect cultures, customs and values in dealings with employees and others who are affected by our activities
- Implement risk management strategies based on valid data and sound science
- Seek continual improvement of our health and safety performance
- Seek continual improvement of our environmental performance
- Contribute to the conservation of biodiversity and integrated approaches to land use planning
- Facilitate and encourage responsible product design, use, recycling and disposal of products
- Contribute to the social, economic and institutional development of the communities in which we operate
- Implement effective and transparent engagement, communication and independently verified reporting arrangements with our stakeholders

This was evident in the Johannesburg Plan of Implementation (JPOI), the internationally agreed action plan adopted at the United Nations World SSD, 2002. Clause 46 of 170 clauses noted that: 'mining, minerals and metals are important to the economic and social development of many countries. Enhancing the contribution of mining, minerals and metals to sustainable development includes ... supporting efforts to address the environmental, economic, health and social impacts and benefits of mining ... enhance the participation of stakeholders ... and foster sustainable mining practices ... through support to developing countries' (UN 2003).

As bland as these statements are, they should not be underestimated for the fact that they represented recognition at the highest level of the international development community that mining has a role to play in a sustainable world. This is still not a view shared by everyone.

Nonetheless, the mining sector had moved into a new era and on to a new agenda. Progress against the statements made at Johannesburg will be critically reviewed at the UN Commission for Sustainable Development (CSD) in 2009–2010, when the sector becomes the focus of the Council for Sustainable Development's work programme of tracking JPOI implementation.

The Extractives Industries Review (EIR)

Arguably the most significant initiative, however, has been the Extractives Industries Review (EIR) (2004), set in motion in 2000 by the then World Bank President, James Wolfensohn. The EIR was commissioned to investigate whether or not the World Bank's investments in the extractives sector (in this case oil, gas and mining) supported, or detracted from, the Bank's mission of poverty reduction and sustainable development. A three-year international stakeholder consultation process, the independent EIR, together with simultaneous internal reviews within the Bank followed (Liebenthal *et al.* 2003).

The EIR eventually concluded that the sector can indeed contribute to the World Bank's mission and that there was a continuing role for the Bank's investments in the sector. The EIR, however, added the rider that this mission could only be achieved if certain enabling conditions were in place. These conditions are:

- good public and corporate governance;
- greater respect for human rights; and
- more effective social and environmental safeguards.

Good governance can generally be taken to include accountable governments, rule of law, absence of armed conflict or the risk of it, respect for human rights and labour standards, protection of indigenous people's rights and the rights of minorities, government capacity to promote sustainable development through economic diversification, and so on.

They further suggested the need to encourage some specific building blocks of good governance: (Extractive Industries Review 2004)

- promote disclosure of project documents;
- develop the capacity to manage fluctuating revenues;
- develop the capacity to manage revenues responsibly;
- help governments to put in place effective and efficient policy and regulatory frameworks;
- integrate stakeholders in decision making; and
- promote the transparency of revenues flows.

The need for good governance: 'the resource curse'

The EIR had been a consequence of, and added to the debate on, a recurring theme in discussions on the role of the mining sector in sustainable development. This is that intuitively one would expect that the prudent exploitation of mineral wealth should be the basis for economic growth, poverty reduction, political stability, and sustainable development. Paradoxically, however, some resource-rich countries remain amongst the poorest and have the highest levels of conflict, poverty and corruption. Of the world's most mineral-dependent states, 11 are heavily indebted and five have ongoing civil wars. This correlation gives rise to the hypothesis that mineral wealth can be more of a 'curse' than a 'blessing'.

"War, poverty, climate change, greed, corruption and ongoing violations of human rights – all of these scourges are all too often linked to the oil and mining industries". *Nobel Laureates for Peace (Jody Williams, Archbishop Desmond Tutu, Rigoberta Mench Tum, Sir Joseph Rotblat, Betty Williams and Mairead Maguire)*

The extractives industries are simultaneously an opportunity and a threat to the development prospects of poorer countries. Extractive industries are important in over 50 developing countries. Mineral resource exploitation represents the potential for many of these developing countries to embark on a more urgent path of economic growth.

Yet it is undeniable that the track record of some resource-rich countries has not been good, and there is no shortage of examples where minerals-derived revenues have fuelled conflicts, corruption, and undermined poverty reduction and sustainable development progress (Goreaux 2001; Montague 2002).

The challenge is to recognize the possibility of the resource curse and work to counter it. The issue of good governance has increasingly framed discussions of the extractive industries and sustainable development. One key element of good governance is transparency, particularly relating to the management of revenues from extractive industries.

The Extractives Industries Transparency Initiative (EITI)

Transparency, applied to revenue flows, enables citizens to hold governments to account for the fate of those revenues received from the exploitation of natural resources. Citizens have the right to know the fate of the revenues government receives from mineral exploitation.

The Extractive Industries Transparency Initiative (EITI) was another outcome of the WSSD (2002). Launched by British Prime Minister Blair, the EITI reinforced the good governance commitments made by G8 leaders (The Action Plan on Fighting Corruption and Improving Transparency, Evian G8 meeting, 2003 with Transparency Compacts being agreed with four countries at the follow-up meeting at Sea Island, USA, in 2004) and African leaders in the New Partnership for African Development (NEPAD). Wiseman Nkhulu – Chairman, Steering Committee of the NEPAD has announced the intention to explore the EITI contribution to the African Peer Review Mechanism).

Transparency is, however, a means to an end, not an end in itself. It is a necessary, but not sufficient, contribution to managing any potential resource curse. The EITI starts from the premise that extractives industries can benefit a country, if managed properly, and that there is nothing inherently wrong with the sector.

At the core of the EITI is the view that if industry pays its taxes and royalties, then government should use these to provide services, rather than have business shoulder the responsibility for aspects of community development that would normally be considered a government's responsibility.

Following the WSSD, a ministerial conference, again addressed by the British Prime

Minister, was held at Lancaster House, London, in 2003. This resulted in the widespread public endorsement of a set of voluntary principles and actions (available on the EITI website, www.dfid.gov.uk) by over 60 participants. Reinforcement and a check on progress in implementing the principles followed at a London conference in March 2005. A number of countries have now moved beyond endorsement and are pioneering the practical implementation of revenue transparency; including Azerbaijan, Ghana, Nigeria (launched personally by President Obosanjo in February 2004), Republic of Congo, Sao Tome e Principe, Timor Leste, Trinidad and Tobago, and Kyrgyz Republic. A Trust Fund and reporting guidelines, and so on, have been prepared to facilitate this process, overseen by a multi-stakeholder steering group (DFID 2005).

Implementation requires all extractives industry companies operating in a particular country to annually, or more often, disclose their payments to government, the government to publish the revenues it receives, the credibility of the data to be verified by independent audit, and civil society to use the disclosed data to hold governments to account for the distribution of those funds in the interests of sustainable development.

Sector specific sustainability reporting

The Global Reporting Initiative (GRI) has developed more general sustainability reporting guidelines for companies in the sector – an exercise that is now routine for large companies www.globalreporting.org.

The Kimberley process

The EITI aims to track payments and receipts and, alongside other efforts to improve public financial management, help build accountability. This differs from another significant initiative, the Kimberley process, which is about tracking a commodity, the origin of diamonds in the market.

The Kimberley process is an initiative in which governments, industry and NGOs joined together to stem the flow of so-called 'blood or conflict diamonds', rough diamonds that have been used to finance conflicts and that have been mostly obtained illegally (Goreaux 2001).

The Kimberley process certification scheme is a voluntary system that requires participants to certify that their shipments of rough diamonds are free from conflict diamonds (sometimes referred to as 'blood diamonds'). It accounts for about 98% of the trade in rough diamonds.

Common interest in improving governance

The benefits of increased transparency are diverse. Governments will benefit from maintaining or increasing inward investment, communities will receive a greater share of the revenues, citizens will be better able to hold governments to account, companies will benefit from more predictable and stable business and investment climates and consumers' will be assured of the origins of their purchases.

There are obvious advantages in tackling the issue of poor governance for an industry that is anchored to the place where the resource it needs to exploit physically occurs and that has limited options to relocate to other countries where better governance prevails.

Improving governance also has importance to the development community. This is because of the changing nature of the way in which international assistance is increasingly being delivered. The development landscape of the past is littered with numerous defunct projects set up by a plethora of, sometimes competing, international agencies, often with their own interests a greater motivation than those of the developing countries that they were supposed to assist. As this project-specific aid sometimes by-passed governments, it is little wonder it frequently proved to be anything but sustainable, rarely surviving long beyond the departure of the 'expatriate experts' sent in to set them up.

Instead, aid is now increasingly being provided more strategically, as direct budget or sector policy support. This has the advantages of greater prospects of country ownership, reduced transaction costs, and greater prospects for harmonizing development agencies' activities and aligning them with the developing country's own priorities not those of the aid agencies (Table 2) (Holman 2003).

For this change to be successful, however, there need to be effective and accountable governments, transparency, and widely agreed goals with targets and strategies to achieve them.

Table 2. *The Millennium Development Goals 2000*

(1) Eradicate extreme *poverty* and hunger
(2) Achieve universal primary *education*
(3) Promote *gender equality* and empower women
(4) Reduce *child mortality*
(5) Improve *maternal health*
(6) Combat *HIV/AIDS*, malaria and other diseases
(7) Ensure *environmental* sustainability
(8) Develop a global *partnership for development*

The above goals are supported by 18 targets.

Heterogeneity of the mining sector: artisanal and small-scale mining

By far the greatest amount of attention to sustainability in mining has fallen on the activities of large-scale mining companies. The mining sector is, however, a heterogeneous entity, ranging from large-scale multinationals, through small-scale enterprises, to the artisanal miners who characteristically are labour-intensive operators using rudimentary tools (Figs 12, 13).

It is estimated that there are now more people directly employed in artisanal and small-scale mining than in larger scale formal mining. Yet the complexity of the artisanal and small-scale mining (ASM) sector is leading to its undermanagement, if not active persecution. Government policies on ASM either do not exist or are poorly developed. The lack of formalization of laws, regulations, rights, fair market prices and safeguards fails to capitalize on the contribution the sector can make to development processes.

Artisanal mining has been described as the "most primitive type of mining characterized by groups and individuals exploiting deposits – usually illegally – with the simplest equipment". The Toronto declaration (2002) recognized that "artisanal and small scale mining... are important and complex (*but*) *beyond the capacity of ICMM to resolve*, and *called on* governments

Fig. 13. Rudimentary milling techniques pose threats to health, safety and the environment, small-scale gold mining, Ghana (photo: Hobbs).

and international agencies *to* assume the lead role in addressing them" (authors italics).

The EIR, in response to the high profile given to this sector at its international consultative workshops, commented on the potential of the artisanal and small-scale mining sectors to 'lessen(ing) the burden of poverty.' It urged the World Bank to help governments to develop policies that also recognize the sector as heterogeneous (in its own right) and to distinguish between community-based and itinerant miners, giving the former clear priority over mining rights.

Artisanal mining is not a new phenomenon. The labours of artisanal miners have laid the foundations for, and their products have adorned, most early civilizations from Angkor Wat to Zimbabwe. The proceeds of artisanal mining opened up early trade relations.

The history of mining itself is rooted in artisanal and small-scale operations and it has frequently been artisanal miners that have pointed the way to mineral deposits for more capital intensive exploitation. For example, the pre-European settler gold mining and ore processing evident in the Francistown/Tati area of Botswana supported the Zimbabwe/Mapungubwe civilizations and opened trade for these civilizations with the Portuguese and Arabs. It was these primitive workings, dating back to AD 900 (Tlou & Campbell 1984) that led early European settlers to deposits that, with increasing mechanization, supported the first European gold mining ventures in southern Africa in the late 1860s, the remnants of which are still evident south east of

Fig. 12. Small-scale emerald mining, Ndola, Zambia (photo: D'Souza).

Francistown in Botswana (Van Waarden 1999) (Figs 14–16).

Today, the ASM phenomenon is widespread and growing throughout Asia, Latin America, and Africa, it features in about 30 countries and, according to ILO estimates, provides livelihoods for 100 million people, although there are obvious difficulties in establishing reliable figures. The ILO also estimates that up to 13 million people are directly engaged in the sector (Figs 17–21).

Some of the unacceptable practices found in the historical roots of mining are still found in today's artisanal sectors. These should have been consigned to the annals of social history and industrial archaeology. The sector has an influence on all of the Millennium Development Goals (Table 2) and other issues at the heart of development policy: HIV-AIDS, child labour, poverty, gender discrimination, environmental sustainability and so on (Figs 17–22).

It is estimated that the artisanal mines of the Lake Victoria goldfields, which engage some 300 000 people, produce nearly 70% of the gold production of Kenya, Uganda and Tanzania. In Mozambique, this is 100% of production. Women comprise 50% of those involved across Africa. It is estimated that gold and gemstones worth US$1 billion per year are produced in sub-Saharan Africa through artisanal mining (D'Souza pers. comm.).

Fig. 15. Pre-European gold milling 'dolly' holes and grinding surfaces in dolerite at Tati, Botswana, from where gold was supplied to the Zimbabwe and Mapangubwe civilizations in present, day Zimbabwe and South Africa.

Fig. 14. Cornish stamp mill; remnants of earliest European gold mining activity in Southern Africa, Vermaak's mine, Botswana (photo: Hobbs).

Fig. 16. Artisanal mining and quarrying provided the materials for many of the enduring legacies of past civilizations: (a) Angkor Wat, Cambodia (photo: Hobbs); (b) the Great Wall of China (photo: Hobbs).

Fig. 17. Artisanal mining and the development agenda. Child labour in Ghana gold mining (photo: Hobbs).

Fig. 19. Artisanal mining and the development agenda. Women in artisanal mining in Rex, Ghana (photo: D'Souza).

The priority need is to integrate the sector into national economies to ensure these miners have access to official markets and get a fair return for their labours and products. Clearly, social, health, safety and environmental safeguards are few and far between in the majority of these activities (Hobbs *et al.* 2003).

To address these issues, the Communities and Small-scale Mining (CASM, www.casmsite.org) initiative was inaugurated at a meeting in London in 2001 under the auspices of the UK's Department for International Development and the World Bank. CASM is now a thriving international network of experts, government and development officials, private sector, NGOs and artisanal miners themselves.

Fig. 18. Artisanal mining and the development agenda. Health and safety issues: artisanal gold mining, Ghana (photo: Hobbs).

Fig. 20. Artisanal mining and the development agenda. Hazardous child labour in Mgusu, Tanzania, hand mixing gold and mercury (photo: D'Souza).

Fig. 21. Artisanal mining and the development agenda. Child labour in Ghana (photo: BGS).

Mirroring large-scale operations, CASM's mission is to move enclave artisanal mining communities to more sustainable, integrated communities pursuing sustainable livelihoods. This calls for a transformation of the sector from the current situation – characterized by self-serving opportunism, violence and conflict, dysfunctional social systems (prostitution and few community institutions), erratic incomes and little or no savings – to communities in which land and other rights are respected, community structures exist, reinvestment takes place, safety, health and environmental management systems are in place, and diverse employment and livelihoods opportunities exist (van der Veen 2003) (Figs 23 and 24).

Fig. 23. Irresponsible legacy of recent small scale mining activity, in the footsteps of earlier generations, small-scale discards, Tati, Botswana (photo: Hobbs).

Fig. 22. Getting the ASM product to market, transporting artisanally mixed copper oxide, Laputo, Democratic Republic of the Congo (photo: d'Souza).

Fig. 24. Irresponsible legacy of abandoned small scale mining activity, hazardous gold mine, exposed shaft, Tati, Botswana (photo: Hobbs).

Conclusion

It is clear that governments, especially developing countries, are often now the weakest link in the momentum behind the drive to 'sustainable mining' that has been led by the private sector. This gulf has become wider as the private sector has devoted considerable attention to improving its contribution to sustainable development.

The role of developing countries is the creation of more conducive business environments and investment climates and the better management of natural resources, renewable and non-renewable. This will enhance sustained economic growth, political stability, and the contribution the mining sector can make to the attainment of poverty reduction, the other Millennium Development Goals (MDGs) and sustainable development.

This paper has tracked recent progress of the mining sector's progress to greater sustainability, highlighting key initiatives. Two further recent initiatives hold out promise of progress in this area. The first of these is the development of the inter-governmental forum on mining and minerals – formerly the Global Mining Dialogue (a Canadian and South African partnership initiative originating from WSSD but only receiving the required 25 member countries to bring it into effect in 2005), which is attempting to provide the necessary forum to encourage more urgent progress in the way governments manage mineral resource endowments and problematic elements such as the artisanal and small-scale mining sector.

Secondly, and at a regional level, the African Mining Partnership was launched in Cape Town in 2004 (initially chaired by Ghana). Twenty-two Ministers responsible for mining in their countries participated (Angola, Burkina Faso, Chad, Republic of Congo, DRC, Djibouti, Egypt, Ethiopia, Gambia, Ghana, Kenya, Malawi, Mali, Mauritania, Namibia, Nigeria, Senegal, Sierra Leone, South Africa, Sudan, Tanzania, Uganda) and stressed the important part that the extractives sector has to play in poverty reduction.

The overall challenge in encouraging sustainable development in mining is to convert what has been described as the *vicious circle* of extractive investments, historically at times characterized by enclave activities that fail to generate indigenous jobs or local investment, operate under a veil of secrecy, have little or no beneficiation or added value, result in little economic diversification and exhibit little compliance to environmental and social standards and laws, to a *virtuous circle*, where jobs are created, revenues collected and managed competently, incomes saved and reinvested, there are forward and backward economic linkages, diversification is encouraged and environmental and social impacts managed and where poverty and unsustainability are replaced by prosperity and sustainability, in other words 'sustainable mining'.

The author is grateful to Kevin d'Souza (Wardell Armstrong), Paul Henney and Andrew Bloodworth (BGS), and Peter van der Veen, Jeffrey Davidson and Gotthard Walser (World Bank, IFC) together with many other colleagues, for sharing their insights and expertise over the past two years of collaboration in the work of CASM. Sincere thanks are also due to Dr. Anne Blackbeard of Blue Jackets Ranch, Tati Town, Botswana.

References

BARR, J. 1970. *Derelict Britain*. Penguin, Harmondsworth, England.

BUSINESS PARTNERS FOR DEVELOPMENT. 2002. *Putting Partnering to Work 1998–2002 Results and Recommendations*. Business Partners for Development, World Bank, Washington, DC.

DANIELSON, L. 2002. *The Role of the Minerals Sector in the Transition to Sustainable Development*. International Institute for Sustainable Development (IISD), London.

DEPARTMENT FOR INTERNATIONAL DEVELOPMENT (DFID). 2005. *Extractive Industries Transparency Initiative*. DFID, London.

ENVIRONMENTAL PROTECTION AGENCY. 1995. *Best Practice Environmental Management in Mining* (26 Modules). Commonwealth of Australia (now known as the 'Sustainable Minerals Series').

EXTRACTIVE INDUSTRIES REVIEW (EIR). 2004. *Striking a Better Balance – the World Bank Group and Extractive Industries – Final Report*. World Bank, Washington, DC.

GOREAUX, L. 2001. *Conflict Diamonds*. Africa Regional Workshop Paper Series no. 13. World Bank. (http://www.worldbank.org/afr/wps/index.htm).

HALLER, R. & BAER, S. 1994. *From Wasteland to Paradise*. Koschany, Munchen, Germany.

HAWKEN, P., LOVINS, A. & HUNTER LOVINS, L. 1996. *Natural Capitalism – the Next Industrial Revolution*. Earthscan, London.

HOBBS, J., BLOODWORTH, A. & HOADLEY, M. 2003. Escaping poverty through artisanal mining – a year down the line and how are we doing? *In*: *'Vision' Endangered Wildlife Trust*, volume 11. Future Publishing, Johannesburg, 118–121.

HOLMAN, M. 2003. Critical issues in developing a mining related Poverty Reduction Strategy for the People's Republic of Tanzania. Minerals and Strategies Research, Cape Town.

HOPE, K. R. 2003. Lessons for Africa. *Business in Africa*, December 2002/January 2003, Johannesburg, 28–32.

INTERNATIONAL INSTITUTE FOR ENVIRONMENT AND DEVELOPMENT (IIED)/WORLD BUSINESS COUNCIL FOR SUSTAINABLE DEVELOPMENT. 2002. *Breaking New Ground – Mining, Minerals and Sustainable Development*. Earthscan, London.

LIEBENTHAL, A., MICHELITSCH, R. & TARAZONA, E. 2003. Extractive industries and sustainable development – an evaluation of World Bank experience. World Bank, International Finance Corporation and Multilateral Investment Guarantee Agency, Washington, DC.

MILLER, G. T. 2000. *Living in the Environment: Principles, Connections and Solutions* (11th Edition). Brooker/Cole Publications (Division of ITP), Pacific Grove, USA.

MONTAGUE, D. 2002. Stolen goods: Coltan and conflict in the Democratic Republic of the Congo. *SAIS Review*, **XX11**(1) (Winter–Spring).

MURSHED, S. 2004. *When Does Natural Resource Abundance Lead to a Resource Curse?* International Institute for Sustainable Development, London.

SCHMIDHEINY, S. 1992. *Changing Course – A Global Business Perspective on Development and the Environment*. MIT Press, Cambridge, Massachusetts.

TLOU, T. & CAMPBELL, A. 1984. *History of Botswana*. MacMillan, Gaborone.

TYLER MILLER, G. 2000. *Living in the Environment*. Brooks Cole Publishing, Pacific Grove, USA.

UNITED NATIONS: World Summit on Sustainable Development; Political Declaration and Plan of Implementation. 2003. United Nations, New York.

VAN DER VEEN, P. 2003. CASM: A force toward sustainable communities. Paper presented at the Communities and Small Scale Mining (CASM) Annual Conference, Elmina, Ghana.

VAN WAARDEN, C. 1999. *Exploring Tati – Places of Historic and Other Interest Around Francistown*. Marope Research, Francistown, Botswana.

WORLD BANK AND INTERNATIONAL FINANCE CORPORATION. 2002. *Treasures or Trouble? Mining in Developing Countries*. IFC, Washington, DC.

The role of minerals in sustainable human development

JEREMY P. RICHARDS

Department of Earth and Atmospheric Sciences, University of Alberta, Edmonton, Alberta, T6G 2E3, Canada (e-mail: Jeremy.Richards@UAlberta.CA)

We humans now have a name for our survival instinct: sustainable development. This means, quite simply, living on this planet as if we intended to go on living here forever.

Porritt 2002, p. 75

Abstract: Sustainable mineral resources development can be seen as the equitable conversion of transient mineral wealth into durable social and environmental capital. In the past, this conversion has not been efficient or equitable, with benefits accruing mainly to First World investors and consumers by externalization of social and environmental costs to local people and places. Modern industry, led by large multinational corporations, is in the process of changing its *modus operandi* to embrace ideas of corporate and social responsibility. The damage from past practices to the developing world is severe, however, and may require measures beyond voluntary or current legal instruments to reverse degenerative trends. Central among these requirements is Third World debt cancellation. However, the mining industry can also contribute by fully internalizing the costs of mineral production, and paying a fair price for the resources it extracts; these internalized costs should be reflected in higher commodity prices. This can be achieved through a combination of financial instruments and incentives, innovation, and best practice, with essential consumer buy-in through increased awareness.

One might ask why those of us fortunate enough to live in the comfortable developed world should worry about issues such as resource depletion and environmental change (anthropogenic or otherwise), when even the worst model scenarios predict little serious effect in our lifetimes. No other living species on this planet is worried, so why should we be? Because, of course, we are the only species conscious of the long-term risks and implications of these potential changes. And therein may also lie our salvation, because we are also the only species that might be able consciously to affect the outcome. Currently, however, we show little inclination, or even ability, to face up to this challenge (the Kyoto protocol, for example, even if adhered to, will not come close to solving the problems that climate change scientists tell us we are facing). The societal transformation required to make global, collective decisions on this scale, if made, will mark the most fundamental change in human thinking and cooperative action since the birth of modern civilization. If not made, however, we and other species may face extinction, or at least a seriously degraded existence.

Although mining is a bit-player in this global drama, the problems and challenges the industry faces are the same as in other human interactions with the natural world, and solutions learned in one field can often be applied more widely. Moreover, although we as individuals cannot solve all the world's problems, if each of us contributes within his or her field of expertise then the collective effort of 6.3 billion people can surely prevail.

This paper reviews, fairly subjectively, the role that minerals have played, and can potentially play, in the development of society. Mineral wealth is seen by some as the starting point for economic growth, but others argue that it has been a curse for many (Sachs & Warner 1995; for opposing views see Manzano & Rigobon 2001; Weber-Fahr 2002; Wright & Czelusta 2003). Our challenge in the twenty-first century is to find ways of converting transient mineral wealth into sustainable development for the greater and equitable good of society, while at the same time allowing realistic returns on investment.

Global impacts of mineral production

It is trite to rehearse the fact that the use of rocks and minerals has been the basis of development of civilization since the Stone Age. Access to a good supply of flint or obsidian from which to make hunting tools was a matter of life or death for Paleolithic man as long ago as 700 000 BP, and increasingly sophisticated use of metals was the basis for the Bronze and Iron Ages, beginning around 5000 and 3000 BP, respectively. Although we now live in the hydrocarbon age, and might be about to enter the hydrogen age, these are still raw materials, and both commodities depend inescapably on mineral products for their extraction, conversion, and usage. Civilization has therefore been tied to mineral extraction from birth, but this relationship has many negative side effects, along with obvious benefits.

As part of the process of articulating the meaning of sustainable development with respect to the minerals industry, the *Mining Journal* recently suggested that extraction of minerals has impacts and dependencies in three main areas: economics, the environment, and society (*Mining Journal* 2000). To this can be added good governance, without which the best-laid plans are ruined (MMSD 2002; Richards 2002, 2003).

Historically, the pursuit of economic benefit, beginning with the first collective farming in the Neolithic age, drove the organized hunt for minerals. This quest both for practical, and later ornamental, riches brought the Romans to Wales and Columbus to the Americas, with little thought other than for personal reward or imperial enhancement. Environmental concerns were almost unheard of, and even the worst social abuses – slavery and forced labour – only began to give cause for widespread concern in the eighteenth century, the Age of Enlightenment. Although slavery was banned first in England in 1774, and the international slave trade outlawed in 1807, human rights abuses continue to this day, and not uncommonly involve the minerals industry (Handelsman *et al.* 2004). Recognition of aboriginal land rights is a late twentieth-century awakening, and the concept of 'prior informed consent', whereby the approval of local peoples must be openly sought before development activities can take place, has only just been incorporated in the World Bank's Extractive Industries Review (2003; to considerable opposition). At the time of writing, two examples of failure or collapse of engagement by mining companies with local communities threaten the very existence of planned mining operations (and millions of dollars of investments) at Tambogrande in Perú, and Esquel in Argentina (e.g. BSR 2003). At issue is the perception that the benefits of mining do not accrue fairly to all stakeholders, but are mostly enjoyed by a select few, often foreigners. Thus, great wealth can be gained by those with the skills to extract ores, but little advantage, and often loss (of land, lifestyle, identity, source of income) is experienced by affected people and sometimes host nations. Even where fair monetary compensation is paid, which may be generous by local standards, local and indigenous people often do not have the skills or capacity to invest that money appropriately to ensure a sustainable future (McMahon & Remy 2001). The result is that problems that companies thought had been addressed (through initial lump-sum cash payments) reappear years later when local people find themselves living in poverty beside a mine that grows ever bigger (McMahon & Remy 2001). An example is the Panguna porphyry copper mine on Bougainville Island, Papua New Guinea, which after 17 years of production was closed by a violent uprising in 1989, and remains closed to this day (Connell 1991; Australian Joint Standing Committee on Foreign Affairs, Defense and Trade 1999).

Although environmental concerns about the impacts of mining have arisen locally at least since Roman times, they first came to prominence on the international stage in the 1960s and 1970s with the appearance of environmental activist groups such as Greenpeace, the World Wildlife Federation, and Friends of the Earth. Today, under the instantaneous scrutiny of the Internet, the environmental performance of the international mining industry is greatly improved. Nevertheless, emissions of gases such as sulphur dioxide from smelting (Fig. 1), and acid mine drainage from waste materials and old workings continue at unacceptable levels, while groundwater contamination by trace metals such as arsenic or mercury, and process wastes such as cyanide, cause international concern (e.g. the Baia Mare cyanide spill in Romania, 30 January 2000).

Realization of the crucial role of governance in sustainable development is again a late twentieth-century/early twenty-first century awakening. It was brought home to me personally when I visited Zaïre (now the Democratic Republic of the Congo, DRC) in 1984, a country endowed with vast mineral wealth (it was then the world's number one supplier of cobalt, and a major supplier of copper and diamonds). But the country's infrastructure was collapsing, the majority of the population was living in poverty, and Zaïre was in debt to the western world to the tune of $4 billion. Coincidentally, the country's President Mobutu Sese Seko (Fig. 2) was amassing a personal fortune of

Fig. 1. The Chuquicamata Cu smelter, photographed in 1993. Modernization of this 1970s plant has reduced SO_2 emissions from 100% prior to 1980 to 20% in 1999, although technology is now available that can capture >99% of emissions (Tilton 2003).

comparable magnitude in foreign bank accounts. Since then, the DRC has collapsed into civil war, with the deaths of at least a million people since 1998 (Oxfam 2000), and is now one of the poorest countries in the world (GDP per capita of $97; WHO 2002). Elsewhere in Africa, the use of diamond mining to fund conflicts in Angola, Liberia, and Sierra Leone is another well-known example of the squandering of countries' mineral endowments for nefarious ends.

Are the impacts from mining always negative? No, of course not, but often the greatest benefits are felt by a very few at the expense of others. This is a form of externalization of costs, whereby profit is made by offloading costs onto others or other people's environments (Crowson 2002; Corson 2002; Ernst 2002;

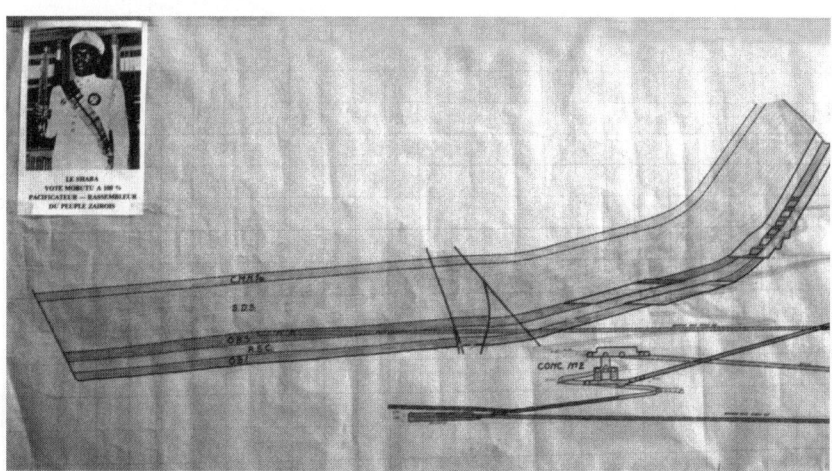

Fig. 2. Cross-section of the Kamoto Cu mine, Democratic Republic of the Congo (formerly Zaïre), in 1984. Note the photograph of then-ruler President Mobutu Sese Seko taped to the map.

Porritt 2002; Tilton 2003; Walker & Jourdan 2003). As argued below, sustainable development must involve the *internalization* of costs, and the paying of fair price to the rightful owners for commodities consumed.

The population and productivity (consumption) explosion

There are two main reasons for this turn-of-the-century global awakening to the impacts of development. First, the Internet now provides an instant window on world events, but, more importantly, the cumulative magnitude of impacts caused by development have increased exponentially as population and productivity increase. Thus, although individual abuses under the Romans' watch 2000 years ago were probably far more severe than today, with a world population of only 300 million (United Nations 1999), globally those impacts were insignificant. (But note that Ruddiman, 2003, has recently argued that agricultural activities involving forest clearances and irrigation may have begun to affect atmospheric CO_2 levels as early as 8000 years ago.) Today, with a world population of 6.3 billion and projected growth to possibly as much as 12.8 billion by 2050 (Fig. 3; United Nations 2003), the cumulative effects are vastly magnified. Couple this with the explosion in productivity rates in the developed world, and the magnitude of human impact really does become visible at a planetary scale. For example, the International Finance Corporation (IFC 2002) estimates that global constant-dollar GDP increased 20-fold in the last 100 years, and 4-fold on a per-capita basis over the last 50 years. However, these numbers conceal the fact that rapid productivity growth mostly occurred only in the developed world, and GDP shrank elsewhere (60% *decrease* in per-capita GDP in Africa since 1900; IFC 2002). Populations did the opposite, however, and whereas growth was moderate in the developed world between 1950 and 1999 (1.3× in Europe, 1.8× in North America), numbers more than tripled in Africa (3.5×) and Latin America and the Caribbean (3.1×; data from United Nations 1999; see also Ernst 2002).

The obverse of production is consumption, and just as production rates have boomed in the developed world, so has consumption. For example, Tilton (2002) estimates that the 20% of world population in developed nations consumes 80% of the world's resources (3% and

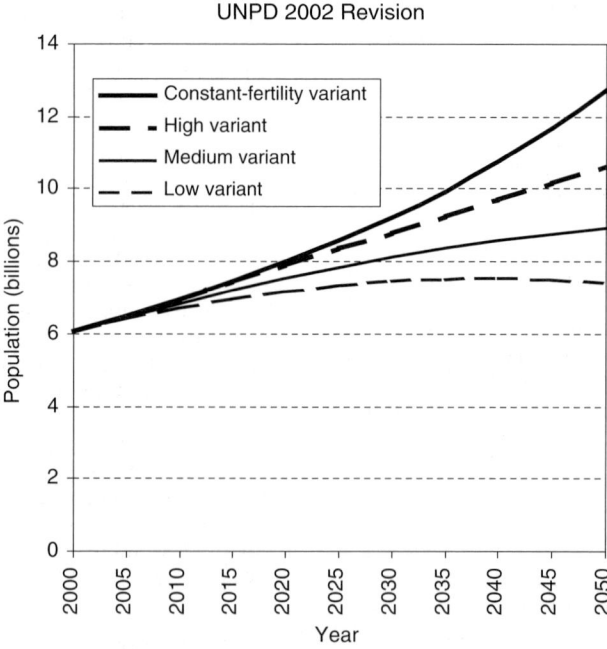

Fig. 3. World population projections 2000–2050 (source: United Nations Population Information Network: http://www.un.org/popin/data.html).

20–25%, respectively, in the United States alone). Increasingly, those resources are being sought from the developing world, and only partly because these less well-explored regions offer new opportunities for discovery. Other less savoury reasons may include lower labour costs, lax environmental regulations, and ineffective or corrupt governments in some countries. Thus, companies can gain competitive advantage by exploiting opportunities in the developing world. It has been argued that this 'apparent competitiveness is actually the result of unfair subsidies extorted from the natural world and disempowered people' (Porritt 2002, p. 80). Meanwhile, First World governments have been known to introduce protectionist measures or subsidies to shield domestic producers and domestic markets from these cheaper sources.

The realization of unsustainable growth

The developed world is slowly coming to realize that the current mode and accelerating rate of consumption is not sustainable. The first indicator that caught global attention was the depletion of stratospheric ozone due to the release of chlorofluorocarbons (CFCs) and hydrochlorofluorocarbons (HCFCs) from refrigeration equipment. Perhaps unfortunately, this problem proved relatively easy to solve, and the Montreal Protocol (1987 and later meetings) has resulted in effective reduction and eventual eradication of CFC and HCFC production. Because of this success, we risk the hubris of assuming we can engineer our way out of all such global environmental threats. However, as is becoming increasingly clear, other challenges such as global warming and the depletion of clean water supplies may be much harder to solve.

More slowly, and more reluctantly, we are also coming to realize that the growth enjoyed by the developed world has been, and continues to be, largely won at the expense of the developing world. This cost is not only measurable in terms of environmental degradation, but also in social impacts such as poverty, conflict, and human rights abuses, which together amount to societal degeneration, the opposite of development.

Who is to blame?

Perhaps no-one is specifically to blame for this course of events – it is just nature's way of survival of the fittest, or in the case of humans, the cleverest and perhaps the greediest (as the *Mining Journal* put it in 2001, 'Equity markets know only two emotions, greed and fear'). However, the natural world also has checks and balances, and no species before ours has managed to sustain unbridled growth for long, because it soon exhausts its life support systems (as argued by Thomas Malthus in 1798). For animals and plants those systems are normally simply food and water, but for humans we are risking our air supply and climate as well.

The issue of blame only applies where there is choice, and because other animals and plants are not conscious of choice, they cannot be blamed for their actions. However, humans *are* conscious of choice, and we are becoming increasingly conscious of the risks we are running. Lawyers will therefore be delighted that blame is a real possibility.

In the case of mineral production, many of the worst environmental and social consequences occur as a result of unregulated artisanal mining (e.g. Kafwembe & Veasey 2001; MMSD 2002) or exploitative state-controlled mining in the developing world (e.g. in Myanmar; MMSD 2002) or under communism, and it would be easy for the developed world to point at its tidy new mines and strict environmental and labour regulations and claim innocence. But this would be disingenuous, because such ruinous activities in the developing world would not be taking place in the absence of poverty and crushing debt, as well as a strong market for the product in the developed world.

Traditional capitalism sees nothing fundamentally wrong with this situation in the international minerals trade: it is a case of supply and demand, and developing countries were presumably not forced to sign on to the generous loans offered by rich nations and institutions (such as the World Bank and the International Monetary Fund) in the 1980s. However, it is unfortunate that a combination of global recession, high interest rates, and declining commodity prices made repayment of these debts increasingly difficult for many countries, who are now trapped in a cycle of poverty.

An alternative view is that the developing world is caught in a poverty trap constructed in such as way that developed world economies grow at the expense of the developing world. For example, the Jubilee 2000 Coalition claims that developing countries paid $13 in debt repayments for every $1 they received in grants in 1998 (http://www.jubilee2000uk.org/jubilee2000/news/imf0904.html).

Who should pay?

It seems an unavoidable truth that the developed world (including large emerging economies such as China, India, and Brazil), as the consumer, should pay for the full cost of its consumption. The problem is that commodity prices do not currently reflect these full costs. They do not, for example, include the long-term costs of environmental damage and remediation, or of social abuses, and fundamentally do not offer a fair price for commodities being taken from other people's land. Furthermore, what income is received often does not benefit those people most affected by the development (McMahon & Remy 2001).

Why is this the case? In a normal marketplace, sellers will cut costs to compete, but not to levels below marginal profitability (or they will soon go out of business and other producers will raise prices again). In the case of raw materials, land owners (if they even own the land) rarely own the mineral rights; these are owned by governments, who under some circumstances can be coerced into selling these rights at unprofitable rates by the necessity to generate revenue at any cost (to service debt, fund wars, and so on), or through greed (bribery and corruption). Thus the real 'sellers', the inhabitants of the land being exploited, cannot dictate the price of their product, and often do not see any of the sale price anyway (taxes and royalties often go straight into central government coffers and are not redistributed to affected populations – this was one of the central grievances of the Bougainville Islanders against the Panguna copper mine; Connell 1991). Again, it is easy to blame poor governance in developing countries for these abuses, but it is often the need to service debts held by the developed world that drives governments to these extremes.

What is a fair price, and how should the developed world pay?

In a market economy, commodity prices should reflect the lowest costs of production, plus a profit margin. Consumers will expect to pay this price and no more; in fact most buyers, be they corporate or private, will seek not only the lowest price but also additional discounts or sales. But in a market where some producers are selling at below cost, or are externalizing costs by disregarding environmental or social obligations, how can a fair price for mineral commodities be set? The answer must be to appeal to a new set of principles, which transcend simple market economics and deal with global citizenship and stewardship. That is not to say that we should throw out the rule books of capitalism, but new financial instruments, such as tax and insurance incentives, and ethical investment funds, can be used to lead the market in more responsible directions (Economist 2001; Corson 2002). Such measures can be used not only to encourage producers to internalize their costs, but also to convince consumers that they should buy these fully costed products, despite the offer of cheaper materials from unsustainable suppliers.

Mineral producers can be separated into two groups: large- and medium-scale (often multinational corporate) producers, and small-scale but widespread and often 'illegal' miners. Economically, these groups can be considered differently and separately, because small-scale mining does not generate a significant proportion of global output, and cannot therefore cause serious market disturbance by undercutting prices.

In the case of small-scale miners, change should come through development aid programs aimed at improving mining methods and conditions, and development of secondary industries that will empower local people in sustainable wealth creation. Development aid should be provided initially from the developed world, but investment should also be made in capacity building in local governments, both to develop and regulate sustainable small-scale mining (Mate 2001; Veiga *et al.* 2001; Pantoja 2003).

In the case of large-scale national and multinational producers, change must come from government and intergovernmental pressure (e.g. taxation, regulation), market pressure (investor responsibility, insurance incentives, corporate responsibility, corporate image), and consumer pressure (NGOs, public sentiment). Some of these issues are already being addressed, and ISO 14001 certification for environmental management is seen as a badge of honour among more progressive companies. For example, BHP Billiton has achieved or is in the process of receiving ISO 14001 certification at all of its major operations worldwide (BHP Billiton 2003). In addition, there are now numerous 'ethical funds' available to investors, many of which set high standards before recommending mining stocks (e.g. EIRIS 1999; the Ethical Trading Initiative: www.ethicaltrade.org). Other measures, such as tax and insurance incentives (e.g. reduced taxation or insurance rates for companies meeting standards such as ISO 14001) have yet to become broadly established, although there would seem to be a strong business case for introducing such schemes in terms of reduced liabilities (IFC 2002).

Measuring and certifying social sustainability has proved harder to implement, however, because of the wide range of social conditions and expectations worldwide (World Bank 1997; IFC 2002). The SA8000 system was developed in 1997 under the auspices of the Council on Economic Priorities Accreditation Agency (CEPAA) to provide an auditable international standard for social accountability (http://www.cepaa.org/SA8000/SA8000.htm). However, SA8000 was mainly developed to safeguard labour rights in manufacturing industries, and has found limited applicability in the mining sector. In the absence of clear rules, many large companies have adopted independent corporate positions on social and environmental responsibility (e.g. Anglo American 2002; Rio Tinto 2002; BHP Billiton 2003), and the US and UK governments recently issued a joint memorandum on voluntary implementation of security and human rights principles (US/UK Governments 2000). Nevertheless, a globally accepted metric for social responsibility (equivalent to ISO 14001) should be an important goal of any accreditation system for holistic sustainable development (Handelsman et al. 2004).

Taking a cue from the agricultural sector, where 'organic' produce successfully competes in many markets despite higher prices, a possibility for the future is the branding of 'green metals' that have been produced from certified mines and processing plants. Branding would enable companies to charge the full internalized costs of metal production while maintaining a competitive advantage. A decision, for example, by a large car-maker to use only green metals may mean a slightly higher unit price, but this could be offset by persuasive marketing appealing to social conscience. Competitors will likely quickly follow suit for fear of being labelled irresponsible. In reality, the costs are likely to be a trivial fraction of the total price of a vehicle. For example, the cost of metals contained in the construction of an average automobile is approximately US$300 (93 kg Al, 18 kg Cu, 13 kg Pb, 985 kg ferrous metals; SAE International 1998). Doubling the price of these metals would only increase the cost of a car by 1–2%, a figure that would be lost in lease or loan repayment rates.

What's in it for industry?

Much of the foregoing discussion has focused on 'sticks', such as taxation, regulation, and external pressure on industry, to coerce it into compliance with society's new and ever changing expectations. However, it is wrong to expect industry to shoulder the costs of these changes alone (i.e. from its profit margin), and ultimately society must be prepared to pay the full internalized costs of the lifestyle it wishes to lead.

Mining companies, like any other industry, are not primarily in business for charitable or philanthropic reasons, and investors will only invest if they can expect a reasonable rate of return (Crowson 2002). It is therefore important to consider the 'carrots' that might be won through the adoption of sustainable development practices.

The first and most obvious carrot is the granting of a licence to operate (IFC 2000; MMSD 2002). The ongoing situations at Tambogrande (Perú) and Esquel (Argentina) show clearly the importance to a company's investment of engaging with communities in order to gain that licence. A dramatic illustration of the impact of failure on the corporate bottom line was the 40% drop in Manhattan Minerals Corp.'s share price on 27 November 2003 amid fears that the Tambogrande project might collapse (Reuters news release, 28 November 2003). If permission to mine is denied, the company has no business, and no development of any kind will take place, so opportunities will have been lost to all.

More positively, however, there are numerous examples of how engaging in sustainable business practices can actually boost a company's competitiveness and business performance. The IFC (2002, p. 72) notes that benefits include 'customer and staff loyalty, product differentiation, resource efficiency, risk reduction, and innovation.' This report also notes that an important aspect of realizing these goals is the move beyond compliance (with standards) to innovation and leadership – or from 'targets to best practice', or 'paternalism to partnership' (McMahon & Remy 2001; MMSD 2002).

In many cases, newly developed technologies, especially for mineral processing and waste treatment, produce positive economic returns through improved efficiency and reduced long-term liabilities (e.g. Chadwick 2001). Some processes may be revenue-neutral, or even negative, but if they reduce the volume of potentially harmful waste then they ultimately will have a positive impact on the balance sheet by reducing long-term liabilities. Innovative uses for waste materials can also result in reclassification as useful product (e.g. McConchie et al. 2002; Harrison 2003; Wiltshire 2003), and proper stabilization of tailings can result in reclassification as inert waste, again reducing liabilities (e.g. Davies & Rice 2002; Vick 2002). Moreover, there are numerous examples of innovative uses of reclaimed mine sites for recreational, ecological, or industrial (brown-site) uses, all of which can return future rent instead of long-term cost and liability (e.g. Tedd et al. 2001).

Many of these benefits are felt in the long term, and may involve actual costs in the short term. This has been a major disincentive in an industry that is driven by quarterly financial reporting, so it was welcome news to read that BHP Billiton has just abandoned this practice. As reported by the *Mining Journal* (2003), the company's chief financial officer, Chris Lynch, stated that 'quarterly reporting ... encourages a short-term focus on what is a long-term business'.

How should 'green revenues' be used?

As noted above, without good governance (at all levels, from local to national and international government, as well as corporate governance), the wealth created by mining will benefit only a select few. In the worst case, the wealth will be stolen by corrupt organizations and public officials (IFC 2000; Economist 2002), but much more commonly, the bulk of wealth accrues mainly to company shareholders. Where some money is returned to governments in taxes and royalties, it is often not reinvested in affected communities (e.g. Newbold 2003). Even where companies adopt their own strict environmental and social policies, they cannot dictate how governments set tax and royalty rates, nor how those revenues are spent. The IFC recommends that companies work closely with governments on these issues, but they can only try to influence decisions (IFC 1998, 2000). Dealing with governments that have strong records of reinvestment into sustainable societies should be one requirement of a 'green metal' certification (see also World Bank 2004).

At issue is the conversion of mineral wealth into other forms of more durable social and environmental capital, which will continue to grow beyond the life of the mine. Thus, Walker & Jourdan (2003) outline mechanisms for investing mineral revenues in value-added 'sidestream' and 'downstream' activities, initially related to the primary minerals industry, but progressively moving out into wider markets. Expansion of activities can support the development of capital goods and service sectors, followed in maturity by 'high-tech' and other knowledge-based industries. Walker & Jourdan (2003) note that a successful transformation to this third phase requires innovation and investment in human resources, with a long-term commitment by government to sustaining development through sound investment of primary revenues (see also Newbold 2003). The authors point to countries such as Sweden, Finland, Germany, Britain, and the USA where this process of technological migration from the original source of wealth (i.e. minerals and other raw materials) has been successful, but suggest that this process is incomplete in other countries such as Canada, Australia, Kuwait, and Saudi Arabia. The latter countries are still operating in an unsustainable mode, with high per-capita income levels supported partly by ephemeral mineral wealth. Far worse off, however, are the majority of developing countries who have never got beyond the very first stage of reliance on natural resource production. As discussed above, debt is much to blame for trapping such countries at this subsistence level of development (Manzano & Rigobon 2001).

Is this a panacea?

Sadly, no – the issues discussed above address only one industry, albeit one that has been held responsible for some of the most acute and dramatic environmental and social abuses in the past. Nevertheless, the minerals industry commonly offers the only realistic opportunity for developing countries to escape the poverty trap. If these countries are allowed, and helped, to grasp that opportunity (e.g. starting with debt cancellation; Hillyard 1998; Jubilee 2000 Coalition), then a win–win situation will result in which both poverty and environmental degradation are reduced. With improved standards of living, reproduction rates will decrease, and consumption rates may become more equitable and even stabilize.

This is the starting point on a road towards a globally sustainable future. But long-term solutions will involve a reassessment of the motivations of human society. Blind competition is nature's way, and it is clear that if we continue to follow that route, natural selection may de-select us as a species. Currently, profits and 'growth' are widely accepted as unquestionable 'goods', and many governments in the developed world also consider population growth to be a necessity (primarily to help pay for future pensions and health care for 'baby-boomers'). Only the anti-globalization community questions these goals, although globalization is potentially the solution, as well as potentially the problem (Economist 2001). The 'goodness' of profits and growth depends on how those benefits are used, and whether they are acquired at the expense of others (e.g. by externalizing costs, that is, making others pay for the environmental or social impact, or through not paying fair value for the resources extracted). If they are simply used to improve the lifestyles of a few, then this must ultimately be viewed as unsustainable. If, instead, at least part of those benefits is invested in the well-being of wider society, then this may herald the start of a 'new deal' for humanity.

A longer term view would ask where humans hope to be in 100 years' time – or 200 or 500 years. And even that timescale is trivial compared with the life expectancy of most species on this planet (measured in millions of years). As Porritt (2002, p. 75) has said, we should be 'living on this planet as if we intended to go on living here forever'. Unfortunately, however, we are in danger, not of becoming the first species to cause its own decline (many species become victims of their own success), but of becoming the first species consciously to *watch* this taking place. Alternatively, we could be the first species consciously to resolve its survival problems, and manage a sustainable future for itself. Moving to another planet is not a realistic solution; instead, we must learn to live within our means and in harmony with our world. This may involve placing collective, global objectives above purely capitalist goals, and making sure that users, polluters, and abusers really do pay.

A number of people have influenced my thinking about sustainable development and the minerals industry, including Ian Thomson, Anthony Hodge, David Evans, and Simon Handelsman. I would particularly like to thank two anonymous reviewers for their constructive criticisms of a draft of this paper, and appreciate and value frank discussions of these issues with Mike Harris, James Macdonald, John Thompson, Sir Mark Moody-Stuart, John Menzies, Richard Morgan, and John Tilton. They did not all agree with all of the points made here, and my acknowledging their willingness to share their views does not imply that they gave their full endorsement to what I have written. The Society of Economic Geologists and the Geological Society, London, are thanked for sponsoring my participation in the 'Sustainable Minerals in the Developing World' conference, and I thank Mike Petterson for his invitation to attend that meeting. This research was supported by a Discovery grant from the Natural Sciences and Engineering Research Council of Canada.

References

ANGLO AMERICAN. 2002. *Towards Sustainable Development*. Anglo American plc, London.

AUSTRALIAN JOINT STANDING COMMITTEE ON FOREIGN AFFAIRS, DEFENSE AND TRADE. 1999. *Bougainville: The Peace Process and Beyond.* http://www.aph.gov.au/house/committee/jfadt/bougainville/BVrepindx.htm.

BHP BILLITON. 2003. *Health Safety Environment and Community Report 2003.* BHP Billiton plc, London.

BSR (BUSINESS FOR SOCIAL RESPONSIBILITY). 2003. *Minera El Desquite Report Esquel, Argentina.* Business for Social Responsibility, San Francisco, CA.

CHADWICK, J. 2001. Benefits of biotechnology. *Mining Environmental Management*, July 2001, 18–19.

CONNELL, J. 1991. Compensation and conflict: The Bougainville copper mine, Papua New Guinea. In: CONNELL, J. & HOWITT, R. (eds) *Mining and Indigenous Peoples in Australasia.* Sydney University Press, Sydney, Australia, 55–75.

CORSON, W. H. 2002. Recognizing hidden environmental and social costs and reducing ecological and societal damage through tax, price, and subsidy reform. *The Environmentalist*, **22**, 67–82.

CROWSON, P. 2002. Sustainability and the economics of Mining – what future? *Minerals and Energy*, **17**, 15–19.

DAVIES, M. P. & RICE, S. 2002. Dry stack filtered tailings. *Mining Environmental Management*, January 2002, 10–13.

ECONOMIST. 2001. Economic man, cleaner planet. *The Economist*, 29 September 2001.

ECONOMIST. 2002. Tragically undermined. *The Economist*, 1 June 2002.

EIRIS (ETHICAL INVESTMENT RESEARCH SERVICE). 1999. *Corporate Environmental Policy, Management and Reporting.* EIRIS, London.

ERNST, W. G. 2002. Global equity and sustainable Earth resource consumption requires super-efficient extraction – conservation – recycling and ubiquitous, inexpensive energy. *International Geology Review*, **44**, 1072–1091.

HANDELSMAN, S. D., SCOBLE, M. & VEIGA, M. 2004. Human rights and the minerals industry: challenges for geoscientists. *Exploration & Mining Geology*, **12**, 5–20.

HARRISON, D. 2003. Industrial mineral products from mining waste. *Mining Environmental Management*, January 2003, 6–7.

HILLYARD, M. 1998. *Cancellation Of Third World Debt.* UK House of Commons, Economic Policy and Statistics Section, Research Paper 98/81.

IFC (INTERNATIONAL FINANCE CORPORATION). 1998. *Doing Better Business Through Effective Public Consultation and Disclosure.* International Finance Corporation, Washington, DC.

IFC (INTERNATIONAL FINANCE CORPORATION). 2000. *Investing in People: Sustaining Communities Through Improved Business Practice.* International Finance Corporation, Washington, DC.

IFC (INTERNATIONAL FINANCE CORPORATION). 2002. *The Environmental and Social Challenges of Private Sector Projects: IFC's Experience.* International Finance Corporation, Washington, DC.

JUBILEE 2000 COALITION. http://www.jubilee2000uk.org/jubilee2000/main.html.

KAFWEMBE, B. S. & VEASEY, T. J. 2001. The problems of artisan mining and mineral processing. *Mining Environmental Management*, November 2001, 17–21.

MALTHUS, T. R. 1798. An essay on the principle of population. Edited and introduced by HIMMELFARB, G., *On Population* (1960). Modern Library, New York.

MANZANO, O. & RIGOBON, R. 2001. *Resource Curse or Debt Overhang?* National Bureau of Economic Research, Cambridge, MA, Working Paper 8390.

MATE, K. 2001. Capacity-building and policy networking for sustainable development. *Minerals & Energy*, **16**, 3–25.

MCCONCHIE, D., CLARK, M., DAVIES-MCCONCHIE, F. & FERGUSSON, L. 2002. The use of Bauxsol technology to treat acid rock drainage. *Mining Environmental Management*, July 2002, 12–13.

MCMAHON, G. & REMY, F. 2001. Key observations and recommendations: A synthesis of case studies. *In*: MCMAHON, G. & REMY, F. (eds) *Large Mines and the Community*. World Bank, Washington, DC, 1–38.

MINING JOURNAL. 2000. Lasting impressions. *Mining Journal*, 17 November 2000, 335, 386.

MINING JOURNAL. 2001. Realpolitik? *Mining Journal*, 27 July 2001, 337, 54.

MINING JOURNAL. 2003. BHPB dumps quarterlies. *Mining Journal*, 5 December 2003, 12.

MMSD (MINING, MINERALS, AND SUSTAINABLE DEVELOPMENT). 2002. *Breaking New Ground (the Report of the MMSD Project)*. Earthscan Publications, London.

NEWBOLD, J. 2003. Social consequences of mining and present day solutions – Region II in Chile highlighted. *Sustainable Development*, **11**, 84–90.

OXFAM. 2000. *A Forgotten War – A Forgotten Emergency: The Democratic Republic of Congo*. Oxfam, London. http://www.oxfam.org.uk/what_we_do/issues/conflict_disasters/forgotten_drc.htm.

PANTOJA, F. 2003. Small scale gold mining in Latin America: Problems and solutions (abstract). Sustainable Minerals in the Developing World, 24–25 November 2003, The Geological Society, London.

PORRITT, J. 2002. From the fringe to the mainstream: The evolution of environmental and social issues in private sector projects. *In: The Environmental and Social Challenges of Private Sector Projects: IFC's Experience*. International Finance Corporation, Washington, DC, 75–84.

RICHARDS, J. P. 2002. Sustainable development and the minerals industry. *Soc. Economic Geologists Newsletter*, **48**, January 2002, 1, 8–12.

RICHARDS, J. P. 2003. The minerals industry as a driver for sustainable human development. *In*: ELIOPOULOS et al. (eds) *Mineral Exploration and Sustainable Development* (7th Biennial SGA Meeting, Athens, August 24–28). Millpress, Rotterdam, 3–6.

RIO TINTO. 2002. *Sustainable Development*. Rio Tinto plc., London.

RUDDIMAN, W. F. 2003. The anthropogenic greenhouse era began thousands of years ago. *Climatic Change*, **61**, 261–293.

SACHS, J. D. & WARNER, A. 1995. *Natural Resource Abundance and Economic Growth*. National Bureau of Economic Research, Cambridge, MA, Working Paper W5398.

SAE INTERNATIONAL. 1998. Life cycle inventory of a generic U.S. family sedan. Overview of results USCAR AMP project: Nr. 982160, Total Life Cycle Conference and Exposition, Graz, Austria, 1–3 December 1998.

TEDD, P., CHARLES, J. A. & DRISCOLL, R. 2001. Sustainable brownfield re-development – risk management. *Engineering Geology*, **60**, 333–339.

TILTON, J. E. 2002. On borrowed time? Assessing the threat of mineral depletion. Unpublished Senate Distinguished Lecture notes.

TILTON, J. E. 2003. *On Borrowed Time? Assessing the Threat of Mineral Depletion*. RFF Press, Washington, DC.

UNITED NATIONS. 1999. *The World at Six Billion*. United Nations, New York.

UNITED NATIONS. 2003. *World Population Prospects, the 2002 Revision: Highlights*. United Nations, New York.

US/UK GOVERNMENTS. 2000. *Voluntary Principles on Security and Human Rights*. U.S. Department of State/U.K. Foreign & Commonwealth Office, London and Washington, DC.

VEIGA, M. M., SCOBLE, M. & MCALLISTER, M. L. 2001. Mining with communities. *Natural Resources Forum*, **25**, 191–202.

VICK, S. G. 2002. Stability aspects of long term closure for sulphide tailings. *Mining Environmental Management*, January 2002, 19–23.

WALKER, M. & JOURDAN, P. 2003. Resource-based sustainable development: an alternative approach to industrialization in South Africa. *Minerals and Energy*, **18**, 25–43.

WEBER-FAHR, M. 2002. *Treasure or Trouble? Mining in Developing Countries*. International Finance Corporation, Washington, DC.

WHO (WORLD HEALTH ORGANIZATION). 2002. *Democratic Republic of the Congo: Country Emergency Profile*. World Health Organization, Geneva.

WILTSHIRE, J. C. 2003. Tailings recycling research. *Mining Environmental Management*, January 2003, 8–12.

WORLD BANK. 1997. *Advancing Sustainable Development: The World Bank and Agenda 21*. Washington, DC, World Bank, Environmentally Sustainable Development Studies and Monographs Series No. 19.

WORLD BANK. 2003. *Striking a Better Balance: The Final Report of the Extractive Industries Review*. World Bank Group, Washington, DC.

WORLD BANK. 2004. *World Development Report 2005: A Better Investment Climate for Everyone*. World Bank Group, Washington, DC.

WRIGHT, G. & CZELUSTA, J. 2003. Mineral resources and economic development. Prepared for the Conference on Sector Reform in Latin America, Stanford Center for International Development, November 13–15, 2003.

Sustainable river mining of aggregates in developing countries

D. J. HARRISON[1], S. FIDGETT[2], P. W. SCOTT[3], M. MACFARLANE[4],
P. MITCHELL[5], J. M. EYRE[3] & J. M. WEEKS[6]

[1]*British Geological Survey, Keyworth, Nottingham, NG12 5GG, UK*
(e-mail: djha@bgs.ac.uk)
[2]*Alliance Environment and Planning Ltd, 276 High Street, Guildford, Surrey, GU1 3JL, UK*
[3]*Camborne School of Mines, Redruth, Cornwall, TR15 3SE, UK*
[4]*Corporate Citizenship Unit, Warwick Business School, The University of Warwick,*
Coventry, CV4 7AL, UK
[5]*Green Horizons Environmental Consultants Ltd, PO Box 137, Bexhill,*
East Sussex, TN40 1YA, UK
[6]*WRc-NSF Ltd, Henley Road, Medmenham, Marlow, Bucks, SL7 2HD, UK*

Abstract: Throughout the developing world, river sand and gravel is widely exploited as aggregate for construction. Sediment is often mined directly from the river channel and makes an important contribution to the national demand for aggregates. However, instream mining, if not carefully controlled, can cause significant damage to the river and its associated biota, and to the adjacent land, as well as creating conflict with other users of the river. The economic and environmental geology of river sand and gravel mining in developing countries is poorly known and there is little knowledge available to inform existing regulatory strategies. Research work on selected river systems in Jamaica and Costa Rica has generated a considerable amount of new information on resources and sediment budgets, on market and supply options, on the physical, biological and social impacts of extraction, and on best-practice legislative and mineral planning issues. A methodology has been developed for effective control of instream sand and gravel mining operations including a Code of Practice, which regulators can use for examining and reconciling the conflicting claims of sand and gravel extraction and the environment.

Sand and gravel is an essential source of aggregate for the construction industry. In developing countries it is often exploited directly from the active channels of river systems, where it is easily extracted and usually requires almost no processing other than size selection. It is often considered a renewable resource. However, in-channel or near-channel mining of sand and gravel inevitably alters the sediment budget of a river system, and may substantially alter channel hydraulics. The impacts of such mining on farmland, river stability, flood risk, road and bridge structures, and ecology are typically severe. The environmental degradation may make it difficult to provide for the basic needs (water, food, fuelwood, communications) of communities naturally located beside the river.

Despite the importance of this extractive industry in most developing countries, the details of its economic and environmental geology are not fully understood and therefore do not adequately inform existing regulatory strategies. The main problem is therefore a need to strengthen the general approach to planning and managing these resources. Compounding the problem is the upsurge of illegal extractions along many river systems. There is therefore a need to foster public awareness and community stewardship of the resource.

This paper summarizes the outputs from a study of river mining in developing countries (Harrison *et al.* 2003) and illustrates how an improved understanding can be used to provide an effective mechanism for the control of river mining operations, to protect local communities, to reduce environmental degradation, and to facilitate sustainable use of the natural resource base.

The research project has involved a multidisciplinary team of researchers from the UK who have carried out field investigations on river systems in Jamaica and Costa Rica. The British

Geological Survey has led the project activities with specialist input from the Camborne School of Mines (University of Exeter), the Corporate Citizenship Unit at the University of Warwick, Alliance Environment and Planning Ltd, Guildford, WRc-NSF Ltd, Marlow and Green Horizons Environmental Consultants Ltd, Bexhill. Overseas partners are principally the Mines and Geology Division of the Ministry of Land and Environment in Jamaica and the Instituto Costarricense de Electricidad in Costa Rica. The project has been funded by the UK's Department for International Development (DFID) as part of their Knowledge and Research (KAR) programme. This programme constitutes a key element in the UK's provision of aid and assistance to less developed nations.

With a better knowledge of available resources, supply options, market and production issues, of the physical, biological and social impacts of extraction, and of best-practice legislative and planning guidelines, a range of solutions are suggested for reducing the problems associated with river mining of sand and gravel. A Code of Practice has been developed, based on best practice guidance, which regulators can use to work within their preferred strategic framework for examining and reconciling the conflicting claims of sand and gravel extraction and the environment.

Aggregate supply issues in Jamaica

In developing countries, aggregate production is often on a small scale, with many small quarries operated by different companies or informal groups. The granting of permissions to extract is less formal, or even non-existent, with illegal mining commonplace. The supply of aggregates is inconsistent and the quality of products may be extremely variable (Alvarado-Villalon et al. 2003).

In Jamaica there are two main sources of aggregates (Scott et al. 2003). These are sand and gravel from alluvium found in, and adjacent to the present rivers, especially the active stream channels, and limestone, which occurs throughout Jamaica, although extraction is mostly in the north and west of the island (Fig. 1). Resources of sand and gravel are located mainly in rivers flowing to the south coast, principally the Rio Minho and Yallahs rivers. Other rocks, which might appear initially to be good resources of aggregates, are either deeply weathered or remote from the major centres of population. Annual aggregate production is estimated to be around 4 million tonnes, made up approximately half from limestone and half from river sand and gravel. There is additional production of marl (a softer limestone) used for fill purposes.

The extraction of sand and gravel in Jamaica is directly from the active stream channel. Extraction from riverbanks, floodplains and older fluvial terrace deposits does not usually occur. The processing varies from a simple screen on an A-shaped frame to remove oversize particles (Fig. 2) to processing plants in which coarse gravel is processed by crushing, washing and screening to make a range of products (Fig. 3). Limestone is extracted mostly by ripping, blasting not commonly being used, and processed by crushing and screening to produce crushed rock aggregate.

The price of aggregates in Jamaica, excluding transport costs, is similar throughout the country. River sand, however, is a premium product, because of its relative scarcity and commands a higher price. Further sand is manufactured by crushing gravel and as a byproduct of crushing limestone, when it is known as 'stone dust'.

With few exceptions, the mobile and fixed plant used in aggregate production in Jamaica are very old; breakdowns are frequent, with consequent loss of production. Considerable overmanning and spare mobile plant capacity is one strategy used to overcome these problems. Finance is available through investment banks for renewal of plant, but the provision of extraction licences for only a short period of time (1–2 years) is a disincentive to re-finance and re-equip.

Production costs for aggregates in Jamaica are very low and a small-scale sand and gravel unit with simple processing is likely to be very profitable. Successful development of an aggregate operation, which requires external finance for a new processing plant, would, however, need a sustained market over a long period of time.

Aggregates must be fit for the purpose for which they are used and should be produced to acceptable quality standards. Although some standards exist in Jamaica, there is often little attention paid to them in the production of aggregates, and aggregate products, such as concrete blocks.

Resource evaluation

In Jamaica, aggregates are extracted directly from the channels of many of the major rivers, but little is known about deposit volume or geometry, size distribution of the material in the deposits, physical characteristics of the material, or the volumes extracted in each river system. In order to determine the quantity and quality of

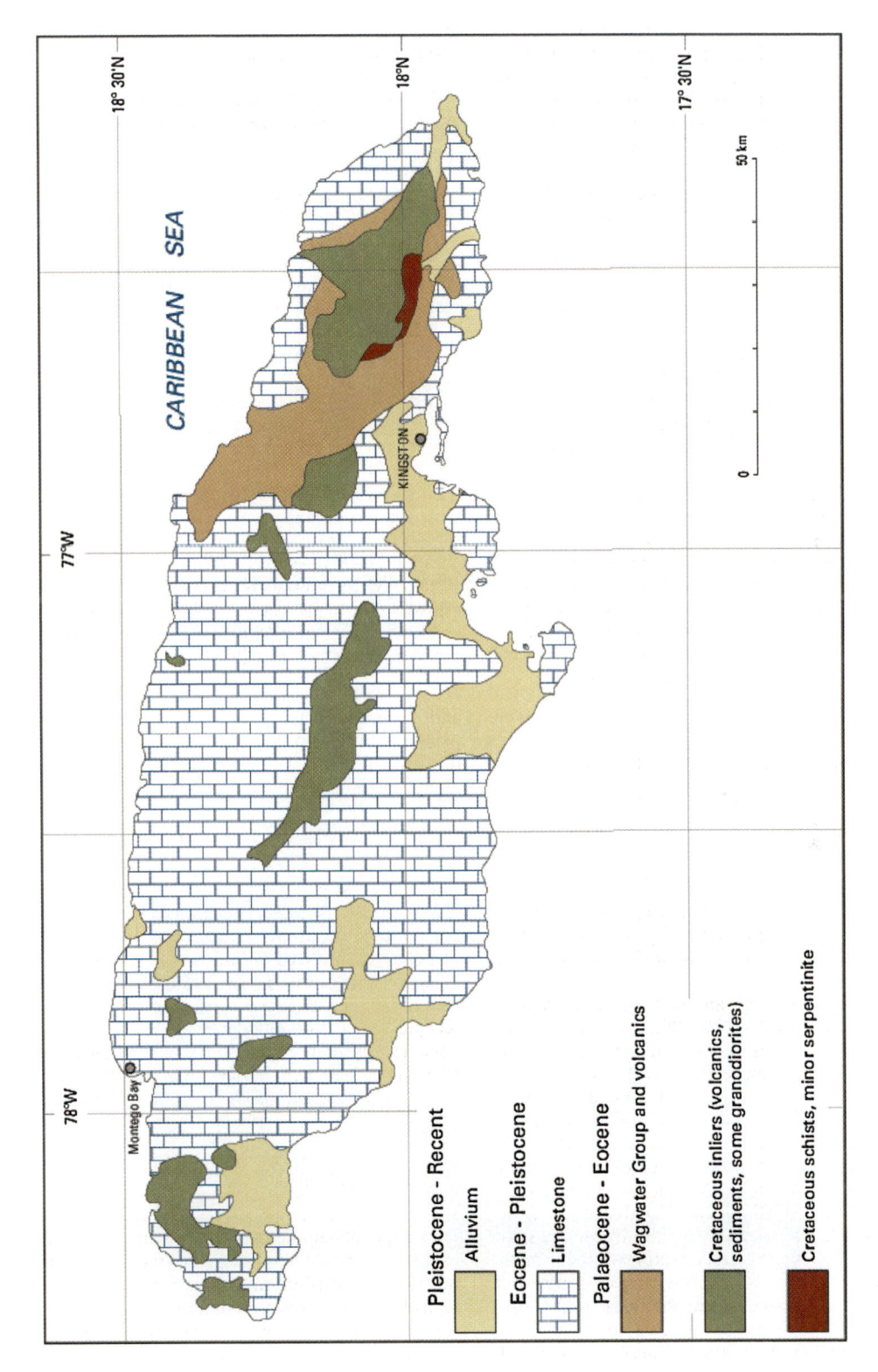

Fig. 1. Simplified geological map of Jamaica.

Fig. 2. River mining, Yallahs River, Jamaica.

sand and gravel available in the rivers chosen as trial sites for the project, resource mapping was undertaken to enable the lower Rio Minho and Yallahs fan-delta to be divided into areas of relative suitability for aggregate exploration (Farrant et al. 2003).

A combination of walk-over geological surveys and studies of aerial photographs were used to show the distribution and relationships of the major sand and gravel bodies and a programme of trial pitting by back-hoe provided ground-truth information and material for particle size analysis. The resulting resource maps form an essential starting point for planning the future supply of aggregates in Jamaica, particularly for sourcing supplies of sand.

In the lower Rio Minho valley, two principal levels of alluvial sand and gravel-bearing deposits were recognized: those forming the broad coastal plain, classified as the Second Terrace, and those related to the meander belt of the Rio Minho incised into the coastal plain and termed the First Terrace. The Second Terrace consists of up to 15 m of silt and clay and is not considered to be a resource, but the First Terrace deposits consistently contain around 5 m of

Fig. 3. River aggregate processing plant, Yallahs, Jamaica.

sand and gravel overlain by up to 2 m of silt. The First Terrace deposits therefore form a major sand and gravel resource, adjacent to the river channel deposits. It is estimated that about 5.25 million tonnes of sand and gravel has been removed from the Rio Minho since 1980, when extraction started. Almost all extraction, has been from river channel deposits (Fig. 4) and the First Terrace deposits are largely unworked.

The Yallahs River drains the southern flank of the Blue Mountains and the river sediments form a lobate fan-delta covering over 10 square kilometres. Coarse poorly sorted sand and gravel resources are inferred to be present throughout the fan-delta. The sediments thicken rapidly downstream and are over 150 m thick near the coast. About 4.4 million tonnes of aggregate has been produced since 1990. Extraction is concentrated in the main channel, which is now incised up to 6 m below the level of the fan-delta surface.

Alternative materials

There is particular concern about the environmental and social effects of in-stream sand and gravel mining and thus there is a need to know to what extent alternative materials can augment or replace natural sand and gravel from this source. In Jamaica, these alternative sources include marine sand and gravel, manufactured sand, river terrace deposits and recycled aggregates (Harrison & Steadman 2003).

Marine sand and gravel

In many countries marine dredged sand and gravel make important contributions to aggregates supply. There is currently no production of marine aggregates in Jamaica, but the environmental concerns of in-stream mining and pressures on land-based development require any marine sources to be seriously investigated.

The Yallahs fan-delta, which forms a major resource of sand and gravel onshore, extends offshore for a further 2 km. It is likely that the sediments offshore are similar in lithology and thickness to the land-based deposits currently extracted at Yallahs. Offshore sands and gravels may, therefore, offer a realistic alternative to the onshore deposits at Yallahs. If adequate resources are proved, and if sufficient financial investment is available, then it is likely that extraction of marine aggregate resources at Yallahs will be viable, provided environmental and development issues are identified and managed effectively.

Manufactured sand

Sand can be manufactured by the crushing and screening of solid rocks, and quarry fines are often sold as fine aggregate for the construction industry. In Jamaica, 'stone dust' is produced during limestone quarrying and is sold as 'sand' for use in concrete and asphalt. Manufactured sand is also produced by crushing river gravel at several of the larger river mining sites. Such crushed rock sands are, however,

Fig. 4. Scour in river bed caused by extraction of river channel deposits, Rio Minho, Jamaica.

largely at the limits, or fall outside of the specifications of fine aggregates for concrete manufacture, and poor quality blocks and other products result. The grading and particle shape of the manufactured sand could be greatly improved by investment in additional processing plant, such as 'Barmac' impactor rock crushers and washing plants.

River terrace deposits

In most countries, fluvial terrace deposits are the preferred source of sand and gravel aggregates as in-stream mining is restricted by environmental concerns and legislation. Resource mapping of alluvial deposits in the lower Rio Minho and Yallahs fan-delta has shown that thick sequences of sand and gravel are developed beneath the agricultural lands bordering the active channels (Fig. 5). These deposits are not currently worked; extraction is largely from the river channels. Development of the terrace deposits would provide good quality aggregate materials and reduce the impacts arising from river extraction. Extraction sites could be effectively restored to nature conservation, tourism or agriculture.

Secondary aggregates

Many mineral wastes fulfil the technical requirements to substitute for primary aggregates and many governments are now encouraging greater use of mineral wastes as aggregates. Concrete, bricks and asphalt, for example, may be crushed and screened to produce secondary aggregates that can be used in construction. Recycling concrete rubble not only reduces the environmental and social impacts of new aggregate production, but also avoids impacts associated with disposal. In Jamaica, the degree of recycling of mineral waste materials into secondary aggregates is currently very small, although some construction and demolition waste is used as fill. The scope for secondary aggregates is often small in developing countries, as regeneration may not produce much useful material and development is mainly on new land.

Environmental impacts of river mining

The extraction of sand and gravel from river channels can have significant physical and biological effects on the river and floodplain environment. The significance and extent of the environmental effects will depend upon a range of factors including the location of the extraction site, the nature of the sediments and the sediment budget, fluvial processes, flood risk, the design, method, rate, amount and intensity of extraction, and the sensitivity of habitats and assorted biodiversity, and other river users in the locality.

To enable the organizations responsible for authorizing extraction to evaluate the nature and scale of the effects and in order to decide whether a proposal can proceed, it is necessary that an adequate assessment of the environmental

Fig. 5. River terrace sand and gravel, Rio Minho, Jamaica.

effects be carried out. A typical environmental impact assessment will include the following stages:

- initial environmental evaluation and screening to determine if a formal Environmental Impact Assessment (EIA) is required;
- definition of environmental issues (the scope of the EIA);
- identification of potential impacts;
- assessment of impacts;
- prediction of impacts;
- identification of mitigation or compensation measures and monitoring requirements;
- consideration of residual impacts and overall conclusions (whether it is acceptable);
- monitoring;
- auditing and review.

Studies in Jamaica (Macfarlane & Mitchell 2003) showed that the most important environmental impacts of river mining activities are physical and aquatic:

- *physical*: dust arising from extraction and processing, riverbank erosion, shifting of river course, flooding and damage to infrastructure such as roads and bridges;
- *water*: contamination with suspended solids and flooding.

The remaining impacts are largely secondary ones that arise from these impacts.

Ultimately, the ability of quarry operators and government agencies to identify, generate, implement and enforce appropriate mitigation and enhancement measures will be affected by a number of factors. The critical factors observed in Jamaica include the legal status of the quarry operators (Fig. 6); regulatory development; government agency coordination; agency funding and corruption. These interrelated factors are in no way unique to Jamaica and are typical of the quarrying sector and the regulatory environment of many developing countries.

Ecological impacts

Sand and gravel extraction in rivers will impact on the aquatic ecosystem and it is important to understand the disturbance resulting from such activity, which may lead directly to impacts on organisms and indirectly to impacts on their habitat.

The effects of sand and gravel mining at the trial sites in Jamaica were investigated (Weeks *et al.* 2003) using indices of biological diversity obtained by biological monitoring of each river (the Rio Minho and Yallahs). Samples, both upstream and downstream of river mining activity, were taken for biological assessment aimed to indicate general changes in biodiversity at each locality. Biodiversity statistics were used to assess species richness or diversity within or between samples.

The results show considerable biological diversity between the samples in the two rivers and indicate major disturbances in the overall biodiversity of the benthic macroinvertebrate fauna as one moves downstream along the watercourses sampled. The greatest change in faunal assemblage occurs in the immediate vicinity and immediately

Fig. 6. Illegal sand and gravel mining, Jamaica.

downstream of sand and gravel mining operations. Biological recovery from these activities is likely to be slow following the removal of the riverbed, which results in massive habitat loss for the benthic fauna. It is anticipated that recolonization of these habitats will be slow, resulting in areas of lowered biodiversity.

Social impacts of river mining

It is recognized that sand and gravel extraction from rivers may cause significant harm to the environment. What is not as widely understood is the effect river mining may have on communities and river users in the locality (Fig. 7). Social impact assessment (SIA) complements the EIA process and is used to promote such understanding. As a result, many countries now require EIAs to address social issues, and impact assessments are now commonly integrated, bringing together environmental and social issues.

There are four principal social 'impact groups': socio-health impacts; socio-cultural impacts; socio-economic impacts; and socio-livelihood impacts. Many impacts are secondary impacts that arise from impacts on the physical and water environment. However, others are a hybrid of environmental, social and economic parameters that can improve social conditions (such as income and local revenue generation) or can reduce it (such as loss of access to clean water).

In Jamaica, the social impacts of river mining were investigated (Macfarlane & Mitchell 2003) at both the local level (by assessing the response from stakeholders such as quarry operators, local residents, customers, local businesses, council and police officials) and the strategic level (through interviews with government bodies and agencies). The most important social impacts were considered to be:

- socio-health: worker, vehicle and pedestrian access;
- socio-cultural: corruption, political collusion;
- socio-economic: road congestion, road degradation, bridge failure (all negative impacts); enhanced services, employment and infrastructure (all positive impacts);
- socio-livelihoods: reduced fishing stocks, loss of land.

Planning guidelines

It is important to ensure an effective and consistent supply of aggregates to meet the needs of developing economies in order to support economic activity and the level of investment in the physical infrastructure, utilities and buildings required to raise living standards. At the same time it is equally essential to ensure that the provision of aggregates does not result in undue environmental and social degradation, such that harm to the population or the environment outweighs benefits derived from such development. The proper control of the effects of river mining is important in achieving improvements in living standards and quality of life.

Fig. 7. Water for drinking and washing, Rio Minho, Jamaica.

The most appropriate forms of regulation are those that allow the operators of quarries to invest on a reasonably secure basis, to put in place proper operational and environmental infrastructure, and over a period of time that reflects the ability to make a reasonable return on the investment. This generally means granting licences for longer periods, but equally means that the level of control exercised by the regulators needs to be carefully considered and individual sites monitored, with effective enforcement of licence conditions to prevent or, when it fails, to remedy harm to the environment. It is clear that at present in Jamaica neither of these basic requirements is met.

Minerals policy

In order to achieve adequate control of river mining, the basic requirements of a system of control needs to be in place. These include:

- an effective legislative framework that allows the exercise of control over river mining;
- an administrative structure that gives clear responsibility and authority to the respective agencies involved in such regulation;
- well-informed policies and proposals for the development and control of river mining and other forms of land use as part of an integrated approach to land use and the environment;
- a shared commitment to the effective operation of the system by government, industry and other stakeholders.

Mineral planning

Effective control of river mining should be based on a planning system that provides:

- a clear and consistent form of forward planning, reconciling policies and proposals for the existing and future use of land determined with reference to carefully considered social, economic and environmental priorities and objectives;
- a system of licensing that requires a consistent level of information to be provided prior to individual decisions on applications to mine or quarry, measured against the policies and proposals set out in the forward planning context (Fidgett 2003);
- a commitment to effective monitoring and enforcement that is adequately resourced and based on clear information on licence requirements and on the occurrence of illegal mining or of activity inconsistent with a licence or permit.

Technical solutions

In addition to the policy and legislative framework that will help shape the future of river mining activity, actions on the ground that help address the environmental and social impacts of river mining are:

- *proactive technical solutions* – these deal with the determination of the best practicable means of undertaking the extraction, processing and restoration of mining or quarrying operations and are focused at the planning stage, prior to the granting of a licence;
- *reactive technical solutions* – these deal with the determination of the best practicable means of mitigating existing mining operations, addressing extraction, processing and restoration. They are focused at the operational stage and during the review or reissue of a licence.

At a practical level, to support the recommendations of this research (Fidgett 2003), a Code of Practice has been developed that is intended to be a guide to each of the potential practical issues raised by river mining, summarizing its key characteristics and setting an objective and performance indicator to help guide the actions of both quarry operators and regulators on the ground. These guidelines are intended to provide a flexible framework that any country can adopt within its own regulatory system.

Code of Practice

The recommended mitigation and enhancement measures within the Code of Practice (Macfarlane & Fidgett 2003; Mitchell 2003) are based on the results of literature review and observation of best practice in river mining and the mining industry generally. The result is an issue-led systematized Code of Practice for the mitigation and enhancement of the respective negative and positive impacts of river mining.

A total of 26 key issues are categorized according to the type of river mining activity with which they are associated, and with the impact group to which they belong. The intervention level for each issue is defined as either principally strategic (policy level) or technical (operational level) and best practice mitigation or enhancement measures are suggested.

Key issues

The 26 key issues of river mining activity covered in the Code of Practice are:

(1) Physical contamination – avoidance of impacts on river quality
(2) Chemical contamination – avoidance of contamination of the water environment
(3) Water supply – avoidance of loss or reduction of water resources
(4) Boreholes – prevention of groundwater depression
(5) Release of wastes – prevention of the release of polluting wastes
(6) Flooding – avoidance of flooding, reduction of flood risk
(7) Waste management – minimize waste production, responsible waste disposal
(8) Dust – dust control and reduction
(9) Fumes – minimize impacts of vehicle fumes
(10) Noise – minimize noise and vibration
(11) Minimize incision of the river channel
(12) Bank erosion – prevention of erosion of the river bank
(13) Livestock – minimize loss of livestock during mining operations
(14) Cultivated land – avoidance of loss of agricultural land
(15) Fishing stock – avoidance of adverse impacts on river ecology
(16) Employment displacement – avoidance of displacement of economic activity
(17) Housing – avoidance of adverse impacts on housing and housing land
(18) Sustainable transport – minimize impacts of traffic
(19) Workplace accidents – minimize worker accidents
(20) Plant and buildings – avoidance of poorly maintained, dangerous sites
(21) External appearance – avoidance of obtrusive operations
(22) Natural resources – avoidance of harm to important nature or environment
(23) Restoration and after-use – promote land restoration to beneficial after-use
(24) Community liaison – promote effective dialogue between operators and communities
(25) Quarry plans – provision of quarry plans for effective resource development
(26) License – avoidance of illegal operations, full compliance with licence conditions

The Code of Practice provides step-by-step guidance on how river mining should be conducted in order to minimize conflicts with other river and land users and optimize the use of natural resources. These guidelines are also intended to provide a basis for assessment to help ensure that sufficient information is produced to enable an EIA/SIA to be carried out, and within a strategic framework that provides a system for examining and reconciling the conflicting claims on land and river.

Conclusion

In many countries, sand and gravel from mining river channels makes an important contribution to the national demand for aggregates. A reduction of instream mining could therefore potentially have a severe effect on the supply and economics of aggregates unless similar material is available from alternative sources. Therefore, the suitability and economic viability of alternatives must be carefully examined, along with an assessment of the environmental, biological and social impacts associated with their production.

Instream mining reduces the pressure to work floodplain terrace deposits of sand and gravel and avoids conflicts of land use (e.g. agriculture and housing). Other benefits include local production of aggregates and reduction in long-distance road transportation from more remote deposits. However, these benefits need to be balanced against the negative impacts of river mining, which may be severe. Aggregate extraction from river channels, if not carefully controlled, can cause significant damage to the river and its associated biota, and to the adjacent land, as well as creating conflict with other users of the river. There will be cases where the social and biophysical environment are too sensitive to disturbance to justify the extraction of aggregate and unless these issues can be satisfactorily resolved, extraction should not normally be allowed. However, the extent to which environmental and social impacts are acceptable is relative, and in part depends on the wealth of the country. In less economically developed countries there may be less funding and little political will available to address such impacts.

A consideration of all factors associated with river mining has resulted in the development of a methodology for effective control of sand and gravel mining operations and a Code of Practice, which provides a basis for determining the potential effects of river mining and for identifying mitigation measures.

References

ALVARADO-VILLALON, F., HARRISON, D. J. & STEADMAN, E. J. 2003. *Alluvial Mining of*

Aggregates in Costa Rica. British Geological Survey Commissioned Report CR/03/50N.

FARRANT, A. R., MATHERS, S. J. & HARRISON, D. J. 2003. *Geology and Sand and Gravel Resources of the Lower Rio Minho Valley and Yallahs Fan-Delta, Jamaica.* British Geological Survey Commissioned Report CR/03/161N.

FIDGETT, S. 2003. *Planning Guidelines for the Management of River Mining in Developing Countries with Particular Reference to Jamaica.* Alliance Environment and Planning Ltd.

HARRISON, D. J. & STEADMAN, E. J. 2003. *Alternative Sources of Aggregates.* British Geological Survey Commissioned Report CR/03/95N.

HARRISON, D. J., MACFARLANE, M. ET AL. *River Mining: Project Summary Report.* British Geological Survey Commissioned Report CR/03/198N.

MACFARLANE, M. & MITCHELL, P. 2003. *Scoping and Assessment of the Environmental and Social Impacts of River Mining in Jamaica.* MERN Working Paper No. 32, University of Warwick.

SCOTT, P. W., EYRE, J. M., HARRISON, D. J. & STEADMAN, E. J. 2003. *Aggregate Production and Supply in Developing Countries with Particular Reference to Jamaica.* British Geological Survey Commissioned Report CR/03/29N.

WEEKS, J. M., SIMS, I., LAWSON, C. & HARRISON, D. J. 2003. *Assessment of the Ecological Effects of River Mining in Two Jamaican Rivers.* BGS Commissioned Report CR/03/162N.

Markets for industrial mineral products from mining waste

PETER W. SCOTT[1], JOHN M. EYRE[2],
DAVID J. HARRISON[3] & ANDREW J. BLOODWORTH[3]

[1]*Camborne School of Mines, University of Exeter in Cornwall, Tremough Campus, Penryn, Cornwall TR10 9EZ, UK (e-mail: P. W. Scott@ex.ac.uk)*
[2]*Wardell-Armstrong International, Wheal Jane, Baldhu, Truro, Cornwall TR3 6EH, UK*
[3]*British Geological Survey, Keyworth, Nottingham NG12 5GG, UK*

Abstract: The composition of mining waste varies according to the nature of the mining operation and many other factors, but where the same mineral is extracted from a similar style of metalliferous or industrial mineral deposit or coal, the waste usually has similar characteristics. There are many potential sources of industrial minerals from mining waste. Waste from one mine may be a byproduct or coproduct in a mining operation elsewhere. Much technical research work on mine waste utilization, for example studies on slate waste, has included a manufacturing process. The waste is invariably an inferior material compared with an industrial mineral from a primary resource for the manufacturing process. Successful markets have not been found. Four scenarios are proposed where an industrial mineral product made from mining waste may be marketed successfully. These are a bulk product for a local market made with minimal or no processing; a low unit value product and a cost-effective alternative source of a mineral for local industry; an industrial mineral commodity traded nationally or internationally; and extraction of a high unit value rare mineral. Making an industrial mineral product from mining waste and successfully marketing it should involve minimal processing of the waste consistent with the value of the mineral product.

The extraction of all solid minerals produces excess unwanted material. This waste occurs as overburden and interburden in open pits, from shafts, driveways and other excavations in underground mines, and as residues from the quality selection of products and mineral separation. Sometimes waste is back-filled in open pits or replaced underground, which may give a positive environmental benefit. Within a mine site, some waste can be used for haul roads, landscaping or for screening the visual impact of the industry. In many mining operations much waste remains in piles or tailings lagoons, taking up valuable space. There is always a positive cost in storage and containment of waste, and its position can sterilize future reserves, requiring its removal to access them. Even with mining operations where the whole rock is potentially useable, such as an aggregate quarry, a significant amount of waste is generated, produced as a consequence of size reduction alone. Extended processing to separate a mineral creates much fine waste. With low-grade ores, virtually all of the material extracted accumulates as waste at and around the mine site.

The waste from many mining activities contains rocks or minerals that have potential as a source of further saleable industrial minerals. The mining and separation of the primary mineral already may have created a concentration of another potential mineral product, or the bulk waste itself may have the properties desirable for an alternative use. This paper reviews possible sources of industrial minerals from mining waste. The importance of establishing markets and making a full economic appraisal are emphasized. Several examples described are from the UK, where much consideration of benefits in the utilization of mining waste has been given. Some others are drawn from Harrison *et al.* (2001, 2002*a, b*). Secondary recovery of the primary mineral, such as occurs when improved technology becomes available to treat an ore, is not discussed, and the uses of downstream products, such as slags, are not included. These activities can also contribute to maximizing the utilization of a resource.

Benefits from waste utilization

Waste accumulates during the life of a mine, and much, if not all, remains after extraction has long since ceased. Often the waste pile requires

From: MARKER, B. R., PETTERSON, M. G., MCEVOY, F. & STEPHENSON, M. H. (eds) 2005. *Sustainable Minerals Operations in the Developing World.* Geological Society, London, Special Publications, **250**, 47–60.
0305-8719/05/$15.00 © The Geological Society of London 2005.

ongoing monitoring for geotechnical stability. The waste may contain potentially hazardous elements or minerals, which could be released into the wider environment through leaching, or erosion by water or wind. Vast quantities of mine and quarry wastes are accumulated worldwide. Even by 1978, a total of 23 billion tons was estimated to have been amassed in the United States with a further 360 000 tons added per day (Aleshin 1978).

The cost of handling and storage of waste represents a financial loss to a mining company, estimated at around 1.5–3.5% of total costs depending on the mineral (Symonds Group 2001). A direct benefit can be made by generating revenue from a saleable product from the waste as long as the expense of making it exceed any marketing, handling, processing and transport costs. Indirect financial benefits can be gained if creating the product from waste enables the size of the waste heap to be significantly reduced in size, rendering it less visible, or in another way providing benefit to the surrounding environment. Space creation in a tailings lagoon through marketing a product made from its contents at marginal cost could be important to a mining company if it offsets the need for extending or building a new structure. An income from a product made from the waste can reduce the financial burden on those left with the responsibility for ongoing monitoring of a waste heap after mine closure.

In recent years, the introduction of legislation and taxation designed to maximize the use of resources, and/or to regulate waste piles, has provided an incentive for mineral producers to consider further the extraction of a mineral product from a waste stream or existing heap. For example, the European Commission has an initiative for the management of waste resulting from prospecting, extraction, treatment and storage of mineral resources (European Commission 2002). It proposes the extension of existing controls on landfill to mining waste. As a result, conditions dealing with permits, upgrading of standards, financial security for closure and aftercare are to be introduced. The landfill requirement for pretreatment is to be considered for adoption to mining waste to promote its minimization, segregation and recovery. In the UK, a levy of £1.60 per tonne on primary aggregates was introduced in April 2002 specifically to encourage recycling and the use of mine waste as a secondary source of aggregates.

For governments in developing countries, there can be a socio-economic advantage in encouraging the utilization of mine waste for economic benefit. A working mine is attractive and an indigenous population often congregates around the site, with many gaining employment there (Harrison et al. 2001) during its active life. However, when the mine closes, the local population often sinks into poverty. The production of a saleable mineral product from the waste, even on a small scale, can provide continuing employment, sustaining at least some of the population for as long as a market for the product exists and the resource remains.

Composition and utilization of mining waste

The composition of mining waste varies with the type of mining operation, the mineral extracted, the nature and structure of the ore body, the overburden, and many other factors related to the geology and mining of the ore. When examined closely, the waste from every mine is unique. However, where the same mineral is extracted from the same or a similar style of deposit, there are some common features in the waste. This applies to many metalliferous ores, coal, and several industrial mineral extractive operations. Table 1 gives details of the nature of the waste, the current practice for disposal, and some potential markets for a large number of different types of mineral deposits, especially industrial minerals. The information is derived from literature and the authors' collective experiences. Some are expanded in Harrison et al. (2001). With several of the types of mineral extraction shown in Table 1, the waste material may already constitute a coproduct, but elsewhere the waste is not used. For example, maximizing the efficiency of extraction, and hence maximizing the revenue, from a bulk commodity such as aggregates or limestone in the UK involves making a wide range of products (e.g. Scott and Dunham 1984). For example, crusher scalpings can be used for bulk fill applications, and limestone fines for agricultural lime. Limestone considered as waste at one site is a product at another. There are well-known examples of dual or multiple industrial mineral products from a single deposit, for example, mica, feldspar, quartz, spodumene from pegmatites, North Carolina (e.g. Kauffman 1994; Tanner & Van Dyk 1994); silica sand and kaolin from a kaolinized arkose (Bristow 1989) from some of the types of mineral operation listed. The same geological process that concentrates one mineral to yield an economically viable reserve, often creates concentrations of other potentially exploitable minerals or further, different grades of the primary mineral

Table 1. Waste in mining operations and potential markets

Type of mining operation	Nature of waste	Current practice for disposal	Potential markets and industrial minerals	Notes
Sand and gravel	(a) Silt and/or clay fines (<63 μm)	Tailings pond	Bricks (ceramic or adobe), tiles, heavy mineral concentrates (e.g. Ti minerals, zircon), asphalt filler	1
				2
	(b) Oversize boulders	Remain in quarry	Armour stone	
Crushed rock aggregate (excluding limestone and dolomite)	(a) Silt and/or clay fines (<63 μm)	Tailings pond	Asphalt filler, bituminous paints filler	2
				3
	(b) Flakey particles	Waste pile	Pipe bedding, bulk fill, filler in paints and plastics	5
				3
	(c) Primary crusher scalpings	Waste pile, or used for haulage roads	Bulk fill	5
Limestone and dolomite aggregates	(a) Primary crusher scalpings	Waste pile, or used for haulage roads	Bulk fill, agricultural lime	5
				6
	(b) Secondary scalpings and other fines	Waste pile or tailings pond	Agricultural lime, asphalt filler, all-weather sports surface	6
				2
				4
Limestone/marble dimension stone	(a) Large blocks (>1 m³)	Waste pile	Marble tiles, terrazzo tiles, armour stone, aggregate	8
				7
	(b) Cut stone waste	Waste pile	As for large blocks, white mineral filler, small-scale lime production	9
				10
Igneous and metamorphic rock dimension stone	(a) Large blocks (1 m³)	Waste pile	Armour stone, aggregate	11
	(b) Other waste, excluding overburden	Waste pile	Aggregate	12
Slate	Substandard slatey rock and processing waste	Waste pile	Aggregate, mineral filler	13
				14
Silica sand	(a) Hydrosizer underflow	Waste pile	Building sand	
	(b) Spiral concentrator residue	Waste pile	Building sand, TiO_2 and/or other heavy minerals	
	(c) Clay/silt fines from washer and attrition scrubber	Tailings pond	Bricks, asphalt filler, all-weather sports surface	1,2,4
Bentonite	Substandard or out-of-specification bentonite	Waste pile or not extracted	Pond/reservoir lining, landfill site lining, animal bedding	

(*Continued*)

Table 1. *Continued*

Type of mining operation	Nature of waste	Current practice for disposal	Potential markets and industrial minerals	Notes
Common clay for ceramic bricks and tiles	(a) Substandard clay	Waste pile or not extracted	Adobe bricks, pond/reservoir/landfill lining	15
	(b) Mis-shaped bricks or tiles	Waste pile	Recycle into clay body	20
Kaolin (sedimentary)	(a) Sand from processing	Waste pile or tailings pond	Silica sand, feldspar, building sand, TiO_2 and/or other heavy minerals	
	(b) Silt/clay underflow from hydrocyclone	Tailings pond	Fine quartz for ceramics mica	20
Kaolin (residual from granite)	(a) Sand from processing	Waste pile	Silica sand, feldspar, mica, Nb/Ta minerals, cassiterite, zircon, other rare element minerals	16
	(b) Silt/clay underflow from hydrocyclones	Tailings pond	As for (a)	16
Pegmatites	In quarry waste from selective mining	Remains in quarry	Other pegmatite minerals (e.g. quartz, mica, feldspar, rare lithophile metals (Li, Be etc.)	17
Phosphate (sedimentary)	(a) Limestone interburden	Waste pile or backfilled	Agricultural lime, aggregate	6,12
	(b) Crusher scalpings	Waste pile	Low-grade direct application phosphate fertilizer	
	(c) Clay/silt fines	Tailings pond	Low-grade direct application phosphate fertilizer	
Phosphate (carbonatite)	(a) Sand-sized material from flotation plant	Tailings pond	Rare lithophile metals (e.g. REE, Nb, Ba), fluorite	20
	(b) Clay/silt fines	Tailings pond	Low-grade direct application phosphate fertilizer	
Coal	(a) Overburden and interburden clay	Backfilled or waste pile	Ceramic/refractory kaolin clay	
	(b) Overburden and interburden sandstone	Backfilled or waste pile	Silica sand, building sand, bulk fill	
	(c) Coarse residue from washing plant	Waste pile	Secondary coal recovery	18
	(d) Fine residue from washing plant	Tailings pond	Clay for bricks and tiles	
	(e) Burnt colliery spoil	Waste pile	Unbound pavement aggregate	

Gold in quartz veins	From heap leach pile	Tailings dump	Quartz powder for ceramics, glass, or low-grade filler for paints, plastics	19
Epithermal gold	(a) In quarry waste from selective mining	Remains in quarry	Alunite, kaolin for white cement	19
	(b) From heap leach pile	Tailings dump	Low-grade mineral filler	16
Alluvial gold	Gravel and sand from processing	Remains on site	Sand and gravel	3
Base metal mining (pyrite absent, carbonate absent)	Fine residue from flotation	Tailings pond	Low-grade mineral filler	16, 19
Base metal mining (pyrite absent, carbonate present)	Fine residue from flotation	Tailings pond	Agricultural lime	19
Base metal mining (pyrite present; e.g. volcanogenic massive sulphides)	Fine residue from flotation	Tailings pond	Sulphuric acid	21
Mississippi Valley-Type Pb/Zn	Fine residue from gravity separation or flotation	Tailings pond	Fluorite, barite or calcite	
Talc (dolomite host)	Fine residue from flotation	Tailings pond	Dolomite (mineral filler) agricultural lime	
Talc (ultrabasic host)	Fine residue from flotation	Tailings pond	Low-grade magnesite, or low-grade mineral filler, agricultural lime	
Cement	(a) See limestone and common clay (b) Waste clinker	Waste pile	Bulk fill, recycle into cement kiln	6
Potash	Residue from flotation	Tailings pond	Halite for road de-icing	
Ti-mineral sands	Sand	Left on site as disturbed ground	Zircon, garnet, monazite, silica sand	
Gemstones in alluvial sands	(a) Sand (b) Gravel	Left on site as disturbed ground Left on site as disturbed ground	Building sand Concreting aggregate	
Bauxite	Below grade bauxitic clay	Waste pile	Ceramic or refractory clay, red/yellow mineral pigment	

(Continued)

Table 1. Continued

Type of mining operation	Nature of waste	Current practice for disposal	Potential markets and industrial minerals	Notes
Iron ore	Below grade ore	Waste pile	Red mineral pigment, secondary recovery of Fe ore	
Fluorite and barite veins	Residue from jigging/tabling/flotation	Tailings pond	Galena, calcite (mineral filler)	22
Uranium in sandstone	(a) Tailings from crushing	Tailings pond	Silica sand	

Note: Every extractive operation is unique. Thus, the waste will by unique in its nature and properties.
The list of potential markets takes no account of practical difficulties or economic viability. In many extractive operations, where a major amount of waste is generated, it may be suitable for backfilling and/or reinstatement of land for agricultural or other purposes. In low-wage economy countries, secondary recovery of the primary mineral product may be possible using hand sorting or other labour-intensive separation or mineral selection procedures.
[1] Blending with more clay may be needed.
[2] Processing needed to remove clay.
[3] Grinding and classification needed.
[4] If sufficient clay is present, it may make a suitable material for making a sports surface after compacting, rolling and levelling.
[5] Will need to be capable of being compacted.
[6] Requires further crushing.
[7] Requires further crushing and screening. Most suitable uses are as exposed aggregate (marble chippings) or as filter stone (e.g. for sewage). May require secondary blasting.
[8] Will require crushing and screening to approx. 20–40 mm.
[9] Source material required to be free from other quarry waste and overburden and to be homogeneous and white or very pale coloured.
[10] Requires a kiln nearby, but stone may decrepitate during burning.
[11] Requires crushing and screening. May require secondary blasting for initial size reduction.
[12] Requires crushing and screening.
[13] Requires crushing and screening, but flakey particle shape is likely to restrict potential market.
[14] Requires crushing and grinding. Only suitable for use where colour is not important.
[15] Requires crushing and grinding.
[16] Depends on original rock composition.
[17] Assumes pegmatite as currently mined is only a source of one mineral. Other minerals can be selected by hand sorting.
[18] Hand sorted.
[19] Toxic metal levels may be high and restrict uses.
[20] Will require much further processing.
[21] Chemical plant nearby required.
[22] Possibly other metalliferous minerals (e.g. sphalerite).

within the same deposit. An example is fluorite, barite and calcite from wastes in Mississippi Valley Type deposits (Harrison et al. 2001) (Table 1).

Fine-grained waste from the separation of industrial minerals from clastic sedimentary deposits, such as silica sand, and sand and gravel operations, may contain concentrations of heavy minerals. Shaffer (1996) showed that the considerable amounts of fine-grained waste discarded (estimated at 15–20%) during processing of sand and gravel from glacial deposits in Indiana, USA, contain heavy mineral concentrations averaging about 2.5 wt%. These are made up of ilmenite, garnet, zircon, magnetite and traces of other economically important minerals, primarily chromite, monazite and gold. The heavy minerals present in waste from processing silica sand used in the float and container glass industries in the UK are dominated by Ti-rich minerals (ilmenite, rutile, 'leucoxene') and zircon, irrespective of the geological setting and age of the deposit. These minerals concentrate in the waste stream, arising from spiral classifiers or hydrosizing.

Successful utilization of mining waste may be prevented if it is likely to create a potential or actual environmental hazard. This could apply to the use of waste from base metal mining, where trace elements, such as cadmium or selenium, might be present and potentially leachable, for example, when the waste is used in bulk fill applications. Waste containing sulphides from tin, copper and other base metal mining cannot be considered an aggregate source because of oxidation and subsequent acid production, causing reactions with cement and concrete decay. Characterization of the waste and assessing all the technical difficulties in its use is an essential prerequisite in addition to establishing a market for the product. A waste pile should be investigated using the same rigour as that associated with the evaluation of a primary mineral deposit.

Allowing for the likely variation within a waste pile, its perception as an inferior material compared with that from a primary source, and the likely absence of waste having the ideal properties for a given use, it would appear obvious that limiting the extent of processing to low-cost, simple procedures, such as size selection or size reduction, is an important factor for sustaining a business using mining waste. An example of slate waste, given below, illustrates this. Further evidence is found in the large volume of literature on mine waste utilization, including papers in a series of biennial conference reports from 1968 to the early 1980s (e.g. Cutler & Nicholson 1970; Heins & Geiger 1970; Mindess & Richards 1970; Collings et al. 1974; Collins 1978; Collings 1980; Rampacek 1980). Many of these studies refer to the use of mine waste in manufacturing processes, for example in ceramics, refractories, concretes, bricks, and glass. These are high capital-intensive activities and success would rely on a consistent and uniform supply of the waste and a market for the products, competing with similar products from primary raw materials. Others relate to usage in road construction and as aggregates. Some of the technical studies are very detailed. Most of these studies show that, although mine waste can be used as a substitute for a primary mineral, the resulting product's properties are either inferior or further additional research is required to modify the properties of the waste or to overcome technical difficulties. Under such circumstances, a direct economic benefit is unlikely to be achieved.

Scenarios for profitable utilization of mining waste

As with all industrial minerals, a product made from mining waste is only of value if there is a market. The consumer will compare a potential product from waste with the quality of an existing industrial mineral extracted from a primary source. Without a favourable comparison, the waste is unlikely to find a market. Other technical, economic or legislative factors may be involved before persuading the consumer to take the product from waste, and these may positively or negatively distort the market. However, waste is often very variable, especially that found on an existing heap accumulated over a large number of years. Even simple processing of the waste will require an outlay of capital with associated risk before a product can be made. Markets for industrial minerals are constantly changing. Industrial minerals from new sources are continually becoming available to replace existing sources, and advancing technology results in changes in patterns of use of raw materials. Thus, any industrial mineral product made from mining waste needs to be considered in the same way as a new source of a primary industrial mineral. The priority is to establish a market for the potential product.

It is proposed that there are four scenarios where an industrial mineral product made from mining waste may be marketed successfully.

(1) The waste becomes a bulk product for a local market with minimal or no processing.

(2) The waste is a low unit value product and a cost-effective alternative source of a mineral for local industry.
(3) The waste is a source of an industrial mineral commodity, traded nationally or internationally.
(4) The waste contains a high unit value, rare mineral for which there is a high demand internationally.

All four scenarios involve minimal processing of the waste consistent with the value of the mineral product. That is, if the mineral is of high unit value (i.e. Scenario 4 and for some minerals in 3), some extended processing can be justified; but, for Scenario 1 only size separation by screening at most, is possible. Note that product manufacture, beyond mineral grain separation, is not part of any scenario. The place value (i.e. the location of the waste) is critical in establishing a market for minerals in Scenario 1. It is also important for those in Scenario 2. The mineral product in Scenarios 1 and 2, and sometimes 3 are likely to be useful in a basic industry of a country (e.g. construction, cement, ceramics). Examples of each scenario are discussed below. In some cases, the free market situation is distorted by other factors, such as indirect financial incentives.

Scenario 1: aggregates from china clay (kaolin) extraction in Cornwall, southwest England

China clay extracted from kaolinized granite is the UK's largest mineral export. It is a major world source of kaolin, with around 2.4 million tonnes annual production. For every million tonnes of china clay, two million tonnes of rock, six million tonnes of sand and gravel sized material, and one million tonnes of feldspar, mica and quartz (mixed fines) and overburden are produced (Kessell 2001). Around 20 to 24 million tonnes of waste are produced annually and there is a stockpile of 450 to 600 million tonnes (Kessell 2001; see also Whitbread et al. 1991). The waste rock is mainly unkaolinized granite, tourmalinized granite, greisened granite, porphyries, and tourmaline and quartz-rich vein material, sometimes with a hydrothermal breccia containing granite, porphyry, metasediments and a tourmaline rich matrix. Sulphides are not present. The sand is mainly quartz, with some unkaolinized feldspar, minor mica and tourmaline. There are some technical constraints in using the waste in concrete, including poor workability and increased cement content to achieve equivalent strengths compared with some primary aggregates (Whitbread et al. 1991).

The waste supplies 45% of the three million tonne local aggregate market in Cornwall (Kessell 2001), the china clay area being situated near the middle of this rural county. The aggregates have acceptable properties for most applications. Further markets for this waste exist only outside southwest England, and although the china clay areas are not directly located at the coast, they are only 10 to 15 km from the port of Par, from where much of the china clay is exported. The considerable demand for aggregates in southern and southeastern England (the area surrounding London) has resulted, since 1999, in the successful marketing of aggregates made from this waste to these regions, transported by ship, and also to Germany (Kessell 2001). The imposition by the UK Government of a tax of £1.60 per tonne on primary aggregate production from 1 April 2002 enables aggregates from china clay waste, which is not subject to the tax, to compete even more favourably with crushed rocks and sand and gravel from primary sources nearer to southern and southeastern England. It is predicted (Kessell 2001) that the market for aggregates transported by ship from Par extends to 400 to 440 nautical miles (i.e. towns on both sides of the whole of the English Channel and southern North Sea), and the target is to achieve sales of five million tonnes per annum within 5 to 10 years. The primary aggregate tax, which represents 35% of the transport costs by sea to the River Thames estuary (London), is predicted to enable a thriving market for aggregates from china clay waste to develop.

Scenario 2: clay waste

The waste from several mineral operations in Table 1 is rich in clay, which will be plastic to a varying degree. Successful utilization of such material would involve selling it to an existing brick or tile manufacturer, who would blend or otherwise incorporate it into the clay body. The major brick manufacturers in the UK are now actively using mineral wastes for blending so that they can achieve a wide range of products and make more efficient use of their primary raw material sources (Smith 2002). In a study of clay-rich waste material from a silica sand operation in Costa Rica, Harrison et al. (2002a) showed that an indirect financial benefit to the silica sand company could be achieved by selling the kaolin clay waste to local brick companies at very low cost. An economic assessment showed that a small negative net present value for the clay waste used in this way is more than offset by the saving of space within the tailings lagoon. A greater demand for the waste would exist if there was a larger brick-making industry in Costa Rica. Unfortunately, the kaolin clay waste is insufficiently white for higher value ceramic uses such as pottery.

Scenario 2: armour stone from dimension stone waste

The production of dimension stone create a large amount of waste, and the presence of the waste is ubiquitous around any of this type of extractive operation. The waste varies in size, but much is often in large (100 kg to 10 tonnes) irregular blocks. It is generally unsuitable for primary use because of its shape, the presence of irregularities in colour or texture, or planes of weakness. A small amount may find use as a secondary source of dimension stone, for dressed stone, masonry or walling stone, or for artisanal sculptures and artwork, the irregularities in colour or texture possibly being of benefit for the latter applications. When suitably located, dimension stone waste can find a market as armour stone or rip-rap, or in the case of smaller blocks (200 mm maximum; Smith 1999, p. 261) for use in gabion cages.

The overall market in the UK for armour stone is in excess of 1.5 million tonnes per year (Rees-Jones & Storhaug 1998), mostly for marine uses. With sea level rise and the associated need to protect high-value coastal properties and amenities, the demand for such stone worldwide is likely to increase. Since 1991, the region around the coastal town of Larvik in Norway has become a major source of armour stone (Rees-Jones & Storhaug 1998), with sizes ranging from 20 kg to 20 tonnes. This rock is waste from the major dimension stone operations in that area, and the market penetration has included providing material for coastal defences in the UK (2 million tonnes since 1990; Woodstock-Deering 2001). One advantage of using waste dimension stone, extracted principally by means of drilling and diamond cutting, is the absence of incipient fractures, which might be present in rock obtained through blasting.

Waste produced in dimension stone quarries may be suitable as crushed stone aggregate, and in special circumstances as a raw material for a higher value industrial mineral. An example of the latter is the successful use of waste Carrara Marble from Italy as a source of white marble for high-brightness, white pigment manufacture. In the UK, using dimension stone waste as a source of aggregates is not normally permitted by the Mineral Planning (Permitting) Authorities (Smith 1999, p. 133) as the large volume of traffic associated with aggregate production, when compared with building stone extraction, is normally an unacceptable factor when granting permits for dimension stone quarries. Legislation, driven by environmental impact, can sometimes prevent the use of mining waste.

Scenario 2: slate waste

A good example of efforts to utilize mine waste is the vast amount of accumulated material from slate mining in the UK (400–500 million tonnes; Whitbread et al. 1991), especially that from North Wales. A good market (40 000 tonnes per year) has developed for pulverized slate waste granule and powder products (Richards, Moorehead & Laing Ltd 1995).

Technical studies on Welsh slate waste have shown that, amongst other uses, it can be used for making an expanded lightweight aggregate, fused with additives for ceramic flooring, autoclaved with lime to make bricks, for the recovery of aluminium metal, and melted, spun and woven as fibrous insulation material (e.g. Gutt et al. 1974; Crockett 1975; Anon. 1989; Harries-Rees 1991). Similar studies have been undertaken on slate waste from North America (e.g. Vermont slate waste; Mackenzie & Horiuchi 1980) These processes all require a major investment of capital, even though the cost of the raw material is very low. Slate waste may not be the most appropriate raw material for these applications, as ideally, manufacturers would seek the most suitable minerals for their process, rather than chose to take something that is inferior, solely because it is a waste material. Slate waste is not the ideal material to make an expanded lightweight aggregate, as its bloating capacity is far lower than that of many clays, and it usually requires a higher temperature for reaction. If aluminium metal could be extracted in large quantities economically from any aluminosilicate mineral, it is likely to have been achieved several centuries ago. A manufacturing plant was established in the 1970s to produce an expanded lightweight aggregate from slate waste in Wales, but it closed down after a short period of time. None of these applications have proved to be commercially viable (Richards, Moorehead & Laing Ltd 1995).

Slate waste is a source of bulk fill for the construction of highway embankments and other similar areas where it is available locally, and a road sub-base product (Type 1) can be made and successfully marketed (Whitbread et al. 1991).

There are other examples of successful use of locally available mine waste in the construction of highways (e.g. Pettibone & Kealey 1972; Emery & Kim 1974) and as crushed rock aggregates and sand (e.g. Anon. 2000). The waste's convenient location and its zero mining cost are likely to be a major consideration in its use. For slate, costly extended treatment is not an economic option for sustained utilization of the waste.

Scenario 3: fluorite and barite from base metal mining

Hydrothermal vein and replacement ore deposits in Derbyshire and the Northern Pennines in England (Dunham 1948, 1974; Ford 1976) were exploited for lead and zinc, and occasionally iron and other metals, through hundreds of years and especially in the 19th and early 20th centuries. Through most of the 20th century until recently there were profitable operations exploiting fluorite and barite from the very dispersed waste piles, supported by further primary extraction of fluorite and barite from veins and replacement deposits, to supply the steel and artificial cryolite, and oil-well drilling industries, respectively, in the UK and Europe (Mason 1978). A further example of the successful extraction of barite from waste is that from the sediment-hosted, stratiform copper–lead–zinc mine at Tynagh, Ireland. Tailings were reprocessed for barite, also for use in drilling muds (Down & Stocks 1977).

The current fluorite market is dominated by extraction from primary sources in China (2.25 million tonnes in 1999; Stockwell 2001), with much less production from Mexico (561 265 tonnes in 1999), South Africa (217 540 tonnes in 1999), Mongolia (153 693 tonnes in 1999) and elsewhere. In each of these countries, production costs are low compared with Europe and North America, which has resulted in a significant reduction in fluorite production from countries in these regions in recent years (e.g. Burger 1991). Fluorite prices (chemical grade or acidspar) are only around US$180 per tonne (*Industrial Minerals* February 2004) delivered from China to US Gulf ports. Any fluorite produced from mining waste could compete with that from the primary producers in an international market.

Scenario 3: feldspar and mica from pegmatites

Although pegmatites are an important source of feldspar and mica in their own right, these minerals are often ignored when metal ores, such as tin and tantalum, and lithium minerals are exploited from this source. The feldspar and mica waste accumulates as sand and silt sized particles from gravity separation processes. The individual minerals are already liberated, and a further processing stage, such as air or froth flotation, can create a product. Quartz may also be present, and available for concentrating as a high-purity silica product. Sulphide minerals, which would require a separate additional processing stage, do not usually occur in significant amounts in most pegmatites.

An example is waste from former pegmatite mining for tin ore in Uis, Namibia (Richards 1986; Harrison *et al.* 2002*a*, *b*). Around 75 million tonnes of waste, made up of quartz, feldspar and mica, with very minor amounts of other minerals, is accumulated from the low-grade primary ore, which had a mill-feed of 1 250 ppm Sn (Richards 1986). Iron oxide bearing minerals are especially low in amounts. It is a significant resource of feldspar and mica, and possibly of a mixed feldspar–quartz product (cf. several feldspar products contain significant amounts of quartz; Scott & Power 2001). Processing using flotation without further grinding yields a feldspar product, and a mica concentrate can be produced by air classification. World feldspar production is around 12 million tonnes per annum, with major amounts from some industrialized nations (e.g. USA, Japan, Italy, France, Germany and Spain) (Stockwell 2001), while others have no commercial feldspar resources (e.g. UK). There is significant international trade, with nearly one-third of world production being exported from the country of origin. Mica production is similarly dispersed through several countries and there are significant exports. Ceramic-grade feldspar is US$112–165 per tonne FOB Durban (*Industrial Minerals* February 2004). Mica is a more expensive product, at more than US$200 and up to US$1000 depending on the extent of processing (grinding). Even though transport costs to a port would be a major expense in supplying feldspar and mica from Uis, as it is around 200 km to the nearest port at Walvis Bay, these comparable prices from the adjacent country of South Africa would appear to give scope for further investigation of the economics of production of feldspar and mica from this waste material. Governments of developing countries such as Namibia are keen to develop further their mining activities and give favourable taxation arrangements. They are also becoming aware of the environmental damage created by past mining activities, especially in areas where there is an emerging tourist interest, and thus may provide additional support for initiatives to recover minerals from waste. Quartz as a source of silica for the glass industry is unlikely to be a potential product from any location remote from a glass works, such as at Uis, as its unit value is too low.

Scenario 4: rare minerals from pegmatites

The significance of this example is in providing a small revenue for a mining company or other organization responsible for the management, treatment or removal of a waste pile, or for supporting an indigenous population remaining after a primary mining activity has ceased. It does not reduce a waste pile significantly in volume. Pegmatites often contain rare minerals and crystal forms worthy of extraction as mineral specimens from large-sized mining waste. Gemstones (e.g. tourmaline, beryl) may also be present. Extraction of colourful minerals found in pegmatites in Brazil takes place in areas of former pegmatite mining. Attractive minerals found in the waste are used in the production of ornaments, jewellery and other artefacts (Rao *et al.* 2002). These are marketed internationally.

Scenario 4: tantalum minerals

This metal is in high demand for use in capacitors in mobile telephones. Its price escalated in 2000 to more than US$300 per kilogram of ore concentrate. It has subsequently returned to its previous much lower price (approximately US$30 per kg). A primary source of tantalum is highly evolved alkali-rich granites and associated pegmatites (e.g. Cerny *et al.* 1986; Cuney *et al.* 1992; Schwartz 1992). Tantalum is also present along with niobium, titanium and tin in the fine silt-sized tailings from the china clay extractive operations in the most evolved granite (topaz granite) in the St. Austell region of Cornwall, southwest England. Columbite–tantalite, Nb–Ta rutile, and cassiterite occur only as accessory minerals in the granite, and become concentrated into one size fraction during processing of the china clay (Scott *et al.* 1998). A tantalum-enriched concentrate can be recovered by gravity separation from the fine-grained waste, and when the price was high, consideration was given to producing it commercially. However, insufficient tantalum and other rare metals are present in the waste in Cornwall to make extraction economic at current prices for the metals. Waste from alkali granite pegmatites (e.g. Uis, Namibia) are similarly concentrated in tantalum, and have been used for the successful extraction of this metal (Harrison *et al.* 2002a).

Other rare metals concentrated in mining waste that might find a demand in the future are rare earth elements concentrated in phosphate wastes from carbonatites. The waste from other mining activities where rocks of a highly fractionated nature (e.g. nepheline syenites) are involved and waste from processing of ore bodies created through extreme chemical weathering (e.g. secondary oxide ores and bauxites) also contain increased concentrations of uncommon elements and minerals.

Conclusions

The waste from many mining operations both for metal ores and industrial minerals contains other industrial minerals that have potential for use. Geologically similar primary mineral extractive operations contain waste of a similar type. Some mining wastes contain large quantities of an industrial mineral and can be considered a resource for that mineral. Detailed technical studies of the potential for utilization of mining wastes, which involve the manufacture of products, do not usually result in a sustainable business, as the capital cost is high, the market usually is not properly established, and/or the waste is not the most suitable raw material for the manufacturing process. There are some good examples of successful uses of mining wastes, especially for bulk products, and there are other scenarios where industrial mineral products and rare metal ore concentrates made from mining waste may find markets.

This research was partly funded through a UK Government, Department for International Development project, entitled Minerals from Waste (Project No. R7416). Camborne School of Mines library staff are thanked for their help in locating relevant literature. This paper is published with the approval of the Executive Director of the British Geological Survey (NERC).

References

ALESHIN, E. 1978. Comments on the utilization of mining and mineral wastes. *In*: ALESHIN, E. (ed) *Proceedings of the Sixth Mineral Waste Utilization Symposium*. U.S. Bureau of Mines and IIT Research Institute, 77–78.

ANON. 1989. Away with slate waste. *Industrial Minerals*, **265**, 105.

ANON. 2000. Make waste pay. *Mining, Quarrying and Recycling*, January/February 2000, 18–20.

BRISTOW, C. M. 1989. World kaolins: genesis, exploitation and application. *In*: CLARKE, G. (ed) *Industrial Clays, a Special Review*. Industrial Minerals Division, Metal Bulletin plc, London, 8–17.

BURGER, J. 1991. Fluorspar producers, fighting for survival. *Industrial Minerals*, **284**, 17–27.

CERNY, P., GOAD, B. E., HAWTHORNE, F. C. & CHAPMAN, R. 1986. Fractionation trends of the Nb- and Ta-bearing oxide minerals in the Greer Lake pegmatite granite and its pegmatoid aureole, southeastern Manitoba. *American Mineralogist*, **71**, 501–517.

COLLINGS, R. K. 1980. Mineral wastes as potential mineral fillers. *In*: VAN NESS, M. (ed) *Proceedings of the Seventh Mineral Waste Utilization Symposium*, 39–48.

COLLINGS, R. K., WINER, A. A., FEASBY, D. G. & ZOLDNERS, N. G. 1974. Mineral waste utilization studies. *In*: ALESHIN, E. (ed) *Proceedings of the Fourth Mineral Waste Utilization Symposium*, 2–12.

COLLINS, R. J. 1978. Construction industry efforts to utilize mining and metallurgical wastes. *In*: ALESHIN, E. (ed) *Proceedings of the Sixth Mineral Waste Utilization Symposium*, 113–143.

CROCKETT, R. N. 1975. *Slate*. Mineral Resources Consultative Committee, Mineral Dossier, No. 12, HMSO.

CUNEY, M., MARIGNAC, C. & WEISBROD, A. 1992. The Beauvoir topaz-lepidolite albite granite (Massif Central, France): the disseminated magmatic Sn–Li–Ta–Nb–Be mineralization. *Economic Geology* **87**, 1766–1794.

CUTLER, I. B. & NICHOLSON, P. S. 1970. Ceramic products from mineral wastes. *In: Proceedings of the Second Mineral Waste Utilization Symposium*, U.S. Bureau of Mines and IIT Research Institute, 149–154

DOWN, C. G. & STOCKS, J. 1977. Positive uses of mill tailings. *Mining Magazine*, September 1977, 213–223.

DUNHAM, K. C. 1948. *Geology of the Northern Pennine Orefield*, Volume 1. Memoir of the Geological Survey of Great Britain. HMSO, London.

DUNHAM, K. C. 1974. Epigenetic minerals. *In*: RAYNER, D. H. & HEMINGWAY, J. E. (eds) *The Geology and Mineral Resources of Yorkshire*. Yorkshire Geological Society, Leeds, 293–308.

EMERY, J. J. & KIM, C. S. 1974. Trends in the utilization of wastes for highway construction. *In*: ALESHIN, E. (ed) *Proceedings of the fourth mineral waste utilization symposium*, 22–32.

EUROPEAN COMMISSION 2002. *The Management of Waste Resulting from Prospecting, Extraction, Treatment and Storage of Mineral Resources*. Working Document, No. 3, D. G. Environment, Brussels, 15 June 2002.

FORD, T. D. 1976. The ores of the south Pennines and Mendip Hills, England – a comparative study. *In*: WOLF, K. H. (ed.) *Handbook of Strata-Bound and Stratiform Ore Deposits*, Volume 5. Elsevier, 161–195.

GUTT, W., NIXON, P. J., SMITH, M. A. HARRISON, W. H. & RUSSELL, A. D. 1974. *A Survey of the Location, Disposal and Prospective Uses of the Major Industrial By-Products and Waste Materials*. Building Research Establishment Current paper, 19/74, 81 p.

HARRIES-REES, K. 1991. Slate market split. *Industrial Minerals*, **284**, 44–55.

HARRISON, D. J., BLOODWORTH, A. J., EYRE, J. M., SCOTT, P. W. & MACFARLANE, M. 2001. *Utilization of Mineral Waste: a Scoping Study*. British Geological Survey Technical Report, **WF/01/03**.

HARRISON, D. J., BLOODWORTH, A. J., EYRE, J. M., MACFARLANE, M., MITCHELL, C. J., SCOTT, P. W. & STEADMAN, E. J. 2002a. Utilization of mineral waste: Case studies. British Geological Survey Commissioned Report, **CR/02/227 N**.

HARRISON, D. J., BLOODWORTH, A. J., EYRE, J. M., MACFARLANE, M., MITCHELL, C. J., SCOTT, P. W. & STEADMAN, E. J. 2002b. Minerals from waste project summary report. British Geological Survey Commissioned Report, **CR/02/228 N**.

HEINS, R. W. & GEIGER, G. H. 1970. Potential utilization of mine waste tailings in the Upper Mississippi Valley Lead-Zinc Mining District. *In: Proceedings of the Second Mineral Waste Utilization Symposium*, 181–193.

KAUFFMAN, R. A. & VAN DYK, D. 1994. Feldspars. *In*: CARR, D. D. (ed) *Industrial Minerals and Rocks*, 6th edition. Society for Mining, Metallurgy and Exploration, Inc., Littleton, Colorado, 473–481.

KESSELL, C. 2001. *China Clay By-Products: Between a Rock and a Hard Place. Mining Waste Management, New Approaches to Environmental Sustainability*. CDrom. Mineral Industry Research Organization, Leeds.

MACKENZIE, J. D. & HORIUCHI, T. 1980. Development of asbestos substitutes from domestic raw materials. *In*: VAN NESS, E. (ed) *Proceedings of the Seventh Mineral Waste Utilization Symposium*, 34–38.

MASON, J. E. 1978. Geological aspects of fluorspar mining in Derbyshire. *In: Discussions and Contributions, Fluorspar. Transactions of the Institution of Mining and Metallurgy*, Section A, **87**, 70–71.

MINDESS, S. & RICHARDS, C. W. 1970. Criteria for selection of mineral wastes for use in the manufacture of calcium silicate building bricks. *In: Proceedings of the Second Mineral Waste Utilization Symposium*, 155–165.

PETTIBONE, H. C. & KEALEY, C. D. 1972. Engineering properties and utilization examples of mine tailings. *In*: SCHWARTZ, M. (ed) *Proceedings of the Third Mineral Waste Utilization Symposium*, 161–169.

RAMPACEK, C. 1980. Progress in mining and metallurgical waste utilization. *In*: VAN NESS, E. (ed) *Proceedings of the Seventh Mineral Waste Utilization Symposium*, 3–15.

RAO, A. B. DE CASTRO, C. & ADUSUMILLA, M. S. 2002. Waste materials from pegmatites and others in ENE Brazil: economic and environmental benefits. *In*: SCOTT, P. W. & BRISTOW, C. M. (eds) *Industrial Minerals and Extractive Industry Geology*. Geological Society Publishing House, 341–343.

REES-JONES, R. G. & STORHAUG, E. 1998. Meeting the demand for rock armour and its grading specification. *In*: LATHAM, J.-P. (ed) *Advances in Aggregates and Armourstone Evaluation*. Geological

Society Engineering Group, Special Publications, **13**, 87–90.

RICHARDS, T. E. 1986. Geological characteristics of rare-metal pegmatites of the Uis type in the Damara Orogen, South West Africa/Namibia. *In*: ANHAEUSSER, C. R. & MASKE, S. (eds) *Mineral Deposits of Southern Africa*, Geological Society of South Africa, Johannesburg, 1845–1862.

RICHARDS, MOOREHEAD & LAING LTD. 1995. *Slate Waste Tips and Workings in Britain*. HMSO, London.

SCOTT, P. W. & DUNHAM, A. C. 1984. Problems in the evaluation of limestone for diverse markets. *Proceedings of the 6th Industrial Minerals International Congress*. Metal Bulletin, London, pws1–21.

SCOTT, P. W. & POWER, M. R. 2001. The influence of geology on the quality of feldspar products. *In*: BON, R. L., RIORDAN, R. F., TRIPP, B. T. & KRUKOWSKI, S. T. (eds) *Proceedings of the 35th Forum on the Geology of Industrial Minerals – the Intermountain West Forum, 1999*. Utah Geological Survey, Miscellaneous Publication **01–2**, 221–229.

SCOTT, P. W., PASCOE, R. D. & HART, F. W. 1998. Columbite–tantalite, rutile and other accessory minerals from the St Austell Granite, Cornwall. *Geoscience in South-West England*, **9**, 165–170.

SCHWARTZ, M. O. 1992. Geochemical criteria for distinguishing magmatic and metasomatic albite enrichment in granitoids – examples from the Ta–Li granite Yichun (China) and the Sn–W deposit Tikus (Indonesia). *Mineralium Deposita*, **27**, 101–108.

SHAFFER, N. R. 1996. Heavy minerals in fine-grained waste materials from sand and gravel plants of Indiana. *In*: AUSTIN, G. S., HOFFMAN, G. K., BARKER, J. M., ZIDEK, J. & GILSON, N. *Proceedings of the 31st Forum on the Geology of Industrial Minerals – The Borderland Forum*. New Mexico Bureau of Mines and Mineral Resources, Bulletin **154**, 215–222.

SMITH, A. 2002. Brick making – the ultimate waste repository. *In*: SCOTT, P. W. & BRISTOW, C. M. (eds) *Industrial Minerals and Extractive Industry Geology*. Geological Society, London, 323–326.

SMITH, M. R. 1999. Stone: Building stone, rock fill and armour stone in construction. Geological Society Engineering Group, Special Publications **16**.

STOCKWELL, L. E. 2001. *World Mineral Statistics, 1995–1999*. British Geological Survey.

SYMONDS GROUP 2001. A study on the costs of improving the management of mining waste. Final Report. Report to DG Environment, European Commission, http://www.europa.eu.int/comm/environment/waste/studies.mining/mining_cost.pdf

TANNER, J. T. JR. 1994. Mica. *In*: CARR, D. D. (ed) *Industrial Minerals and Rocks*, 6th edition. Society for Mining, Metallurgy and Exploration, Inc, Littleton, Colorado, 693–710.

WHITBREAD, M., MARSAY, A. & TUNNELL, C. 1991. *Occurrence and Utilization of Mineral and Construction Wastes*. HMSO, London.

WOODSTOCK-DEERING 2001. *Larvik Armourstone*, http://www.woodstock-deering.co.uk/larvik.html

Sustainable small-scale gold mining in Ghana: setting and strategies for sustainability

P. A. ESHUN

Mining Engineering Department, University of Mines and Technology, PO Box 237, Tarkwa, Ghana (e-mail: arrojaeshun@yahoo.co.uk)

Abstract: In Ghana, small-scale/artisanal gold mining has been on-going for more than a century. Artisanal mining has been the support for the rural people who more often than not are forced to sacrifice their farmlands and means of livelihood for large-scale mining operations. In order to reduce the activity of small scale/artisanal during the colonial era, laws were passed to bar indigenous operators from dealing in gold ore, amalgam, bullion, retorted gold, slags, concentrates and mercury. In recent years, however, under the auspices of the German non-governmental agency, *Gesellschaft für Technische Zusammenarbeit* (GTZ) and the World Bank, the Ghana government has undertaken a number of initiatives to formalise and regulate small-scale mining operations. Unfortunately, small-scale mining activities are characterised by lack of capital and minimum use of appropriate technology in the mining and treatment of the minerals into finished products. In addition, the industry is associated with land degradation and water pollution. This paper explores possible strategies that aim to make small-scale gold mining in Ghana more sustainable (i.e. more efficient, less destructive to the environment and more meaningful to the operators and the country as a whole). The roles of stakeholders in the small-scale mining industry in Ghana are also identified. It concludes that, for sustainable small-scale mining, a pragmatic synergistic approach must be adopted by all stakeholders in the organisation, regularisation, training and support of small-scale mining operations in Ghana. Mineable lands need to be delineated, illegal operators should be organised and brought under a responsible umbrella, small-scale mining operators should be supported with funds, technology and education, and alternative livelihood programmes must be pursued in mining communities.

Historical perspective of small-scale mining in Ghana

Gold was mined in Ghana on a small scale long before the arrival of colonial powers. This is demonstrated by several ancient workings in the gold-bearing areas of the country. Legislation drafted during the Colonial era in Ghana led to a disincentivization of small-scale mining in favour of large-scale mining. The passing, in 1905, of the Gold Mining Products Protection Ordinance (CAP. 149) prevented indigenous Ghanaians ('Gold Coasters') from dealing in gold ore, amalgam, bullion, retorted gold, slags and concentrates. In 1932, the Mercury Ordinance was added, making it illegal for Ghanaians to own mercury. In recent years, under the auspices of the German non-governmental agency Gesellschaft für Technische Zusammenarbeit (GTZ) and the World Bank, the Ghanaian Government has undertaken a number of initiatives to formalize and regulate small-scale mining operations. This article examines the current state of smale-scale mining in Ghana within this setting.

Regulatory framework

Legal framework

In 1989, the following three laws were passed, which enabled Ghanaians above the age of 18 years to engage in small-scale mining operations and to purchase and use mercury required for their operations.

- *The Small-Scale Gold Mining Law (PNDCL 218)*: This provides for the registration of activity; the granting of gold-mining licences to individuals or groups; the licensing of buyers to purchase product; and the establishment of district-assistance centres.

From: MARKER, B. R., PETTERSON, M. G., McEVOY, F. & STEPHENSON, M. H. (eds) 2005. *Sustainable Minerals Operations in the Developing World*. Geological Society, London, Special Publications, **250**, 61–72.
0305-8719/05/$15.00 © The Geological Society of London 2005.

- *The Mercury Law (PNDCL 217)*: This legalized the purchasing of mercury (for mineral processing purposes) from authorized dealers.
- *The Precious Minerals Marketing Corporation Law (PNDC Law 219)*: This transformed the Diamond Marketing Corporation into the Precious Minerals Marketing Corporation (PMMC), which was authorized to buy and sell gold.

Administrative framework

For effective administration of the Small-Scale Gold Mining Law, seven small-scale mining district centres have been established by the Minerals Commission within some of the gold-mining communities in Ghana (Fig. 1) and are adequately staffed (two mining engineers and a secretary) and equipped with logistics to carry out the following functions:

- educate prospective small-scale miners and the public on the need for the acquisition of a small-scale mining licence, prior to engaging in small-scale gold mining;
- appeal to illegal operators to regularize their status by acquiring a small-scale gold mining licence;
- register prospective small-scale miners;
- inspect sites of prospective small-scale miners and assist in the preparation of requisite documents;
- monitor operations of small-scale gold miners and offer technical advice in appropriate mining and processing methods, health and safety and environmental issues; and
- send reports on small-scale mining activities to the Minerals Commission.

The Small-Scale Mining Department of the Minerals Commission coordinates the activities of the District Centres.

Definition of small-scale gold mining in Ghana

In Ghana, the definition of small-scale gold mining, as given in the Small-Scale Gold

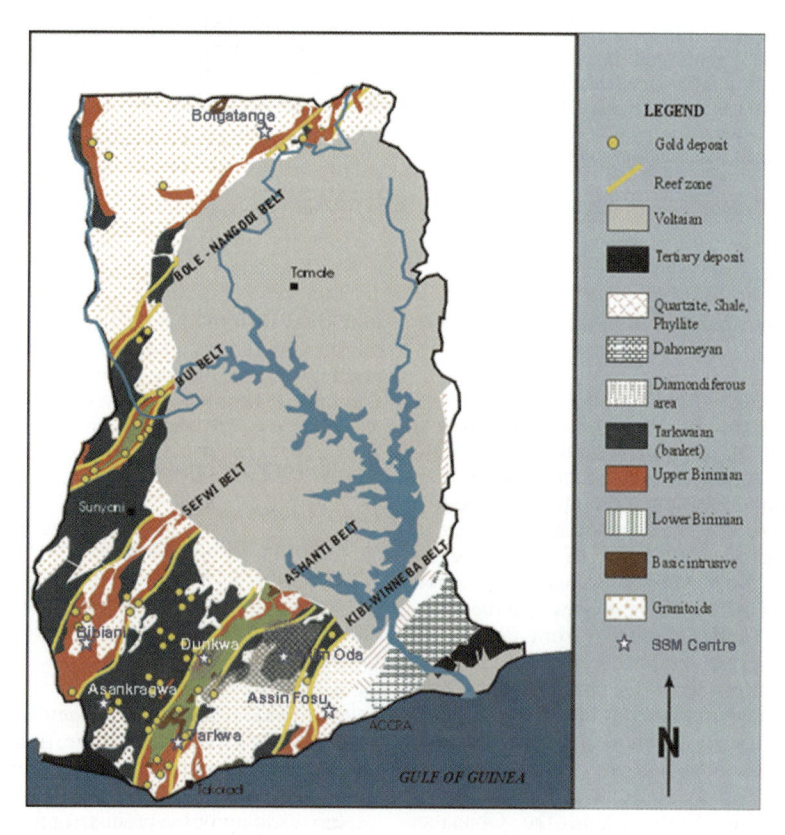

Fig. 1. Map of Ghana showing gold deposits and the seven small-scale mining centres.

Mining Law, 1989, PNDCL 218, Part III, Section 21, is:

> The mining of gold by any method not involving substantial expenditure by an individual or group of persons not exceeding nine in number or by a cooperative society made up of ten or more persons (Anon 1989).

The following provisions in the law amplify the definition of small-scale gold mining in Ghana:

(1) The size of the area in respect of which a licence may be granted under this law shall not exceed 10 hectares (25 acres), even in the case of a grant to a cooperative society of ten or more persons; and
(2) No small-scale miner shall use any explosives in his operation.

It can therefore be concluded that, in Ghana, small-scale gold mining refers to the mining of gold in a concession not exceeding 10 hectares (25 acres) in an area in which no explosives are used.

Mining methods

Initially, small-scale miners directed their attention to mining alluvial ores as the gold existed in a free state and therefore could be extracted in a relatively easy manner. In recent years, however, small-scale miners work on all deposit types. Although mining may take place underground or on the surface, the mining techniques being used have not changed much despite the passage of time (Amankwah & Anim-Sackey 2003).

Underground operations

Underground mining by small-scale miners usually takes place in abandoned pits, adits and shafts. However, the level of technology does not allow mining of ore at great depths. Access into shafts is by the help of old metal structures and through footholes made on the shaft walls. The miners carry torchs and hurricane lamps to illuminate their work places. Figure 2 shows an illegal miner with a torch strapped to his head descending into an underground working (Harkinson 2003). In order to prevent flooding, pumps may be sent underground to get rid of water from the stopes when necessary. If ventilation is poor, the exhaust gases from the pumps can, at times, lead to suffocation of the miners. Loosening of rocks is by means of hand-held hammers and chisels, mattocks and shovels. In rare cases, very daring miners use explosives to loosen the rocks, thereby acting in contravention of the small-scale mining law. The loosened chippings are shovelled into sacks and buckets and carried up the pit to a convenient place where the material undergoes further comminution. Some miners spend days and weeks underground, grinding the ore and recovering gold before returning to the surface.

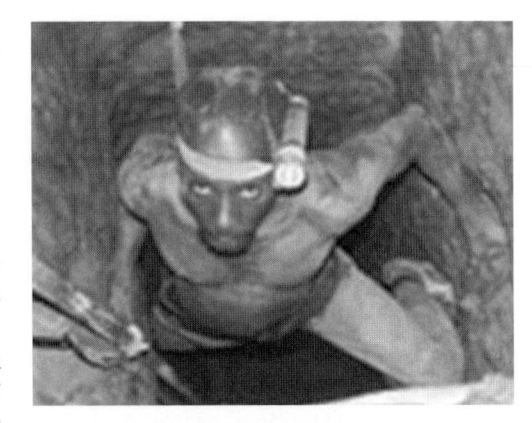

Fig. 2. An illegal miner descending into an underground working area (*source*: Harkinson 2003).

Surface operations

Small-scale miners use surface mining methods on alluvial and other deposits that are close to the surface. Miners usually search for outcrops, which serve as doorways to the mineralized zones below. Upon locating a vein or other mineralized zones, mining may take place to great depths.

In mining alluvial deposits, small-scale miners dig out gold-bearing materials from old river courses and terraces. At times, streams and rivers may be diverted and their banks mined out to depths reaching 10 m and beyond. The overburden removed is used to build dams around the pits to prevent water from flowing into them. Materials mined out from the pits are carried in headpans to a suitable location for processing. In rare cases, miners dive into the rivers to scoop loose materials from the riverbeds, a manual dredging method locally termed as *anomabu*.

In mining outcrops, the vegetation is cleared using cutlasses and hoes. The humus soil and sand are then dug up with pickaxes and hand-shovelled away to expose the ore. The rich material is dug out and loaded into small nylon sacks for washing on wooden sluice boxes.

Processing methods

According to Amankwah & Anim-Sackey (2003), the processing techniques currently in use by small-scale miners are significantly more advanced than previously used artisanal techniques. More appropriate comminution equipment and better recovery units have been introduced and are run in a more professional manner. Unfortunately, however, most illegal operators still employ crude processing techniques and at best send their ore to improvised comminution sites dotted around the mining communities.

Comminution

Comminution is only relevant when consolidated and slightly consolidated ores are being treated. Small-scale miners begin size reduction as soon as rock is loosened using hand-held hammers and chisels, and other simple implements. At the processing site, the material is dried and size reduction is carried out using a combination of jaw and rocker crushers, and hammer and stamp mills (metal mortars and pestles). The stamp mills and rockers are manually operated, while the others are powered by diesel or electricity. These units are now cheaper to use because most of the machine parts are produced in local foundries. Disc mills, used locally in grinding corn and other foodstuffs, have now been modified and equipped with harder grinding surfaces that are well suited for grinding ores (Fig. 3).

Scrubbing and screening

The main preconcentration processes performed on alluvial material are scrubbing and screening. Scrubbing is carried out mainly on alluvial gold ores to remove obscuring dirt and slime. The extent and mode of washing usually depend on physically inspecting how strongly the clay adheres to the surface of the gold. If the clay content is low and adhesion to the mineral surface is weak, scrubbing and screening are done simultaneously using a trommel or by pulsating a hand-held screen (with the material in it) in water.

Ores with high clay content usually demand the use of harsher scrubbing techniques. Some miners carry this out in rotating drums used locally for mixing concrete. Others perform this operation by pounding and shearing the material with their feet in basins after soaking with water. In both cases, the added water is decanted from time to time and fresh water is introduced. The process continues until the gravel and sands are relatively clean.

Concentration

The types of concentration units available to the small-scale miner include the sluice, strake and shaking tables, pans, jigs, gold savers and centrifugal concentrators. The most common one in use is the sluice board, because it is relatively easy to build and use. The sluice board is built locally using hard wood boards or metal sheets (with sides slightly curved inwards) or split bamboo (Fig. 4). The board is lined with a corduroy, jute material, miner's moss or astro turf to obstruct the flow of slurry during the washing of ore on it.

Knelson concentrators are generally too expensive for individual small-scale miners; however, they are available for hiring for a fee.

Fig. 3. Corn grinding mill modified for ore grinding.

Fig. 4. Washing of sand and gravel on a sluice in the Akim Oda mining district.

Fig. 5. Gold–mercury amalgam being heated in a can on an open fire.

Some legalized alluvial gold miners usually pile up dug-out material for several days and then hire a Knelson concentrator to concentrate the ores for a few hours. This practice, though faster, is more expensive and is recommended for relatively high-grade ores only.

Recovery of gold from concentrates

Gold is recovered by winnowing, amalgamation, retorting, and direct smelting. Although many of the miners are becoming environmentally conscious, some still engage in amalgamation and heating without using mercury retorts (Fig. 5) causing health hazards to the miners and the neighbouring communities.

Impacts of small-scale gold mining in Ghana

Positive impacts

Revenue generation. Prior to the promulgation of the Small-Scale Gold Mining Law and the establishment of the Small-Scale Mining Department under the Minerals Commission and the Precious Minerals Marketing Corporation (PMMC), gold mined by small-scale miners was smuggled out of the country. The capturing of the potential of small-scale gold mining into the mainstream of the economy (among others) has led to an increase in gold output and raised the foreign exchange earnings of the country. It is therefore not surprising that from 1992 onwards, gold replaced cocoa as the nation's leading foreign exchange earner. From Table 1, it is seen that the contribution of small-scale gold production increased from 9272 oz in 1989 to over 173 000 oz in 2001, an increase of about 1800%. In comparative terms, this represented 2.2% of large-scale production in 1989 as compared to 7.8% in 2001. It is also seen that the relatively low gold prices between 1999 and 2001 affected large-scale gold mining by a drop in production of 13.45%, while the production of gold from small-scale mining increased by 32.7% within the same period. This trend is also depicted by the revenue graph shown in Figure 6.

Social benefits. Labour force engaged in small-scale gold mining in Ghana is normally quoted at being around 30 000–50 000 (ILO 1999; Hilson 2001). Considering the fact that most of these people operate in rural areas, the problem

Table 1. *Gold production in Ghana, 1989–2001*

Year	Small-scale miners (oz)	Large-scale miners (oz)	Total Ghana (oz)	% small-scale to total Ghana
1989	9272	420 204	429 476	2.2
1990	17 234	517 818	535 052	3.2
1991	15 601	825 114	840 715	1.8
1992	17 297	976 223	993 520	1.7
1993	35 145	1 222 344	1 257 489	2.8
1994	89 520	1 338 491	1 428 011	6.7
1995	127 025	1 581 506	1 708 531	7.4
1996	112 349	1 474 746	1 587 095	7.6
1997	107 097	1 677 911	1 785 008	6.3
1998	128 334	2 244 819	2 373 153	5.7
1999	130 833	2 358 423	2 489 153	5.5
2000	145 662	2 168 802	2 314 464	6.3
2001	173 610	2 041 129	2 214 739	7.8
13 years	1 108 979	18 847 530	19 956 509	5.6

Source: Minerals Commission.

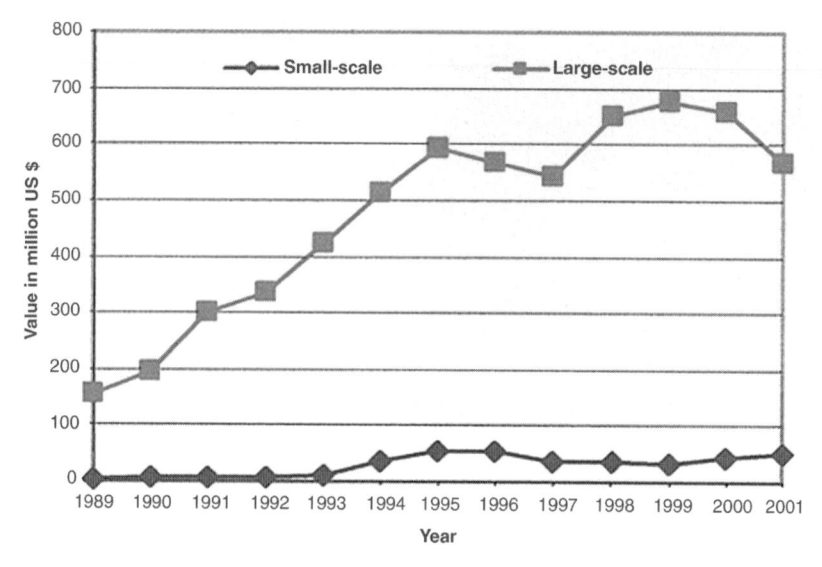

Fig. 6. Revenue for large- and small-scale gold mining in Ghana, 1989–2001 (*source*: PMMC).

of rural–urban migration seen in other parts of Africa has been addressed by the operation of small-scale mining. The net effect has been a relative increase in the purchasing power of rural people and an improvement in their socio-economic lives. Allied industries such as local goldsmiths and international jewellery companies are also flourishing in the country as a result. Furthermore, the sector also provides indirect employment to food and drink sellers, clothing sellers, boutiques, hardware stores and many more. Most of these businesses are found at mining sites, as well as in the towns and villages where mining is active.

Small-scale mining as a precursor to large-scale mining. Often, prospectors for large-scale mines depend on local artisans and the mining history of an area to commence their exploration programmes. As a result, local artisans have been instrumental in the discovery of several of the large-scale mines in Ghana. Because of this, some local artisans feel they have the moral justification to mine illegally even after concessions have been taken over by large-scale mines. One illegal operator at Dumasi, a small-scale mining community close to Bogoso Gold Limited, recalled how his ancestors mined freely: 'They were working in peace, they were not having any problems, but now, we are being harassed' (Harkinson 2003).

Mining of what otherwise might be on uneconomic resource. A valuable contribution of small-scale mining that is often not acknowledged is that the operations typically exploit what might otherwise be considered an uneconomic resource. It can be seen from Table 1 that between 1999 and 2001, when the gold price was in decline, production from large-scale mines decreased, but that of small-scale mines increased. This is attributed to the use of simple equipment and technology in small-scale mining operations and corresponding low operating costs. This scenario should perhaps motivate large-scale mines to shed some of their concessions, which could then be worked on by legalized small-scale mining operators.

Negative impacts

Land degradation. Small-scale gold mining operations are widespread and not quite localized; hence, their effects on the land are extremely varied. The operations have resulted in the unnecessary destruction of forests, food and cash crops. Stirring and exposure of the soil to the mercy of the weather renders the soil very susceptible to various forms of erosion and leaching of plant nutrients, to the extent that it might no longer sustain crop production. The haphazard placement of overburden material and tailings in mounds and ridges between the pits makes the land even more difficult to till for food cropping. The present state of the mined lands, as can be seen in Figure 7, is also hazardous to the health and safety of nearby inhabitants. People, in particular farmers and

Fig. 7. An unclaimed mined-out area in the Akim Oda mining district.

hunters, may easily fall into these pits and get injured or even drown as the pits are often filled with water. The pits also serve as breeding ground for mosquitoes, which are agents for causing malaria, hence posing a health hazard.

Air pollution. A major source of air pollution in the case of small-scale gold mining, commonly takes the form of solid suspension as a result of ore crushing and sieving (Fig. 8). Gold ore from hard rocks are pounded and sieved in enclosed areas and sometimes in congested habitats. In addition, grain mills are sometimes used in milling the ore (Fig. 3). This causes serious air pollution and poses potential health hazards to the communities. During heating of the gold amalgam in the open (Fig. 5), mercury fumes are released into the air and this is also a potential health hazard.

Water pollution. Small-scale miners introduce solid suspensions and mercury into the water courses in which they operate. For example, tests carried out at Dumasi, a small-scale mining community close to Bogoso Gold Limited (BGL), to determine the levels of mercury in the bloodstream of the inhabitants, livestock, fish and sediments from the Apopre stream and settling ponds established strong evidence of mercury exposure among the Dumasi population (UNIDO 2001).

There are instances where small-scale gold miners have effectively diverted the courses of rivers and streams to make alluvial gravel accessible. Diversions are normally performed during the dry season, resulting in some rivers drying up, thus denying populations living downstream of traditional sources of drinking water. In addition, slopes can become unstable and collapse during the subsequent rainy period, leading to rapid silting up of many streams.

Social problems. The rapid growth of small-scale gold mining in Ghana has encouraged migration of people seeking jobs to such communities, resulting in an increase in population. The vibrant economic activity as a result of this population increase and the competition for food and other basic commodities has drastically increased the cost of living in such communities, with the Tarkwa district being a prime example.

In some cases, it has affected education to such a degree that some parents do not have enough money to pay for school fees. In addition, increase in loss of teachers and in truancy of school children, as well as juvenile delinquency has resulted. The quest for money forces children to play truant. At Akuntasi, a village near Tarkwa, the case was so serious that the primary school had to be closed down (Eshun & Mireku-Gyimah 2002).

The growth in population of people from different backgrounds brings with it different, and in some cases, conflicting customs and traditions. This has increased active social life, resulting in prostitution and sexual promiscuity and as a result the increased spreading of communicable diseases such as syphilis, gonorrhoea and AIDS. There are also cases of drug abuse among small-scale gold miners. This is especially prevalent among illegal operators.

Land use disputes between large-scale and illegal small-scale (galamsey) operators. The issue of land disputes between small- and large-scale operators who are in constant competition over the plots of land overlying the country's rich gold deposits is particularly of concern. Several brutalities and clashes have been reported, sometimes resulting in the deaths of illegal small-scale operators who encroach on concessions held by large-scale mines. Most of these conflicts have occurred

Fig. 8. Dust from sieving.

Fig. 9. Extensive *galamsey* operations on and at the outskirts of the Dadieso reclaimed pit of Ayanfuri Mine.

within the Western and Ashanti Regions of Ghana. Notable among these are the following.

- At Ashanti Goldfields Company's Obuasi mine, violent clashes erupted in 1996 between *galamsey* operators and local security forces, resulting in over $1 million in damages at the site, and in December 2000, vandalizing *galamsey* operators set ablaze the poultry farm at Obuasi, and stole valuable livestock. In addition, at its smaller Bibiani property, located slightly west of Obuasi, management reported needing police assistance from Kumasi to control resident *galamsey* operators. Figures 9 and 10 illustrate the extent to which *galamsey* operators stalled the reclamation programme of Ashanti's Ayanfuri mine before final mine closure. Figures 11 and 12 show similar invasions of mined-out pits of Bogoso Gold Limited by illegal small-scale miners.
- In 1990, Abosso Goldfields Ltd (AGL) applied for a prospecting licence for an area in the Tarkwa mining district, which was occupied at the time and mined by seven groups of small-scale cooperatives. Shortly after AGL began extensive prospecting, over 600 miners migrated from other parts of the concession onto the company's trenches, where explosives were being used. After failing for months to prevent the conflict with these resident illegal miners, management decided to pursue a different strategy. Following successful communication with the cooperatives, it identified, registered and regrouped the small-scale

Fig. 10. Pictures of active *galamsey* operations at the Bokitsiso mined-out pits of Ayanfuri Mine.

Fig. 11. Illegal miners (*galamsey*) in the floor of Chujah East Pit, Bogoso Gold Limited.

miners, and equipped them with Abosso Goldfields' identification cards. Initially, 700 miners were registered, most of whom were using small picks and shovels to work alluvial and consolidated gravel deposits. A commissioned agent periodically visited the site to purchase product from these miners (Hilson 2001).

- The present expansion programme of Bogoso Gold Limited (BGL) to cover Prestea, an old mining town, has resulted in serious confrontation between the company and the local community. Illegal small-scale operators have invaded the Bogoso concession at Prestea and have threatened to blow up the only bridge over the River Ankobra that connects Prestea to Bogoso and in fact to the rest of the country. This threat was to prevent BGL from transporting their equipment required to continue its mining operations in the town. The company, in a gesture of harmony, has allowed the illegal miners to continue working on the concession until their operations advance to where the illegal operators are working. A more pragmatic solution is therefore required.

Fig. 12. Illegal mining working showing acidic water in Dumasi Pit, Bogoso Gold Limited.

The following reasons could be assigned to the disputes that exist between the large-scale operators and the illegal small-scale miners.

- The illegal operators have a cultural relationship with the land, which results in their reluctance to relocate, even after the land is leased out to large-scale operators.
- Some human rights activists advocate that compensation rates paid to communities affected by mining are inadequate and that communities are forced to accept inadequate resettlement packages. However other practical investigations reveal that monies paid to land owners, which appear adequate at the time of payment, soon reduced in value, probably due to inflationary pressures and/or misplaced priorities on the part of the beneficiaries.
- Lack of expertise of the illegal operators dramatically reduce their employment potential in more specialized areas of mining or in different industries.
- Illegal operators lack both capital and education to regularize their operations should they have access to land.
- Lack of willingness by the illegal miners to be trained as they are used to working illegally to achieve quick subsistent incomes.
- Lack of commitment on the part of the government and traditional authorities to commit taxes, mineral rents and royalties paid by the large-scale mines to solve local problems.
- Lack of commitment on the part of large-scale mines to address such conflicts pragmatically by instituting more permanent structures such as alternative livelihood projects for affected communities.
- Communities affected by the operations of large-scale mines normally do not have

structured development plan for the large-scale operators to help them execute.
- Lack of delineated lands for people who are willing to mine legally.

Strategies for sustainability: recommendations

Since it appears that small-scale mining has come to stay in Ghana, the challenge is not how to eradicate it but how to assist it to become more productive, less polluting and socially beneficial in the most cost-effective way. However, experience has shown that interventions, whether undertaken by government or the private sector, such as technical assistance, appropriate training or even access to equipment or cash credits, in order to demonstrate positive results, must occur within a coherent organizational framework.

In other words, any attempt to assist small-scale mining cannot easily or effectively be delivered to an unorganized sector operating anarchically or independently. That is why a pragmatic synergistic approach is proposed to be adopted by all stakeholders in the delineation of mineable lands, organization, regularization, training and support of small-scale mining operations in Ghana; to make small-scale mining a tool for sustainable rural and community development.

The role of the government

The government of Ghana has shown the goodwill to support small-scale mining operations by establishing a legal framework to control the regularization, marketing and further set-up of an administrative framework for supervision of small-scale mining activities in the country. However, there is leniency on the part of the government in dealing with illegal small-scale operators as a result of humanitarian and, at times, political reasons. This results in extreme frustration for the large-scale operators, who are sometimes left with no option but to administer initial threats of brutalities. However, they usually succumb to pressure and allot portions of their concessions temporarily to the illegal small-scale miners who at the end of the day mine irresponsibly. The government should therefore be prepared to take serious actions in streamlining and enforcing the regulations against illegal operations to save the environment, health and safety of the operators and others in the communities.

In support of sustainable small-scale gold mining in Ghana, the government should commit funds to explore mineable lands, which will be administered by the Small-Scale Mining Department of the Minerals Commission to regularized investors. The government is also encouraged to create a special fund out of the Poverty Alleviation Fund, to assist small-scale miners to acquire improved tools and equipment for mining and processing of the ore.

The role of the District Assembly

The current democratic dispensation of the country has seen the decentralization of authority and responsibility of the government, making the District Assembly an important arm of government. The District Assemblies have the prerogative to see to the peace, development and progress in the various communities in the country. To achieve sustainable small-scale mining, chief responsibility has to be placed on the District Assembly. The District Assembly needs to bring all stakeholders on board by, first, organizing into groups the locals involved in illegal operations. The Assembly should further secure funds to allow areas outside large-scale concessions to be explored and demarcated. These sites will then be allocated to organized groups, who will pay taxes to the Assembly. In instances where large-scale mines are prepared to release part of their concessions, the Assembly should take custody, and in collaboration with the District Small-Scale Mining office, allocate them to organized cooperatives. The Assembly should channel proceeds from mining into other alternative livelihood projects such as fish farming, planting of trees of commercial value and other ventures that the communities can recommend.

The role of regulatory and supporting institutions

The main players include the Small-Scale Mining Department under the Minerals Commission, Environmental Protection Agency, Geological Survey Department, the Police Service, the Precious Mineral Marketing Company and other legal gold buyers, mercury dealers and the Electricity Company of Ghana. All these institutions should come together to find solutions to sustainable small-scale gold mining in Ghana, especially focusing on empowering the rural communities. The law enforcement agencies such as the Police Service and the Environmental Protection Agencies need to be loyal to the tenet of their authority and exercise it impartially when the need arises. The Minerals Commission should dialogue with the government and other agencies for funds to enable the Geological Survey Department to demarcate mineable lands for small-scale mining. The Small-Scale Mining Officers of the Small-Scale Mining Department of the Minerals Commission should be transparent in their guidance and licensing of investors. They need to be empowered to arrest illegal operators for onward prosecution. The Minerals Commission should revise the definition of small-scale mining in the Law to possibly allow the use of a certain amount of explosives as proponents hide the use of explosives. If the use of explosives is legalized, small-scale miners can then be guided and educated. The Electricity Company of Ghana should deny illegal operators access to electricity and prosecute people who connect illegally to the national grid.

The role of the Mining University and other academic institutions

Ghana is privileged to have the only mining institution in sub-Saharan Africa; the University of Mines and Technology, Tarkwa, formerly KNUST School of Mines, Tarkwa. The staff are capable of finding practical solutions to small-scale problems. The strength of the institution should be exploited in the education, training, coordination and search for appropriate technology in the mining, processing and handling of environmental problems concerning small-scale mining. Other academic institutions are also available to contribute in finding workable solutions to the problems of achieving sustainable small-scale gold mining in the country. The Small-Scale Mining Department and the other regulatory bodies should collaborate with the academic institutions in educating the small-scale mining operators. As indicated in UNIDO (2001), 'young illiterate *galamseys* at Dumasi showed strongest exposure (several indicators in class 3) to mercury'. The role of education in sustainable small-scale mining can therefore not be overemphasized. This assignment lies in the purview of the academic institutions.

The role of non-governmental organizations and development partners

Small-scale mining in Ghana has received much support from international agencies in the financing, technological transfer and development of the sector. Other non-governmental agencies (NGOs), both locally and internationally, have supported the country's efforts by criticizing bad practices and periodically taking practical steps to help solve some of these problems. This is the time to call all these agencies together for pragmatic solutions to sustainable development in the small-scale gold mining sector of the Ghanaian economy. Non-governmental organizations should make funds available for prospective small-scale miners. They should criticize constructively any ineptitude from the government and any of the other stakeholders in reaching the common goal to achieve sustainable small-scale gold mining in Ghana.

The role of traditional authorities

Although there have been several chieftaincy disputes marring the authority and integrity of these local authorities, they still hold their place in the democratic governance of the country. The best role traditional authorities can play to achieve sustainable development using the rich mineral deposits of the land include, first, stopping disputes among themselves and displaying to the youth that they stand for peace. Secondly, they should channel the royalties from mining into other alternative livelihood projects. Such projects would reduce illegal mining activities and also provide sustainable employment in the communities after mine closure. They need to ensure that immigrants do not unnecessarily violate the good custom and values of the land.

The role of large-scale mines

The first responsibility of large-scale companies is to plan carefully and pay realistic compensations to communities affected by their mining operations. The companies should, where possible, allocate parts of their concessions to small-scale operators as part of their corporate social responsibility. This would provide a way to coexist with the indigenes who still want to mine but cannot be employed in the large-scale mines. The large-scale mines should make a conscious effort to give local communities preference in regular employment where they qualify. The large mining companies should resort to improved communication and community assistance projects in the area of infrastructure development, education and training to foster good relationships in cohabitation. Lastly, reclamation of pits should be done concurrently with mining so that mined-out pits do not serve as attractive grounds for illegal operations.

The role of the community

The time has come for people living in mining communities to realize that the land is a national asset and that they are privileged to be custodians of such mineral wealth, the benefits of which should be spread across the nation. In this way, they may act more responsibly by indulging in sustainable activities. Illegal mining should be seen as indiscipline and robbery, especially when provisions have been made to organize, regularize and set up alternative livelihood projects for communities having been denied rights to farm or mine on their lands.

Applying the synergy principle

The synergy principle states that, 'every part of an organisation enables every other part', (Hall 2003). It is obvious from the outlined strategies for sustainability (recommendations) that no single stakeholder can shoulder the responsibilities for attaining sustainable small-scale gold mining in Ghana. It therefore behoves on all stakeholders to come together to find workable solutions to this problem. Small-scale mining committees should be formed in every small-scale mining district (Fig. 1). The membership should include representatives from as many of the stakeholders as possible. The District Small-Scale Mining officers and the District Assemblies should form the core with other stakeholders joining from their respective locations. Regular, structured meetings, perhaps on a quarterly basis, will ensure that strategies put in place to ensure sustainable small-scale gold mining are continuously reviewed, modified and updated.

Conclusions

Small-scale mining of gold is now a legally acceptable and important industry in Ghana. Impacts of the industry are both positive and negative. The key challenge is to develop strategies to make small-scale gold mining in Ghana more sustainable, that is, more efficient, less environmentally damaging, in conjunction with greater social benefits. The positive impacts are that it makes a major contribution to foreign exchange earnings of the country; it provides substantial employment to rural communities; it helps stem rural–urban migration; it has been a precursor to large-scale mining; and it enables the exploitation of what otherwise might be uneconomic resources. The negative impacts are that the industry is associated with land degradation; air pollution; water pollution; and social problems such as sexual promiscuity, drug abuse, high cost of living, truancy on the part of both pupils and teachers, land use conflicts between illegal operators and large-scale mines.

Strategies for sustainability will involve delineating mineable lands, organizing illegal operators under a responsible umbrella, supporting small-scale mining operators with funds, technology and education, providing alternative livelihood programmes in mining communities, and adopting a pragmatic synergistic approach by all stakeholders in the search for workable solutions in achieving sustainable development in the small-scale gold mining sector of Ghana.

References

ANON 1989. *Small-Scale Gold Mining Law, PNDCL 218*. Ghana Publishing Corporation, Accra, 1–5.

AMANKWAH, R. K. & ANIM-SACKEY, C. 2003. Some developments in the small-scale mining of precious minerals in Ghana. *Ghana Mining Journal*, 7, 39–42.

ESHUN, P. A & MIREKU-GYIMAH, D. 2002. Small scale mining in the Tarkwa District – A review of its impacts. *Proc. 7th International Symposium on Environmental Issues and Waste Management in Energy and Mineral Production*, Cagliari, Italy, 825–865.

HALL, J. 2003. *The Synergy Principle*, http://www.users.globalnet.co.uk/-iohnone/Enabling/Supporting_version1.htm, 1.

HARKINSON, J. 2003. *Confession of a Dangerous Mine*, http://www.gristmagazine.com, 1, 2.

HILSON, G. 2001. *A Contextual Review of the Ghanaian Small-Scale Mining Industry*, http://www.iied.org/mmsd pdfs/asm ghana.pdf, 9, 18.

ILO 1999. *Report for Discussion at the Tripartite Meeting on Social and Labour Issues in Small-Scale Mines*, International Labour Organisation, Geneva, http://www.ilo.org/public/english/dialogue/sector/techmeet/tmssm99/tmssmr.htm, 6.

UNIDO 2001. US/GHA/99/128 – *Assistance in Assessing and Reducing Mercury Pollution Emanating from Artisanal Gold in Ghana – Phase I, Part I – General Introduction and Assessment of Human Health*, 1–20. Part II – *Conduct of Surveys on River Systems and Overall Conclusions*, http://www.natural-resources.org/minerals/CD/docs/unido/sub2igoatt6part_1.pdf, 1–29.

Approaches to sustainable minerals development in Zambia

IMASIKU A. NYAMBE[1] & VINCENT M. KAWAMYA[2]

[1]*Geology Department, School of Mines, University of Zambia, PO Box 32379, Lusaka, Zambia (e-mail: inyambe@mines.unza.zm or nyambes@zamtel.zm)*

[2]*UNU/INRA – Mineral Resources Unit, School of Mines, University of Zambia, PO Box 32379, Lusaka, Zambia (e-mail: vkawamya@mines.unza.zm or inramru@zamtel.zm)*

Abstract: Zambia is a land-locked country occupying an area of 752 614 km^2, with the geology dominated by Archaean to Neoproterozoic age rocks that contain significant mineral resources. Economically, the most important of these are the Neoproterozoic Katangan rocks, which yield the copper and cobalt ores exploited in the Zambian Copperbelt. Copper and Cobalt exports account for over 80% of Zambia's foreign exchange earnings. Coal-bearing rocks of Karoo age (Permo-Carboniferous to Early Jurassic age) occur in rift valley basins such as the mid-Zambezi Valley in the south of the country. Issues such as a lack of capital investment led to privatization of the Zambian large-scale mining industry. Smaller scale gemstone mining is becoming increasingly important. For example, in 1998, the Ministry of Mines and Minerals Development granted over 200 gemstone-mining licences, 30 small-scale mining licences and prospecting permits, and over 70 artisanal mining rights for various minerals. A policy Framework Paper (1999–2001) published by the Zambian Government encourages the formalization of small-scale mining activities through the provision of licences for small-scale mining and gemstone trading, and the establishment of four regional mining bureaus for licensing and other services to the mining community. Furthermore, the Government has embarked on a Mining Sector Diversification Project with the support of the European Union, with the objectives of increasing export earnings through economic diversification, generating employment opportunities and contributing to poverty alleviation. In addition, the new policies and legal framework encourage private ownership of medium- and large-scale mining operations, and development of new mines. Currently, all former Government-owned mines that were under Zambia Consolidated Copper Mines Limited have been privatized. These activities are slowly generating much needed new investment to the mineral industry in Zambia. For many years, copper mining has supported the social and economic development of Zambia, accounting for around 90% of all Zambia's foreign exchange earnings in 1991. Copper reserves are steadily declining, with total reserves remaining estimated at just over 2 billion tonnes at an average grade of 2.51% total copper. There are no recent economic mineral deposits discoveries to replace depleting copper reserves. Lead and zinc mining at Kabwe finished in 1994 due to reserve depletion. The Maamba Coal Mine in the Zambezi Valley is in need of re-equipment and modernization. Gemstone mining and marketing needs a complete overhaul. This paper reviews the major issues that have contributed to decline in sustainable development of the mineral industry in Zambia and highlights initiatives, both private and public, currently undertaken to revitalize the sector.

Historical perspective

Significant mineral exploration started during the 1920s when the British South African Company (BSAC) offered prospecting rights to large multinational companies. This resulted in the discovery and development of the Luanshya and Nkana Mines (1931), Mufulira (1933) and Nchanga Underground Mine (1936) (Fig. 1). In 1957, Konkola Mine and Nchanga Open Cast Mine (Fig. 1) were brought into production as demand for metals rose in the post World War II era.

A year before Zambia gained independence, in 1964, Chibuluma Mine produced its first ore, followed by the Chambeshi open pit in 1965 (Fig. 1). The underground mine at Chambeshi opened in 1972 and Bwana Mkubwa, near Ndola, opened in 1971 (Fig. 1).

During this time the main companies were Anglo American Corporation (AAC) and Roan Selection Trust (RST). Annual Zambian copper production peaked at over 755 000 tonnes in 1965, and in 1969 the Zambian government

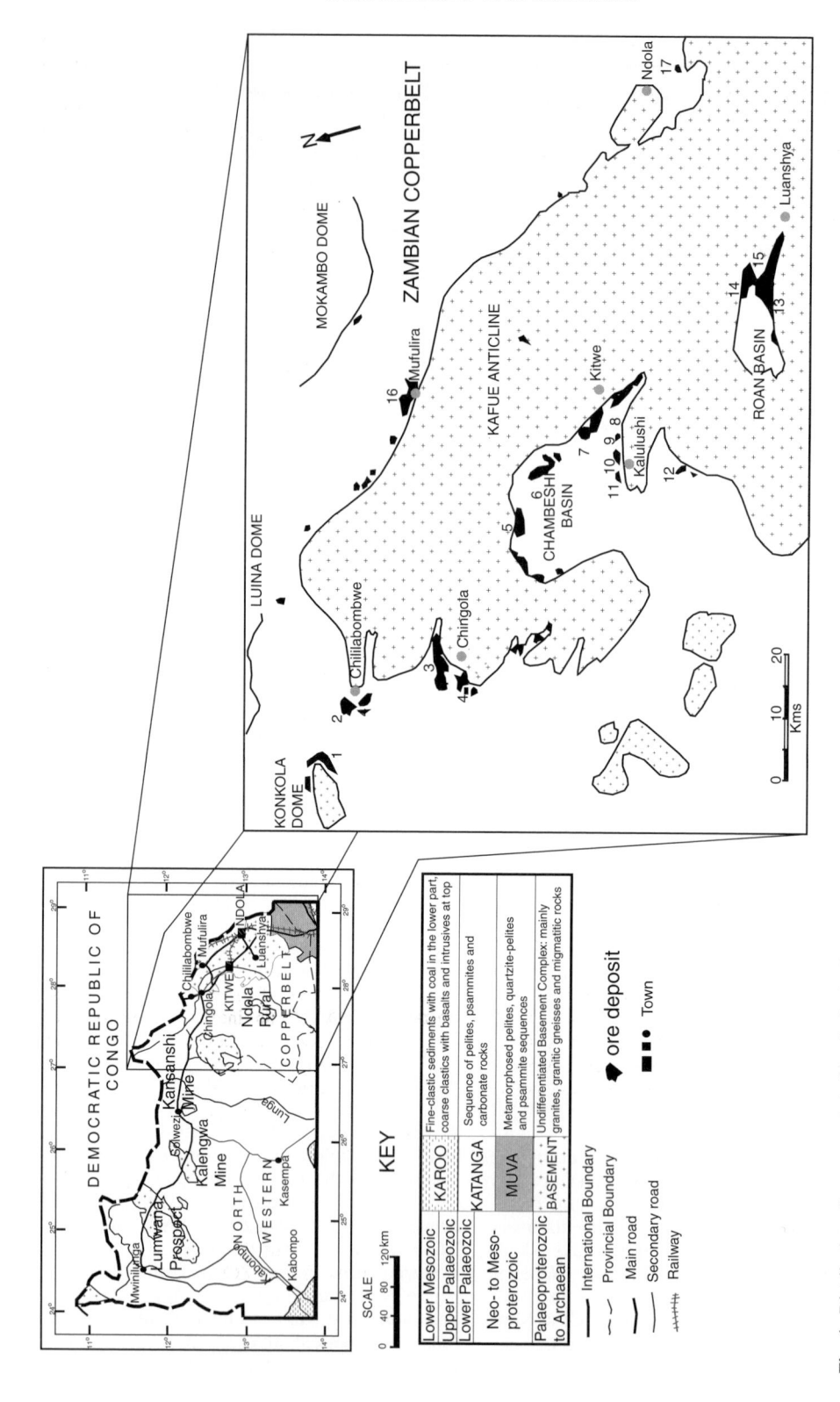

Fig. 1. Distribution of copper–cobalt deposits in the Zambian Copperbelt and Northwestern Province. Names on the map and mentioned in the text include 1, Konkola; 2, Chililabombwe; 3, Nchanga; 4, Chingola Orebodies; 5, Chambeshi; 6, Chambeshi South East; 7, Mindolo; 8, Nkana; 9, Nkana South Limb; 10, Chibuluma; 11, Chibuluma West; 12, Chibuluma South; 13, Muliashi; 14, Baluba; 15, Baluba East; 16, Mufulira; 17, Mbwana Mkubwa. (Note that other named locations in the text are in Fig. 2.)

acquired a 51% stake in all mining utilities and reorganized them into Nchanga Consolidated Copper Mines Limited (NCCM) and Roan Copper Mines Limited (RCM). In 1979, this stake was further increased to 60.3%.

Rapidly falling copper and cobalt prices, coupled with lack of re-investment capital and unfavourable government fiscal policies began to have an adverse effect on total copper production, which rapidly began to decline from the late 1970s.

Mineral resources potential

Zambia is a land locked country located between latitude 8° and 18° south and longitude 21° and 38° east. The country covers an area of 7 52 614 km^2 and is underlain predominantly by Archaean to Neoproterozoic rocks, which host a wide variety of minerals (Fig. 2).

Late Proterozoic rock assemblages of the Lufilian Arc, a large arcuate, northward-convex geological structure, dominate the geology of northwestern Zambia and extend into the southern part of the Democratic Republic of Congo (DRC). These are Katangan-aged rocks that are known for containing copper–cobalt (Cu–Co) deposits that are exploited in the Zambian Copperbelt. Lead and zinc occurrences have been historically exploited to the west of Kabwe. Manganese, silver and gold are mined in small quantities at various localities around the country. In addition to this, some gold, selenium, nickel and silver are produced as byproducts of copper–cobalt mining in the Copperbelt. The Copperbelt area is also host to renowned emerald-bearing pegmatites of the Mesoproterozoic Muva schists (Fig. 2).

In the eastern part of the country, pre-Katangan schists, gneisses, granulites and migmatites of probable Early Proterozoic age occur. These rocks are characterized by numerous post-tectonic pegmatites of various tonnages and grain sizes and are host to a variety of semi-precious minerals such as aquamarines, tourmalines and garnets. Similarly, pre-Katangan gneisses, granitic gneisses and granulites in the southern part of the country are host to good quality amethyst that crystallized in fissures and fractures ranging from centimetres to a metre in width and probably developed during the Permo-Carboniferous to Early Jurassic Karoo times (Fig. 2).

Coal-bearing rocks of the Karoo Supergroup (Permo-Carboniferous to Early Jurassic) are found in rift valley basins, the main ones being the mid-Zambezi and Luangwa valleys in the southern and eastern parts of the country (Fig. 2).

Occurrences of precious and semi-precious minerals in Zambia are not confined to these areas and are scattered sporadically throughout the country (Fig. 2). Areas of occurrences include Lundazi (Eastern), Mkushi (Central), Itezhi Tezhi, Kalomo (Southern), Ndola Rural (Copperbelt), and western and northwestern Zambia (Fig. 2). Various deposits of uranium have been located in Northwestern, Eastern, Southern and Copperbelt provinces (Fig. 2). Prospects for exploitation of large iron deposits exist in Northwestern, Central, Luapula, Southern and Lusaka provinces (Fig. 2). Petroleum and gas deposits are yet to be found, but may occur in Karoo basins of Zambia such as the Barotse (western Zambia), mid-Zambezi Valley and Luangwa Valley (Fig. 2).

Minerals-dependent economy

Mining continues to be a driver of economic development in Zambia. Mineral exports have been contributing between 60 and 90% of total national foreign exchange earnings for over 70 years. The mining industry currently employs over 35 000 people out of 4 75 000 who are in formal employment. The sector generates between 9 and 15% of the gross domestic product (GDP).

Zambia's copper–cobalt production, and hence earnings, has been steadily declining from the peak of 7 55 193 tonnes in 1969 (Fig. 3). Peak production in the 1960s was largely due to the continued development of new mines during the period 1936 to 1970 and the installation of state-of-the-art metallurgical plants. Production began to decline due to declining ore reserves, declining average grades (Fig. 4), increasing mine depth and lack of new ore discoveries. A chronic lack of reinvestment dealt perhaps the heaviest blow to the entire mining industry in Zambia during the two decades that followed peak production.

The shareholders of the large copper–cobalt operations showed no interest in reinvesting in the mines and the Zambian Government, the main shareholder, found itself unable to re-equip the mines. Similarly, mining and marketing of other precious and semi-precious minerals was also in disarray with mostly unaccounted production.

In 1982, in an attempt to redress falling production trends, NCCM and RCM were merged into the Zambia Consolidated Copper Mines Limited (ZCCM), but this did not restore productivity in the sector. ZCCM, which became a large conglomerate, involved itself in non-core business ventures such as transportation, tourism and agriculture, and was heavily indebted by 1990.

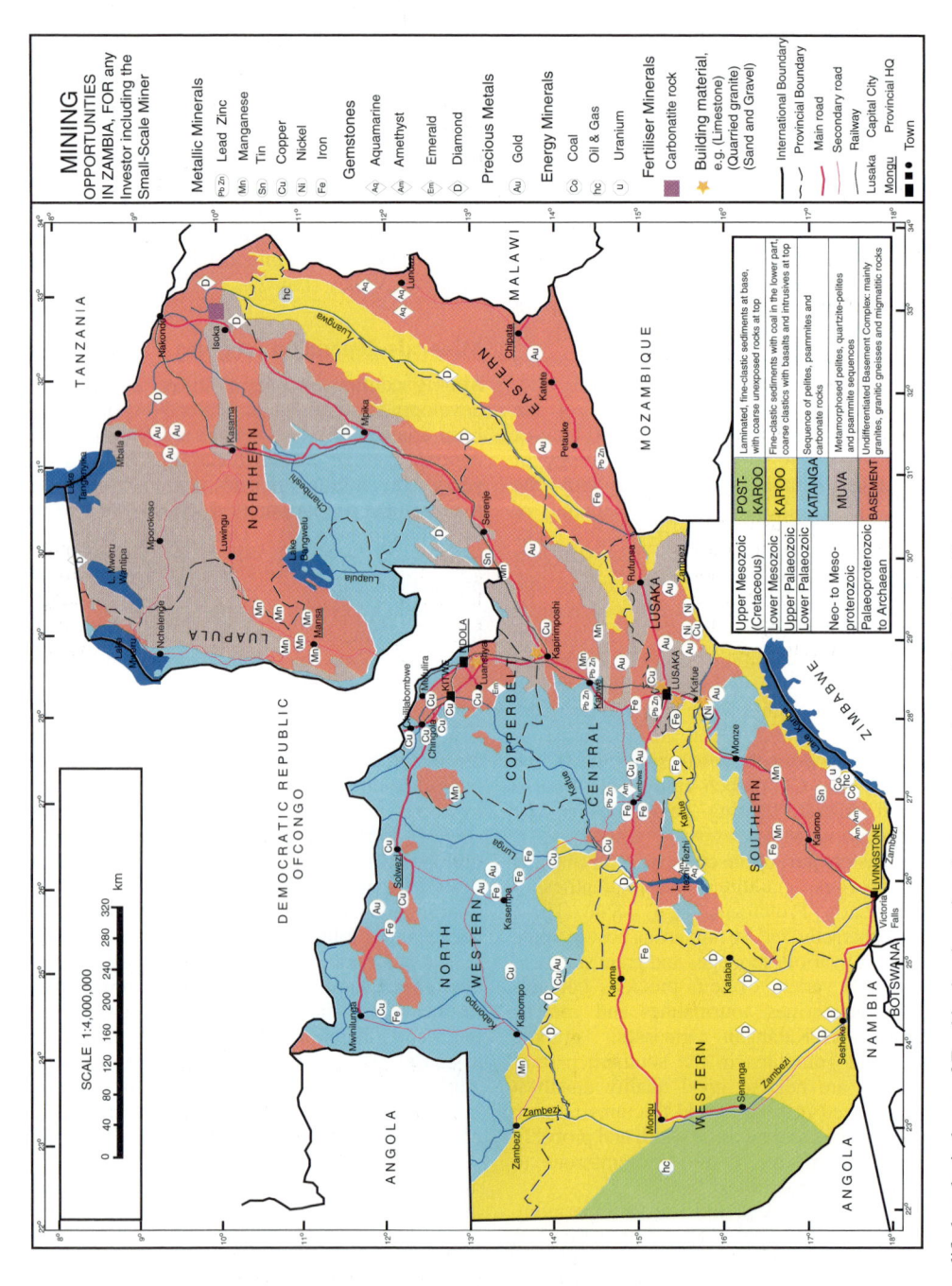

Fig. 2. Simplified geological map of Zambia showing mineral occurrences and named locations in the text.

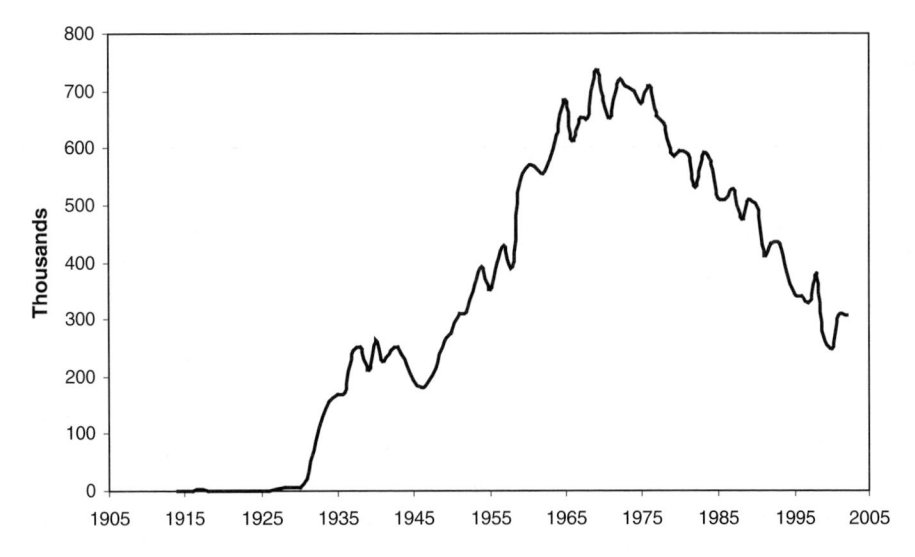

Fig. 3. Copper production 1914–2002 (*source*: Bonel).

In 1991, the Zambian Government embarked on a structural adjustment programme with a prime objective of liberalizing the economy. Policies to promote economic growth, move towards a sustainable balance of payments and encourage the participation of the private sector in economic activities were created.

Sustainable minerals development

In spite of numerous efforts for diversification away from mining, the industry continues to be the country's main source of foreign exchange earning, with exports accounting for more than 60% of total earnings. In 2002, metal exports

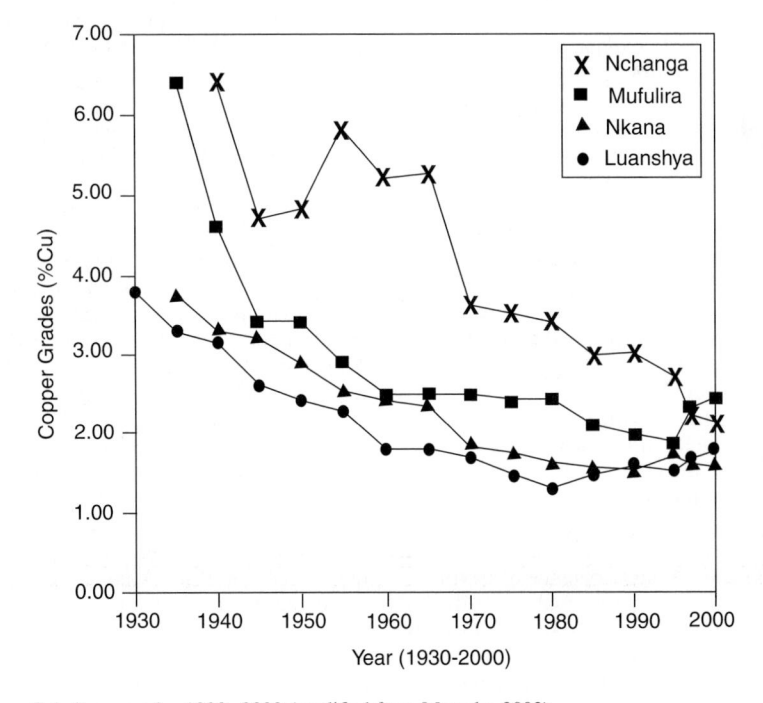

Fig. 4. Copper–Cobalt ore grades 1930–2000 (modified from Maambo 2002).

were around US$600 million of the national's total export earnings of US$917 million, representing a contribution of 65%.

The revenue from copper–cobalt relative to the total value of exports reflects the dominant position of mining as a major foreign exchange earner for the country (Table 1).

Coal production at Maamba in the Southern Province fell by 70% from 211 000 tonnes in 2001 to 64 000 tonnes in 2002 (Fig. 5). Production of magnetite, which is exploited alongside coal at Maamba Collieries in the Southern Province, fell from 6342 tonnes in 2001 to 927 tonnes in 2002, a drop of 85% (Fig. 5).

Pyrite, which is mined at Nampundwe Mine near Lusaka and used for the manufacture of sulphuric acid on the Copperbelt, showed improved performance from 83 700 tonnes to 94 900 tonnes, an increase of 13%.

Earlier, in 1991, with the advent of multiparty politics, the new Government had created a manifesto for a more market-driven economy. The main aims and objectives were to attract private investment for the development and rehabilitation of large-scale mines.

Almost immediately the industry began to enjoy the benefits of the Government's efforts to create an enabling fiscal environment for private investment. Several measures were put in place, which included the abolition of several taxes on mining profits and imports of mineral commodities, divesting of non-core mining activities by ZCCM, abolition of exchange controls and payment of duties on mining equipment imports for new mines.

The industry, however, continued to feel the effect of fluctuating mineral commodity prices (Fig. 6). For example, copper and cobalt prices during this period fell suddenly from record highs of US$3000 per tonne in 1995 to a low of US$1700 in 1999 for copper, and for cobalt from US$29 per pound in 1995 to US$7 in 2002. Austerity survival plans were created and loss making operations at Kabwe and Luanshya were closed in 1994. Chambeshi Mine had closed much earlier in 1988.

During 1991, faced with issues of declining reserves, dilapidated infrastructure and a lack of capital investment, the Government of Zambia embarked on a structural adjustment programme to liberalize the economy. Economic policies were created to promote growth and move towards a sustainable balance of payments position, with the ultimate goal of reducing poverty and raising living standards of all Zambians. In the mineral sector, the Government decided to revive the mining sector, with the aim of again making it the main contributor to Zambia's GDP. In 1995, the Government adopted a new Mining Policy, contained in the Mines and Minerals Act of 1995, and a revised fiscal regime in 1997. The new policy was aimed at encouraging private investment in exploration and promoting the development of new large-, medium- and small-scale mines to exploit metals, energy reserves, industrial minerals and gemstones. The overall objective of the policies included making the private sector the main producer and exporter of mineral products, promote small-scale mining (especially of gemstones) and reduce the environmental impact of mining operations. The main requirements of the policies were to expeditiously privatize the large-scale international mining companies and encourage the development of new mines.

By March 2000, ZCCM was privatized and fresh investment from the new owners promised a revived mining industry in spite of some localized problems.

Private sector initiatives

Large-scale copper–cobalt mining

The privatization process resulted in some large mining companies acquiring majority share holdings in the large-scale mining operations on the Zambian Copperbelt.

Konkola Copper Mines Limited (KCM), controlled by Anglo American Corporation (AAC), acquired majority shares in the Chingola-based Nchanga underground and open pit mines, Konkola mine in Chililabombwe and Nampundwe pyrite mine near Lusaka. In addition to these, KCM also obtained managerial control of the Nkana Smelter, and the acid plant and refinery in Kitwe.

AAC was by far the largest post-privatization investor and invested more than US$300 million before withdrawing in March 2002. This investment, mainly in infrastructure refurbishments, triggered a sharp reversal in total production from 269 000 tonnes in 1999 to over 300 000 tonnes in 2001, with 222 000 tonnes coming from KCM alone (Table 1). The eagerly anticipated US$569 million Konkola Deep project never started due, allegedly, to plummeting copper prices.

Negotiations between the Zambian Government and Sterlite Mining, a subsidiary of Vendenta, as an equity partner to replace AAC have been concluded. Sterlite Mining formally acquired 51% in KCM with management control in November 2004. KCM had, however, not been idle, in spite of the AAC divesture,

Table 1. *Copper–cobalt production and sales in Zambia, 1999–2003*

Mineral	1999 Production (tonnes)	1999 Sales ('000 US$)	2000 Production (tonnes)	2000 Sales ('000 US$)	2001 Production (tonnes)	2001 Sales ('000 US$)	2002 Production (tonnes)	2002 Sales ('000 US$)	2003 Production (tonnes)	2003 Sales ('000 US$)
Copper	265 879	371 653	259 573	425 201	298 150	506 659	377 743	509 703	345 054	637 914
Cobalt	3 761	96 525	4 373	72 174	4 376	82 914	3 984	49 958	1 600	24 705
Total Cu–Co	269 640	468 178	263 946	497 375	302 526	589 573	381 727	559 661	346 654	662 619

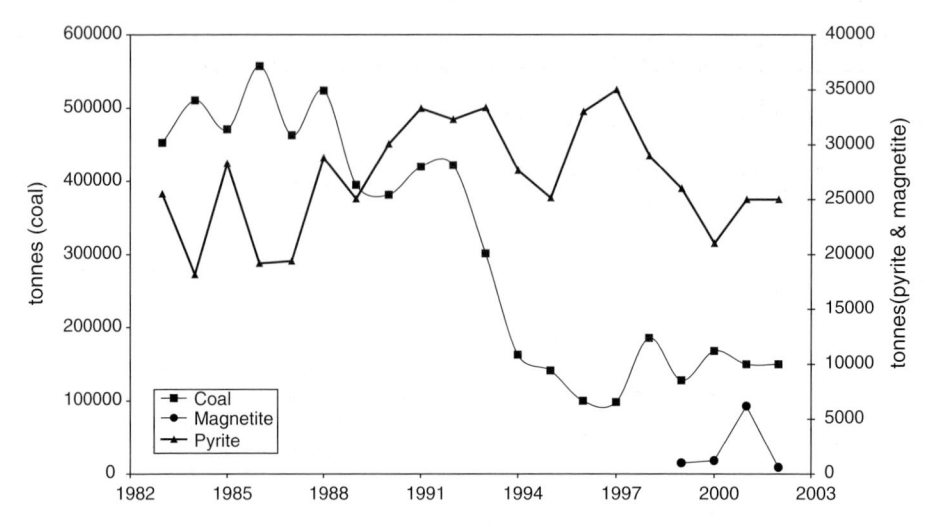

Fig. 5. Production of coal and pyrite (1983–2002) and magnetite (1999–2002) (coal and pyrite data from Bonel).

and had continued producing. During 2003 it produced 220 000 tonnes of copper and around 1000 tonnes of cobalt at a total production cost of slightly more than US$0.65 per pound.

The second largest investor on the Copperbelt is Mopani Copper Mines Limited (MCM), a company owned by Swiss-based Glencore (about 80%) and First Quantum Minerals (less than 20%). MCM owns the Nkana and Mufulira mines, the Nkana concentrator and Cobalt plant, as well as the concentrator, smelter and refinery at Mufulira.

MCM achieved all its own targets for the year 2002 and invested well over US$240 million in

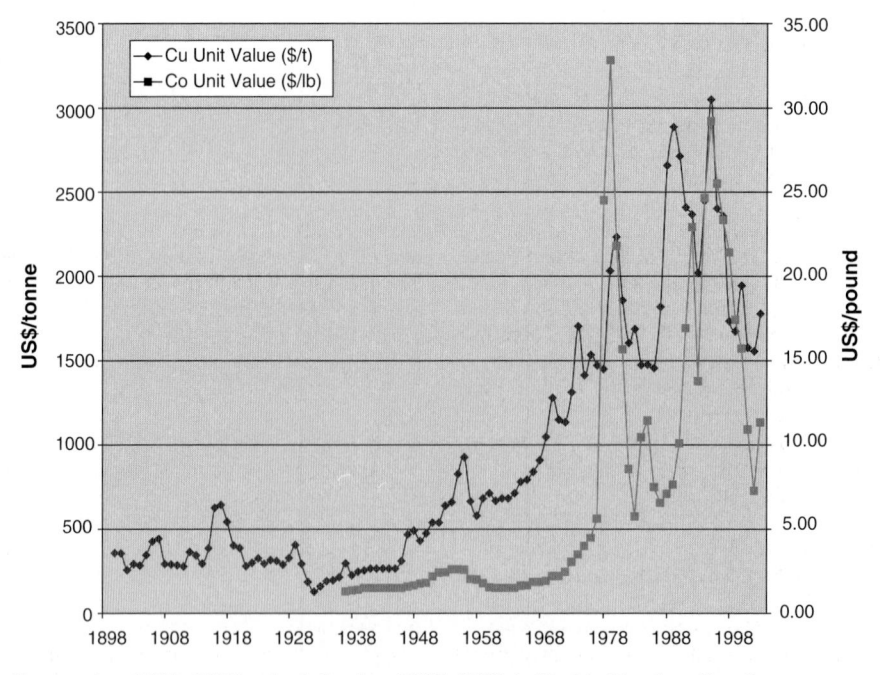

Fig. 6. Copper prices (1900–2003) and cobalt prices (1937–2003) in Zambia (data from Bonel).

both capital and operational expenditures. The company aims to produce 180 000 tonnes by 2005 at a total production cost of about US$0.65 per pound and expects to go into profitability from 2004.

Long- and short-term plans for MCM include deep mining at Mufulira Mine to a level of 1800 m below ground level and the reclamation of copper left in earlier stopes using *in situ* leaching technology. This will add another 15 to 20 years to the life of Mufulira Mine and produce in excess of 400 000 tonnes of finished copper. The life of Nkana Mine could also be extended by another 20 years by exploiting ore reserves in a synclinorium between the Central and South Ore Body shafts at Nkana (Tassell 2003).

The controversy over Luanshya Mine ended when Luanshya and Baluba mines were handed over to J&W Investment Group of Switzerland in January 2004. The operations at Luanshya were initially sold to an Indian company known as Binani in 1999 for nearly US$30 million and re-named Roan Antelope Mining Corporation of Zambia (RAMCOZ), only to be later placed into receivership and re-advertised.

Anglovaal Minerals (Avmin) acquired the surface properties at Chambeshi and a slag dump at Nkana in 1998. The company spent over US$120 million in the acquisition, refurbishment and construction of cobalt and acid plants, as well as a smelter at Chambeshi, for the extraction of copper and cobalt from concentrates and slag. Avmin's main interest at Luanshya was Baluba Mine, which would have offered an alternative source of concentrate for its plant. Unfortunately for Avmin, J&W Investment Group won the Luanshya mines bid, under the new name of Luanshya Copper Mines Plc with a purchase price of only US$7.2 million and a pledge of US$32 million investment over the next five years. With the Baluba bid lost, Avmin decided to withdraw from the Copperbelt and sold its metallurgical plants to J&W.

The acquisitions by J&W did not apparently include various undeveloped oxide ores in Luanshya, including the Muliashi North deposit, which could be worked by open cast mining coupled with advanced metallurgical technology to treat the oxide ores. These near-surface deposits offer the main potential for future expansion in this area.

Adjacent to J&W's metallurgical plants is Chambeshi underground mine, which has now been revived by a Chinese firm known as NFC Africa Mining. The mine was abandoned in 1983, but with substantial capital investment from NFC the mine has restarted, with production expected to be in the range 35 000 to 45 000 tonnes per annum.

New large-scale mines

The most exciting developments in metal mining in Zambia are perhaps the prospects of the industry to expand into the northwestern part of the country where reserves are estimated at more than 1.0 billion tonnes of low-grade oxide copper mineralization.

The Lumwana Joint Venture Copper Project, now owned 100% by Equinox Copper Ventures Limited, completed feasibility studies in October 2003 and has already applied for a large-scale mining licence on the Lumwana prospect.

First Quantum Minerals (FQM) and ZCCM Investment Holdings (80% and 20%, respectively) commenced mine construction in September 2003 at Kansanshi Mine near Solwezi, which will have the capacity of about 102 000 tonnes of copper and 25 000 ounces of gold per annum. Operations began in December 2004 and the mine was officially opened on 10th August, 2005.

FQM also operates the promising Bwana Mkubwa Mine (near Ndola) and Lonshi Open Pit mine across the border in the DRC. Lonshi is reported to have resources in excess of 3 50 000 tonnes of copper and Bwana Mkubwa is capable of producing up to 15 000 tonnes of copper per annum. FQM also produces sulphuric acid in excess for other users.

Other undeveloped copper–cobalt ore bodies include large reserves at Kalengwa in Northwestern Province and at Mkushi in the Central Province.

Future expansions

The most advanced of these is the re-start of operations at Chibuluma Mine, not far from MCM's Nkana Mine. Chibuluma comprises Chibuluma South (open pit) and Chibuluma West, which is an underground mine. Chibuluma South is the target for a planned investment of more than US$15 million by the shareholders of Chibuluma Mines Plc, who are Metorex and the Industrial Development Corporation (IDC) of South Africa.

Having installed a new metallurgical unit at the site, it is expected that Chibuluma South, which will become an underground operation, will produce more than 15 000 tonnes of copper per annum by mid-2005. Chibuluma is reportedly the world's highest grade copper mine at more than 8% copper.

Other envisaged expansions include mining at depth at Mindolo by MCM and the well-publicized Konkola Deep Mine Project in Chililabombwe by the new KCM owners.

Exploration and prospecting

Soon after the adoption of more economically liberal policies in the Zambian mining industry to encourage private sector participation during the period 1991–1992, several international mining houses obtained exploration and prospecting rights over large areas of land. Companies such as BHP-Billiton, Anglo American exploration, Equinox Resources and Anglovaal Minerals performed exploration work for various mineral commodities in different parts of the country. Although exploration activities have generally decreased, companies such as BHP Billiton and Equinox have reached very advanced levels in exploration and many others still hold prospects for copper, gold, cobalt and diamonds.

The Zambian government through the Ministry of Mines and Minerals Development (MMMD) continue to issue mining rights in various categories, especially for small-scale mining of precious and semi precious minerals. As at August 2005, a total of 2539 licences and permits had been issued (Table 2).

The large number of small-scale mining licences issued to date is due to the proliferation of small-scale copper mining activities in the Northwestern Province, which has attracted a number of miners following MCM's purchase of copper ores on the open market to feed its concentrator and smelter. Further, the Government has removed the duty on imported mineral materials, which has allowed MCM and KCM to import ores and concentrates from DRC. In addition, the boost in production is also attributed to wider market availability for production offered by small-scale metallurgical plants in Ndola and Kabwe, which were constructed recently.

Table 2. *Artisanal and small-scale licences and permits issued in Zambia as at June 2003*

Type	Number
Small-scale mining licences	241
Gemstones mining licences	735
Prospecting permits	200
Artisanal mining licences	876
Reconnaissance licences	487
Gemstones selling certificate	1136

Small-scale mining

Small-scale mining in Zambia exploits a variety of gemstones (Fig. 1), including emeralds, amethysts, aquamarines, tourmalines, garnets and citrines. Emeralds are by far the most predominant. These gemstones are mainly mined in five areas: Ndola Rural (emeralds), the Lundazi, Mkushi and Itezhi Tezhi areas (aquamarines and tourmalines) and Kalomo (amethysts).

This variety offers great potential for providing the necessary resources for financing development and poverty reduction. However, gemstone deposits are generally located in undeveloped and very remote parts of the country. Their exploitation requires the development of roads, telecommunications, and other facilities such as schools and clinics for miners. Small-scale mining, if well planned and developed, is one area that could lead to poverty reduction through employment and household income generation. In addition, downstream industries such as carpentry and brick laying are expected to increase. There is also the possibility of small-scale exploration of phosphate deposits in the eastern and northern parts of the country. These may lead to the provision of essential inputs of agro-minerals to agricultural communities.

Although actual gemstone production and sales are difficult to quantify, official figures indicate that total gemstone exports showed a robust rise of more than 50% between 1999 and 2000. From 2000 to 2001, exports increased only marginally by 9% from 1 080 000 kg to about 1 172 000 kg. Between 2001 and 2002, for example, the industry reported a decrease in sales of 6% (Figs 7 and 8). This is probably due to unfavourable conditions prevailing in this sector. For example, in 2002 the Mineral Royalty Tax of 5% was reintroduced for gemstones as opposed to 0.6% negotiated by the large-scale miners in their development agreements. This has not been well received by small-scale operators.

Other minerals that are being worked at low levels include industrial minerals such as building aggregates and agro-minerals. Agro-minerals, such as phosphate rock and lime, are important as they contribute to household food security by providing an affordable source of plant nutrients. Untapped mineral resources include large deposits of tantalite and emeralds in the Northwestern Province, diamonds in Mfuwe in the Eastern Province, sodalite in the Northwestern Province, chromium and gold in the Lusaka Province, lead, zinc, gold and silver in the Central Province.

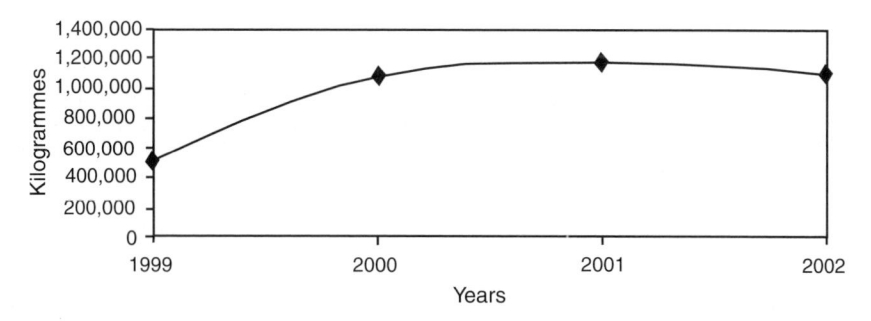

Fig. 7. Total gemstone production (1999–2002) of tourmaline, beryl, garnet, emerald and aquamarine.

National strategies for sustainable minerals development

In October 2002, the Government of Zambia inaugurated a three-year Transition National Development Plan (TNDP) with the goal of achieving 'Sustainable Economic Growth, Employment Creation and Poverty Reduction'. Since the early 1990s national planning has been negated in the name of liberalization. This has had an adverse effect in sectoral planning and implementation. The TNDP heavily draws from the Government's Poverty Reduction Strategy Paper (PRSP) and also covers areas not fully addressed in the PRSP such as science and technology, regional integration, and law and order.

The prime importance of the mining sector to the goals and objectives of the TNDP is recognized and two main objectives are expected to be achieved in the fiscal period 2002–2005. The first objective is to attract private investment for rehabilitation of the large-scale mining infrastructure to increase their contribution to gross domestic product. The second is to revitalize small-scale mining in order to realize its potential as a principal industry for rural poverty reduction.

The Government, through its Ministry of Mines and Minerals Development (MMMD), has identified issues that need to be addressed in order to promote sustainable development in the minerals sector in Zambia.

Promotion of investment in mining

These activities are meant to increase the mining sector contribution to the national economy. These include packaging, production and

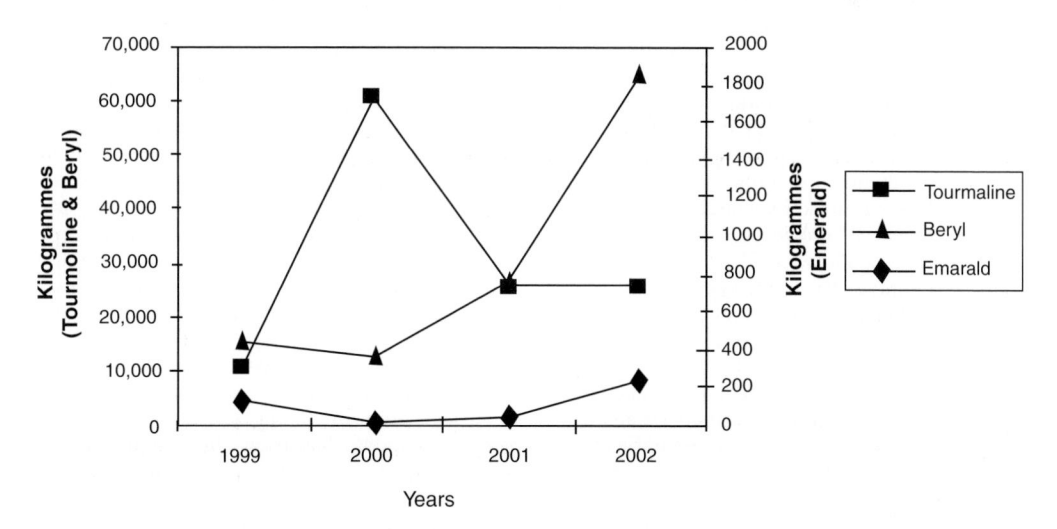

Fig. 8. Production of tourmaline, beryl and emeralds (1999–2002).

dissemination of information on resources, including the existing investment opportunities to both local and foreign investors and the streamlining of the licensing process. Two programmes have been suggested:

Small-scale mining programmes. The following programmes, in order of priority, are ongoing or will be undertaken:

(1) *The Mining Sector Diversification Programme (MSDP).* This programme, which is supported by the European Union Commission, has already started. The programme intends to broaden the focus of production from gemstones to include agro- and industrial minerals. It aims to make adequate geological data available to small-scale miners and prospectors, enable small-scale miners and processors to manage their businesses better; improve access to adequate capital and equipment for small-scale miners; equip small-scale miners with adequate mining safety, valuation and processing skills; enable small-scale miners and processors to obtain fair market prices for their products and ensure that effective and fair mining regulations and taxation policies are in place. Financing mechanisms include the introduction and implementation of the following:
- *Enterprise Development Fund (EDF).* Two lines of credit, namely, the investment credit and the export pre-shipment facility will be made available to small-scale mines operators.
- *Trade and Enterprise Support Facility (TESF).* This is intended to support feasibility and market studies, business re-engineering and corporate recovery and support preparatory activities for companies seeking to list on the Lusaka Stock Exchange.
- *Creation of Mining Community Development Fund (MCDF).* This is a programme that is intended to ensure that owners of productive mines contribute directly to the development of communities in areas where they operate. This programme will be financed and sustained from contributions made by mine owners. Local communities themselves will manage it with appropriate participation of traditional rulers (Chiefs), to ensure that these resources are applied towards development and poverty reduction efforts.

(2) *Reintroduction of the Gemstone Exchange Scheme.* There is a lot of tax revenue lost in the illicit trafficking of gemstones. The marketing methods for gemstones have been unsatisfactory and intermediaries have taken advantage of the marketing situation, by obtaining gemstones directly from workers at the expense of the mine owners. Therefore, the reintroduction of the Gemstone Exchange Scheme will serve as a forum for producers and buyers of rough and processed gemstones to conduct auctions and routine transactions. This will revitalize the sector and increase the incomes of the communities around the mining operations. There will be more production because of the availability of, and accessibility to, the market for these products.

(3) *Introduction of a plant hire scheme.* Capital equipment is a main requirement in the gemstone industry. Most small-scale miners use basic tools such as picks, hoes and shovels that are not effective, resulting in low production levels. The introduction of the plant hire scheme will encourage efficiency and promote the development of new mines. The scheme will target mainly new operators as well as those that may not be able to access the facility under the MSDP programme. The scheme will also make use of newly acquired Zambia National Service equipment. The Zambian Government, through a Chinese grant fund, acquired the equipment in order to spearhead rural development through the provision of such services as land clearance and dam construction.

(4) *Creation of a revolving fund for provision of working capital.* Under this programme, small-scale miners and prospectors who may not be eligible for loans from the bank and the MSDP facility will be able to get money to start their operations. The small-scale miners themselves will manage the fund through a privately owned institution. So far, only one non-traditional sector miner has accessed part of the €16.5 million meant for the credit facility under the Mining Sector Diversification Programme. The performance of the programme in terms of loan disbursements has been sluggish due to the stringency of the conditions attached. Submissions have been made to the European Union delegation to revise the terms of this credit in order to make it more accessible to a majority of miners.

Large-scale mining programmes. The main emphasis will be on marketing 'green fields' for new mines and the rehabilitation of the infrastructure of existing mines by offering an attractive environment for private sector investment. The development of Kansanshi and Lumwana copper deposits would have a positive impact on economic growth and social benefits in terms of improvements in infrastructure and employment opportunities in Northwestern Province.

Establishment of an integrated information system for provision of geological and mining information

There is still much to be done to encourage significant participation of the private sector in the minerals sector. This requires, among other things, good basic geological data in the form of maps, and improved infrastructure, full implementation of new mining policies and legislation, and services to support investment.

Only 55% of Zambia has been geologically mapped at 1:100 000 scale. Lack of comprehensive geological data has greatly affected investment promotion of the mining sector. The Government will, therefore, endeavour to establish and maintain an integrated information system in order to make available comprehensive geological and mining information to interested investors.

Formulation of an appropriate policy and legal framework to guide the mining sector

The objective of this initiative is to provide an appropriate policy and legal frameworks in order to guide the operations of the mining sector. This will involve the amendment of the Mines and Minerals Act of 1995 and subsidiary legislation, as well as revising the Mines Policy.

The Mines and Minerals Act has been diagnosed with a lot of shortcomings, ranging from inadequacy in addressing environmental concerns to the inability to fully safeguard investor interests. In addition to identifiable flaws, the mining legislation lacks regulations with which to implement it. These deficiencies require immediate attention, such as the finalization of the Mines and Minerals Act review, which originally took place from 24th to 30th September 2003.

Human resource development and capacity building

This initiative aims at managing and developing human resources in the minerals industry in order to improve the quality of service delivery. This will involve the recruitment of qualified and competent staff in the industry (small- and large-scale) and measures put in place for their professional development and retention.

Promotion of research and development in the fields of mining

Research and development in the areas of mining will greatly increase the knowledge base and technological development of the sector. It is the intention of the Government to undertake and/or commission a number of research and development activities to ensure that appropriate mining operations and techniques are developed and made available to mining investors. Academic and research institutions will be heavily involved in this activity.

Other strategies

Other strategies will involve the establishment and upgrading of the seismic network to facilitate appropriate interventions and the mitigation and monitoring of the environmental impact of mining activities.

Costs

The outlined intervention measures are expected to cost the government US$36.9 million in the Transition National Development Plan (TNDP), with the Mining Sector Diversification Programme (MSDP) alone costing about US$23.0 million. The setting up of a plant hire scheme and the reintroduction of a Gemstone Exchange Scheme will cost around US$2.0 million and US$6.0 million, respectively.

Concluding remarks

The minerals industry in Zambia will continue to play a key role in fuelling national economic development for many years to come.

Particular emphasis should be placed on supporting, developing and harnessing the small-scale mining of gemstones and other precious and semi-precious minerals. These offer perhaps the greatest potential for advancing both development and poverty alleviation in rural communities. The current intervention measures being focused into this sector are therefore crucial, as they will render the small-scale mining industry viable.

The large-scale mining sector is also on the upswing due to the renewed private participation in both the development of new mines, and in the expansion and rehabilitation of existing operations. It is anticipated that if currently favourable metal prices and high demand, coupled with good fiscal policies, persists, then total copper–cobalt production from large-scale mining activities will continue to positively contribute substantially to national foreign exchange earnings and the promotion of downstream industries.

The sustainable development of resources requires well-qualified personnel in technical fields with managerial administrative skills in order effectively to formulate appropriate policy frameworks, and to implement and monitor the developmental process of minerals industry.

Overall, mining in Zambia has contributed to wealth creation and poverty alleviation and will continue to do so. The industry still continues to sustain the economy through income generation, employment creation, accelerated industrialization and the development of the infrastructure such as roads, power lines and railways.

The authors would like to thank the conveners of the Sustainable Minerals in the Developing World Conference (Brian Marker (ODPM), Mike Petterson (BGS) and Fiona McEvoy (BGS)) without whose assistance, support and encouragement this paper would not have materialized. The authors would also like to thank K. Bonel of BGS and P. E. J. Pitfield for the critical comments and reviews that led to the improvement of the manuscript. Bonel is further thanked for the provision of some of the figures. P. Zimba, S. Nyambe and D. C. W. Nkhuwa are thanked for re-drafting of some of the figures.

References

MAAMBO, P. H. 2002. Overview of mining in Zambia: Past, present and the future. *Zambia Journal of Applied Earth Sciences*, **15**(1), 1–15.

TASSELL, A. 2003. The Copperbelt – new mines planned but uncertainty still in the air. *African Mining*, **8**(3), 44–61.

Sustainable mineral development: case study from Kenya

T. C. DAVIES & O. OSANO

School of Environmental Studies, Moi University, PO Box 3900, 30100 Eldoret, Kenya (e-mail: daviestheo@hotmail.com)

Abstract: The last two to three decades have witnessed a rapid growth in the mining industry in Kenya. The suite of minerals includes metals such as gold, silver, copper, zinc and titanium, and industrial minerals ranging from talc and gypsum to dolomite and gemstones. Methods of exploitation, processing and beneficiation of these mineral resources can have diverse effects on the country's socio-economic position, its varied ecosystems and general environment. The Mining Act (Cap 306) of 1940, which has been the principal Act for regulating minerals use and mining, lacks clear provisions on environmental management. This paper discusses mining guidelines set forth in a new regulatory framework known as the Environmental Management and Coordination Act of 2000, and evaluates their effectiveness and applicability by examining the proposed titanium mining project in the Kwale district. Through a major mining project, the Environmental Impact Assessment exposed the difficulties often encountered in implementing regulatory controls in all new mining ventures in Kenya, both large- and small-scale. It is submitted that the challenge of sustainability should be a major concern of the mineral industry to demonstrate not only profitability but also benefits to society and preservation of environmental integrity.

The known suite of minerals in Kenya range from metals such as gold, silver, copper and zinc to industrial minerals such as talc, gypsum and dolomite. Other minerals commercially exploited include fluorite, trona (sodium sesquicarbonate) and various building stones. Potential for petroleum deposits exists, but its investigation is still only at the prospecting stage. The search for new mineral deposits continues, as do ways of making viable the exploitation of known resources. Consequently, wider and wider stretches of the country are being opened up to development (Davies 1993, 1996). The so-called 'artisanal and small-scale mining' (ASM) is particularly significant in this regard. It comprises the mining of gold by panners in riverbeds and along riverbanks, small-scale mining of gemstones and marble, as well as industrial and constructional materials. It plays a vital role in job creation and poverty alleviation, but it is accompanied by a number of social and environmental consequences.

As a result of increasing public awareness of environmental matters, there has been a corresponding need for tighter environmental legislation and planning controls on all future mining projects, both large- and small-scale. The Mining Act of Kenya (Cap 306) (Government of Kenya 1940), which is the principal Act for regulating minerals use and mining, is currently undergoing revision and will take into account environmental management matters.

The protracted dialogue on both the environmental and socio-ecological issues concerned with the development and operation of the titanium-mining project in Kwale district of Kenya has only just been concluded. This was done within the framework of an Environmental Impact Assessment (EIA) conducted according to the requirements of the Kenya Environmental Management and Co-ordination Act (EMCA) (Government of Kenya 2000) and in compliance with World Bank standards. An EIA document submitted to the enforcement authority, The National Environmental Management Authority (NEMA), has now enabled the issue of an EIA Licence and a Mining Licence.

This paper considers several challenges facing the minerals industry in Kenya, particularly the ASM sector, and describes the manner in which these challenges are being met. The inclusion of a case study illustrates how undesirable environmental and socio-ecological consequences during large-scale operations can be mollified and rehabilitation procedures addressed if a proper EIA is conducted prior to mine inception.

The mineral potential of Kenya

The major rock types constituting the geology of Kenya include the Archaean-Palaeoproterozoic

From: MARKER, B. R., PETTERSON, M. G., MCEVOY, F. & STEPHENSON, M. H. (eds) 2005. *Sustainable Minerals Operations in the Developing World.* Geological Society, London, Special Publications, **250**, 87–94.
0305-8719/05/$15.00 © The Geological Society of London 2005.

granite-greenstone terrain found in western Kenya around Lake Victoria, the Neoproterozoic Kisii Group (formerly the Kisii Series), the Mozambique Belt and the Upper Palaeozoic to Mesozoic Karoo sediments of coastal and northeastern Kenya (Fig. 1). Younger rocks are represented by Tertiary and Quaternary sediments in eastern Kenya and the coastal strip bordering the Indian Ocean.

The greenstones mainly comprise two rock groups, the Nyanzian Group (formerly Nyanzian System), which is unconformably overlain by the younger Kavirondian Group (formerly Kavirondian System). The Nyanzian Group produces gold, copper, silver and other base metals.

The Kisii Group is alternatively called the Bukoban, particularly in Tanzania, where it is extensively developed. In Kenya, the group is represented by a volcano-sedimentary sequence that occurs in the Kisii area south of Lake Victoria (Fig. 1). It is within these sediments that a kaolinized feldspathic unit is mined for the so-called Kisii soapstone, which is extensively used in making handicrafts that are renowned for their beauty. Other mineral resources within the Kisii Group include iron ore, the potential of which is yet to be fully investigated.

The Neoproterozoic Mozambique Belt yields gemstones including ruby, blue sapphire, beryl, amethyst, garnets and tourmaline. Fluorite and marbles are the major industrial minerals currently mined within this belt. Kenya's fluorite deposits at the Kerio Valley represent one of the world's largest known fluorite deposits and it constitute an important foreign exchange earner for the country.

Granites, gneisses, migmatites and quartzites in this belt are extensively used in the construction industry, both as dimensional and decorative stones. Marble is also mined for use in the manufacture of cement and the production of lime.

Groundwater resources are potentially available or exploited in the various aquifers between the underlying Mozambique Belt and its overlying Cenozoic volcanic rocks and sediments, as well as with the cover sequence.

All the Palaeozoic–Mesozoic sedimentary rocks in Kenya occur as north–south oriented belts in the coastal and northeastern regions of the country (Fig. 1). Recent oil exploration work is also reported to have encountered large deposits of these sedimentary rocks at depth in the Turkana Basin of the Rift System.

The Tertiary in Kenya and East Africa is notable largely for its igneous activity associated with rift faulting. The Rift System Volcanics are a source of construction materials such as road gravel, dimensional materials and clays, and in addition host the geothermal fields. Heat from the geothermal fields is currently used to generate more than 200 MW of electricity.

Quaternary sediments, largely terrestrial in origin, are widespread in Kenya, especially in the eastern side of the country and in the Rift System. These sediments are formed of thick lacustrine and fluvial deposits, including evaporates. Diatomite deposits at Kariandusi near Lake Elmentaita and the trona deposits of Lake Magadi are currently being exploited. Gypsum deposits found in several localities in Kenya, together with the lignite occurrences in the Kitui area, are all of Quaternary age.

Current mining activities in Kenya are concentrated on soda ash, crushed refined soda, fluorite, diatomite, crude salt, vermiculite, kaolin, limestone products, corundum, carbon dioxide, gold, lead, gem garnet and other gemstones.

Environmental considerations in exploration, exploitation and processing (ASM)

Among the minerals and materials of interest to ASM are gold, gemstones, marble, as well as industrial and construction materials (Fig. 2). The mining of gold by panners in riverbeds and along riverbanks has a long history in Kenya. Of the various factors associated with ASM, the common denominator is the abject poverty within the sector. One body of opinion supports the individuals who operate in this sector because they have no other options for survival. Another view is that the panners are destroying the country's river systems through poor mining practices. There are also a number of other social and environmental consequences borne by other groups in society, especially inhabitants downstream. These impacts, according to Muhongo (2003), include:

- scarification;
- other landscape degradation such as deforestation;
- siltation;
- soil erosion;
- inappropriate disposal of tailings;
- acid mine drainage;
- heavy metal pollution (e.g. mercury, zinc and copper);
- cyanide pollution;
- radioactive emissions;
- dust;
- mine safety (accidents such as cave-ins);
- security and immigration problems; and
- child labour.

GEOLOGY EXPLANATION

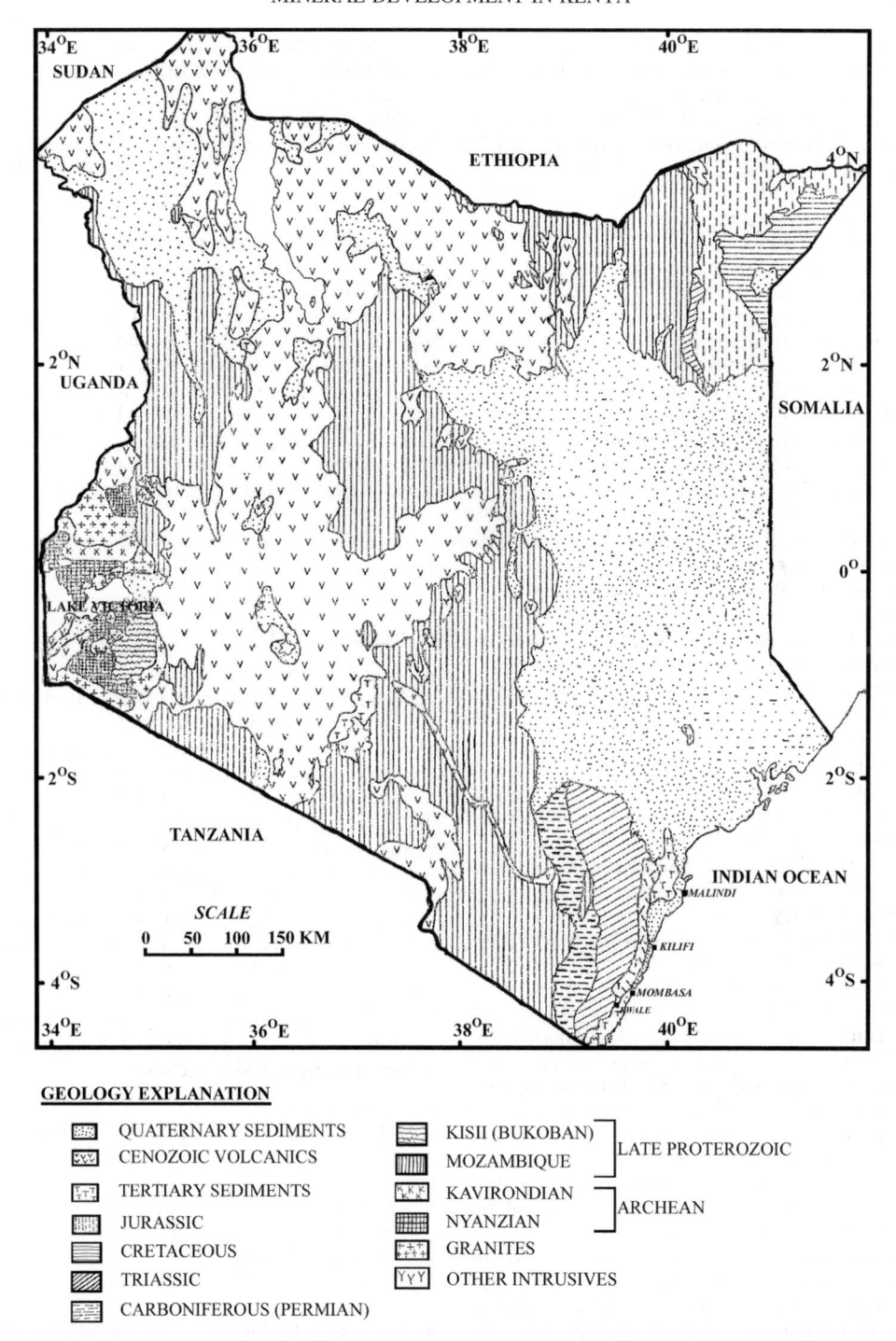

- QUATERNARY SEDIMENTS
- CENOZOIC VOLCANICS
- TERTIARY SEDIMENTS
- JURASSIC
- CRETACEOUS
- TRIASSIC
- CARBONIFEROUS (PERMIAN)
- KISII (BUKOBAN) ⎱ LATE PROTEROZOIC
- MOZAMBIQUE ⎰
- KAVIRONDIAN ⎱ ARCHEAN
- NYANZIAN ⎰
- GRANITES
- OTHER INTRUSIVES

Fig. 1. Simplified geological map of Kenya showing the major rock units (modified from East Africa Educational Publishers Limited 1991).

Fig. 2. Poor mining method along a riverbank.

Major problems faced by ASM include:

- lack of capital;
- lack of cheap and rapid response consulting experts;
- lack of basic knowledge on identification of hosting minerals;
- inability to give reliable estimation of amount of mineral reserves;
- lack of knowledge of most appropriate technology for optimum yields;
- lack of processing facilities;
- lack of training in the understanding of Earth processes;
- lack of reliable and well-paying markets;
- loss of mineralized land to big companies; and
- ignorance of mining regulations and laws.

Information, education and technology

A comprehensive information system is necessary that covers the acquisition, storage, processing and dissemination of information on the locations of mines and quarries. In Kenya, there is a paucity of data on the ASM sector, including the numbers of panners, their socio-economic characteristics, and the economic costs and benefits of gold panning. Effective policy development requires sound information to support decision making. A statistical survey of panners and assessment of key environmental impacts as well as a cost-benefit analysis should be carried out.

For sustainability, it is important to raise the educational standing of the panners. In the developing countries, miners' knowledge acquisition is mainly informal (Styles 2001; D'Souza & Barney 2003; Harrison 2003). From the perspective of an individual Kenyan panner, the mining of gold is profitable. But a broader economic analysis would probably show that in many instances, the economic, social and environmental costs of panning far outweigh the benefits.

- *Benefits* – measured by determining the income from gold sales at world prices and permit fees.
- *Economic costs* – estimate to include capital and labour.
- *Social costs* – include the cost of providing clean water and better sanitation facilities (water and sanitation facilities are often poor with very few panners having access to clean, safe water and proper sanitation facilities).
- *Environmental costs* – include cost of reducing siltation and undertaking riverbank rehabilitation as well as addressing health risks.

Mining methods often show an almost complete disregard for the long-term sustainability of the river systems and the environment, in general. Gold is commonly worked from riverbanks, mainly by horizontal tunneling, which leads to undercutting, collapse and siltation.

Although many techniques are used for gold recovery, few are really suitable. The efficiency of gold recovery can be notoriously poor (reputedly around 30–40%), as evidenced by the fact that tailings are reworked many times. Sluicing, which is widely used, can be relatively efficient, but only where the gold grains are not too small and the gold trapping mechanism (sluice box) (Fig. 3) is efficiently designed. But it is very difficult to obtain accurate information on grain-size distribution of gold, and testing has to be a vital factor before recommendations on mining techniques can be made. Furthermore, the level of mercury used to separate gold is often too low for optimum recovery.

Training and education of panners and local councils need to be improved regarding environmentally sustainable mining practices and use of better technology to improve gold production. At the same time, training in ore testing and sluice box assessments should be incorporated in any training schedule.

A participatory approach

New policy and regulatory instruments can best be developed through a broad consultative process involving key stakeholder groups. Artisanal and small-scale mining in Kenya involves business establishments: companies, cooperatives, individuals and families, and is affected by:

- human resources – knowledge, skills, experience, contacts;
- capital;
- mining and processing technology;
- production; and
- business transactions.

From earlier Knowledge and Research Programmes by the Department for International Development, UK, and interaction with different stakeholders in the minerals industry, it has become increasingly evident that there is a need for stakeholders to work together to address issues of common concern. Attempts to address problems in the past have been piecemeal and *ad hoc*, with little research or consultation with the groups that are supposed to benefit from these actions (Sridharan 2003). The observed dissatisfaction and breakdown of trust in mining regions is a result of company actions as well as of government inaction. Combined inputs from many stakeholders will allow for a more robust and sustainable mineral development plan.

Legal and regulatory framework

Environmental regulatory framework

The Mining Act (Cap 306) of Kenya (Government of Kenya 1940) emphasizes matters of occupational health and safety, but lacks clear provisions on environmental management. In light of this omission, another framework law on environmental management known as the National Environmental and Coordination Act (Government of Kenya 2000) was enacted for the following purposes:

(1) to provide a framework for sound environmental management in Kenya;
(2) to provide a framework for improved legal and administrative coordination of the diverse sectoral initiatives for the management of the environment;

Fig. 3. Typical sluice box arrangement used in artisanal and small-scale mining.

(3) to be the principal instrument of Government in implementation of all policies relating to the environment.

This Act embodies the principles of sustainable development and also stipulates that all EIA procedures should engender the management tools inherent in Sections 58 to 67 of this Act. The projects to be subjected to EIA are specified in the Second Schedule of this Act.

At present, full mining licences can be granted on the basis of:

(1) Mining Location;
(2) Exclusive Prospecting Licence;
(3) Special Licence;
(4) Mining Lease;
(5) Special Mining Lease.

The Exclusive Prospecting Licence is granted on approval of the mining location, allowing exploration work to proceed. This is then be followed by the issue of a Special Licence, upon which an approved EIA must be undertaken before obtaining a mining licence, which is not commodity-dependent.

In Kenya, the need for the development of an effective policy framework for the ASM sector has been realized. Current mining legislation and institutional structures are being evaluated to determine how they affect the operation of the ASM sector. Further work should involve in-depth consultation with the local government and other stakeholders on the services within the legislative arrangements.

Environmental management guiding principles

Two environmental management requirements, amongst others, are fully recognized in Kenya:

- sustainability;
- provision of ecosystem services within a wholesome environment.

New policies and strategies for Kenya's resources and environmental management are beginning to emerge in line with the innovative approach of Agenda 21 set by the countries of the world during the UN Conference on Environment and Development in Rio in 1992. However, it is evident that some policy and regulatory changes under the Mining Act need to be made to legalize small-scale gold panning and shift the burden of administration to local government. Existing policies and the underlying legal/regulatory framework must be reviewed to develop incentives and penalties for more sustainable mining practices. A ban on gold panning is doomed to failure given the high costs of monitoring and enforcement. A more pragmatic approach might be to establish policies and incentives for panners to work in groups or cooperatives.

It is recommended that under existing regulations, local governments can raise permit charges significantly, with a portion of this revenue held in trust and returned to the permit holder after the mined site has been inspected and approved as having satisfied the laid-down environmental guidelines. To improve the income of panners operating under these new conditions, opportunities to market gold at the official or world market prices must be improved. The Reserve Bank and its agents must establish more accessible marketing outlets. This would reduce the role of roving dealers and loss of gold on the black market.

Case study: the proposed titanium mining project in Kwale District, Kenya

A vigorous campaign to support sustainable development in the mineral sector in Kenya that began during the 1990s was given renewed impetus in the last three years following the discovery of large deposits of titanium minerals in the coastal strip of Kenya (Abuodha 2002; Fig. 1). Heavy mineral reserves in coastal Kenya total some 5300 million tonnes (Table 1). The titanium deposits account for 40% of the world's known unexploited titanium reserves and are claimed to be worth trillions of US dollars on the world market (Abuodha 2002). The socio-economic benefits are clear: for example, construction of a smelting plant will add value to the minerals and the multiplier effect will contribute to more jobs and revenue for the Government. Iron that is present in

Table 1. *Summary of heavy mineral reserves in coastal Kenya (from Abuodha 2002)*

District	Location	Reserve in millions of tonnes
Kwale	Nguluku/Maumba	200
Kilifi	Vipingo	500
Kilifi	Sokoke	1700
Mailindi	Sabaki	400
Malindi	Mambrui	700
Malindi	Ras Ngomeni	1100
Tana River	Simiti Island	400
Tana River	Ras Kitua	300
Total	Coastal Kenya	5300

ilmenite in appreciable amounts could be extracted as a byproduct, thereby avoiding environmentally sensitive disposal of iron oxide.

The Kwale mining project is located approximately 65 km from Mombasa and 10 km inland. The deposit (approx 5 km^2) is located in an agricultural area with coconut and cashew nut as the main crops. An EIA for mining of titanium in Kwale district was carried out by Coastal and Environmental Services of South Africa on behalf of Tiomin Kenya Ltd., the local subsidiary of Tiomin Resources Inc., a Canadian multinational. The EIA plan did not, however, address adequately the impact of released toxic substances, including radioactive emissions from the mining operations on the coastal ecosystem, nor the need for continuous rehabilitation of the mining area. Limited water, energy and land resources in Kwale will mean further deprivation of the people with respect to these resources, related infrastructure and new settlements. Also envisaged is the disruption of major tourist attractions, including marine fauna and cultural and historical properties. Because the EIA was commissioned by Tiomin, who will be working the resources themselves, this provoked charges of bias, and the ensuing wrangles have forced the parties concerned to engage in protracted legal proceedings.

Public participation and adequate consultation, including as many interested parties as possible, would educate the hostile public on anticipated environmental impacts and economic benefits expected from the mining venture. The Tiomin project revealed many conflicts in the national laws, which need to be harmonized with the EMCA for effective implementation.

Conclusion

Mining projects should go ahead, but all efforts should be made to limit damage to the environment and society. This of course entails the conduct of genuine environmental impact assessment programmes and rigid plans for rehabilitation and reclamation.

In view of the vital role that the ASM sector plays in job creation and poverty alleviation, it is important that due recognition is given to this sector by central government and local authorities. There needs to be a move towards a culture of wealth generation and to draw the ASM sector into the legitimate social fabric of the country. Not only will this enhance the quality of life, but it will also demonstrate the benefits of working within the legal framework rather than outside it. The aim should be to inspire the miners to be entrepreneurs, not to constrain and hinder them. The best inducement will be to encourage them to act within a legal structure, with due regard to health, safety and the environment. The approach to sustainable mineral development in Kenya should therefore be all-encompassing, dealing with each of the relevant issues that are fundamental to the mining sector as a whole.

References

ABUODHA, J. O. Z. 2002. Environmental impact assessment of the proposed titanium mining project in Kwale District, Kenya. *Marine Georesources and Geotechnology*, **20**, 199–207.

DAVIES, T. C. 1993. The impact of mineral development in Kenya. *In*: OPIYO-AKECH, N. (ed) *Geology for Sustainable Development*. UNEP/UNESCO Publications, Nairobi, 23–26.

DAVIES, T. C. 1996. Mineral development in Kenya. *In*: RAI, K. I. & SINGH, G. (eds) *Mineral Development and Environment: Recent Researches in Geology Series*. Hindustan Publishing Co., New Delhi, **15**, 40–43.

D'SOUZA, K. & BARNEY, I. 2003. A livelihoods analysis and policy framework for the small-scale mining sector: a preview of two new KaR projects. *Earthworks*, **16**, 7.

EAST AFRICAN EDUCATIONAL PUBLISHERS LIMITED 1991. *Phillips EAEP Atlas*. EAEP, Nairobi, 136.

GOVERNMENT OF KENYA 1940. *The Mining Act (Cap 306)*, Government Press, Nairobi.

GOVERNMENT OF KENYA 2000. *Environmental Management and Coordination Act*. Government Press, Nairobi.

HARRISON, D. 2003. River mining – managing aggregate extraction in alluvial systems. *Earthworks*, **16**, 2–3.

MUHONGO, S. 2003. The future of artisanal mining in Tanzania. 10th Workshop of the Geological Society of Kenya on Mineral Resource Development and Environment, 27–28 March 2003, Nairobi, Kenya.

SRIDHARAN, P. V. 2003. Planning sustainable regeneration in mining areas using tri-sector partnerships. *Earthworks*, **16**, 8.

STYLES, M. 2001. Recovering the lost gold of the developing world. *Earthworks*, **13**, 2–3.

Artisanal and small-scale mining in Africa: the poor relation

K. P. C. J. D'SOUZA

Wardell Armstrong LLP, Lancaster Building, High Street, Newcastle-Under-Lyme ST5 1PQ, UK (e-mail: kdsouza@wardell-armstrong.com)

Abstract: The artisanal and small-scale mining (ASM) sector in sub-Saharan Africa is a sector usually associated with conflict minerals, fatal diseases, smuggling, criminal activity and civil war. Throughout Africa the ASM sector is unfortunately viewed in a negative and distorted manner with little appreciation or understanding for the realities and hardship of miners, their families and communities. However, the sector is burdened and plagued with issues ranging from child labour, gender inequality, the spread of HIV/AIDS, environmental devastation, poor health and safety, migrant workers, lack of capital and fair markets, and conflict with the private large-scale mining sector. The paper discusses the many overlapping and complex drivers, challenges, constraints and issues that characterize the sector and considers the potential solutions through the adoption of appropriate best practice, hopefully leading to sustainable livelihoods in the ASM sector and overall poverty alleviation. Key issues, such as institutional capacity, governance, assistance schemes, legislation, miners' organizations, gender mainstreaming, child labour, health and safety, environmental protection, mineral trading and marketing, adding value, finance and credit, and the co-existence with the large-scale mining sector are all detailed. The paper also highlights ideas of what individual countries can do to help formalize and provide assistance to this vulnerable sector. Some of what is discussed concurs with the findings of the Mining, Minerals and Sustainable Development (MMSD) Project, the multi-donor Communities & Small-Scale Mining (CASM) initiative, the UN Economic Commission for Africa (UNECA) and UN Department for Economic and Social Affairs (UNDESA) – Yaoundé Seminar on ASM in Africa, and most recently the African Mining Partnership (AMP). The bulk of this paper, however, has come from the experience gained by the author while managing and working on numerous projects commissioned by the UK's Department for International Development (DFID), the World Bank, United Nations, various African Governments, private mining companies and NGOs in over 20 African countries.

Artisanal and small-scale mining (ASM) definition

There is a lack of an internationally agreed definition of artisanal and small-scale mining (ASM), not unsurprising given the diversity within the sector. However, country-specific definitions do exist, reflecting locally relevant situations and development processes (Ali 1986; Ghose 1994). Characteristics used in country-level definitions include level of employment, annual production output, amount of capital investment, size of claim or concession, and depth of mining operations. A commonly made distinction, although not always specified, is that between small-scale and artisanal miners. Artisanal miners are often defined as those who employ manual, low-technology mining, conducted on a minor scale and usually involving individuals or families (D'Souza 2000, 2002). They are also often considered illegal or informal and have various local colloquial names ranging from *Galemsey* in Ghana, *Panners* in Zimbabwe, *Cresseurs* in the République Démocratique du Congo, *Nyonga* in the Tanzanite miners of Tanzania to *Orpailleurs* in many West African Francophone countries and *Garimpeiros* in the Lusaphone countries in Southern Africa. Small-scale miners, on the other hand, can have some degree of mechanization, have a legal licence and/or are organized in some form of mining workers' group. For convenience, however, most people now discuss the sector collectively using the acronym 'ASM' and it is widely acknowledged that the sector is 'typically practised in the poorest and most remote rural areas by a largely itinerant, poorly educated populace, men and women with few employment alternatives' (MMSD 2002).

At the national level, in different countries, criteria for identifying artisanal and/or small-scale mining are usually tied to the legislative system. For example, in Côte d'Ivoire, the key criterion is the level of mechanization; in Zambia the size of the concession area; in Ghana the amount of capital invested and the number of participants; in Guinée the type of mineral exploited; in Ethiopia the depth of working and a ban on the use of explosives; similarly, in Sénégal, ASM is recognized in accordance with the depth of working, crude production and the methods applied (UNECA 2002). What tends to be absent from most of these definitions is an understanding of the way in which artisanal and small-scale mining can be associated with a highly differentiated sector in socio-economic terms. However, despite the commonly used terminology and now popular acronym, it is critical to differentiate between artisanal and small-scale mining when implementing legislative amendments since they actually represent quite distinct activities.

ASM in Africa

Since there is no clear definition of ASM, and because many work casually, seasonally or informally, it is impossible to count the total number of miners. Research conducted back in 1999 by the International Labour Organisation (ILO 1999) suggests that throughout Africa, ASM activity involves between 3.2 and 4 million people directly, and that through a social multiplier it affects the livelihoods of a further 16 to 28 million, the majority of whom live in abject poverty on less than U$1 a day. However, many now claim that this figure is probably a gross underestimate and that the number of people seeking to work in the sector in many parts of Africa could triple over the next 10 years (D'Souza 2002). This view is based largely on the continued underperformance of many national economies, and the expectation that the formal sector will not be able meet job creation demands.

Asking the question 'Why does this sector exist in Africa and what forces people to work in the sectors?' may help focus assistance. Artisanal and small-scale mining (ASM) is not a modern phenomenon in Africa; indeed, historically, most mining on the Continent could be considered within this category. From 4000 BC, Egyptians, Nubians and Cushites mined placer deposits between the Nile and the Red Sea in modern-day Sudan, Eritrea and Ethiopia. After 300 AD, many of the African kingdoms and old empires such as Ghana, Mali and Songhai in West Africa and Monomotapa and Great Zimbabwe in Southern Africa used ASM to exploit mineral deposits for building materials, base metals, and of course, gold. Indeed, all over Africa, prosperous and rich kingdoms thrived, including Kongo, Kilwa, Luba, Baganda, Kanem Borno, Benin, Ife, Asante and many others who all had a tradition of ASM to exploit minerals (D'Souza 2000).

However, what about more recent times, under the influence of post-colonial politics and global economics? ASM activities, influenced by these new factors, started to impact in Africa during the 1970s and 1980s (AGID 1980; UNECA 1992). A combination of decreased demand for minerals, drastic falls in minerals prices, plummeting mineral investment, retrenchment from the nationalized large-scale mining companies, the impact of the Structural Adjustment Programmes (SAPs implemented by the World Bank), deep economic recessions, and widespread drought, triggered an attempt in a number of countries to diversify mineral production and reduce dependence on a single mineral for export. Indeed, wide-scale ASM was encouraged in many countries, as it required minimal technological and economic investment for start-up. But why does this sector still exist in Africa in the 21st century? The simple answer is abject poverty and dwindling livelihood choices in an increasingly marginal environment. With over 40% of Africans (UN Development Programme (UNDP) 2003) living below the US$1 per day poverty line, and vulnerable to a great variety of natural and man-made forces and shocks, increasingly more rural communities are seeking a livelihood within the ASM sector.

What are the macro factors in modern Africa that make people turn to the ASM sector? Seasonality is a contributing factor as subsistence farmers often mine in the dry season when there is less agricultural work, in order to supplement their meagre incomes, or during periods of national economic recession. With over 25 armed conflicts in Africa since 1963, affecting around 20% of the Continent's population, many people have also been forced into the ASM sector as a last-resort survival strategy. This has been the case especially in countries like Sierra Leone, Liberia, Sudan, the République Démocratique du Congo (RDC) and Angola. Many other people can suddenly be drawn into mining following the discovery of new mineral reserves. This has been seen with gold or gemstone 'rushes', during which thousands of migrants hope to make their fortunes. The number of miners also fluctuates with the international demand for (and thus the price of) a particular mineral. A recent example was the rush for coltan that took place in the Kivu region of Eastern RDC in 2000 (D'Souza 2003). People are also drawn to the ASM sector following natural disasters or environmental shocks, such as occurred in the severe droughts that affected numerous countries in Horn and Eastern and Sahel regions of West Africa, or the floods and cyclones that devastated Moçambique in 2000.

Based on the diversity of ASM in Africa, there have been recent attempts by development experts to try to differentiate specific types of ASM. The four main identified groups appear to be permanent, seasonal, rush and poverty-driven (Weber-Fahr et al. 2002). Such a distinction may prove to be useful when formulating policies as these groups have very different characteristics, again, depending on the country and region where they are taking place. There can also be considerable overlap between the four categories, for example, permanent, seasonal and gold-rush mining can each be poverty driven.

At present, African ASM operations are as diverse as the minerals they produce, and engage in the extraction of more than 40 different mineral commodities (Carman 1985). One could crudely state that ASM in Africa falls into two broad categories. The first is the mining of high-value minerals, notably gold and precious stones, which are generally for export and are by far the most important group economically. In many African countries gold and diamonds (together accounting for around 60% of all ASM activity) are particularly attractive as they have the advantage of being relatively simple to extract, refine, and transport. In fact it has been estimated by the ILO that around 18% of the Continent's gold production is from ASM and in some countries, such as Burkina Faso, Sénégal, Uganda, Congo, Moçambique, Cameroun and Niger, nearly 100% of gold production is from ASM. The second category comprises mining and quarrying for base metals (D'Souza 2000, 2004), industrial/agricultural minerals and construction materials. This is mostly for local markets and exists in all African countries, although this mineral sector is often overlooked and ignored. Indeed, diversification into high-volume, low-value industrial minerals should be encouraged and viewed as a key priority, especially for countries whose mineral sectors are currently dominated by a single commodity, such as Ghana, Tanzania, Zambia and Mali. Throughout Africa, this mineral subsector has a huge revenue-generating potential and could also lead to import substitution, improved self-sufficiency and the establishment of numerous and diverse associated local mineral-using added-value industries.

The wide variety of minerals exploited in Africa means that there are also many different mining and processing techniques employed within the sector. However, despite this heterogeneity, there are a number of characteristics of the ASM sector that can be found throughout Africa (D'Souza 2000, 2002; Traore 1994) (Table 1).

Figures 1 to 9 illustrate many of the typical negative characteristics of ASM in Africa, including the widespread use of child labour, gender inequality, environmental destruction

Table 1. *Key characteristics of ASM*

- Labour-intensive
- Semi- or un-skilled workforce
- Low production levels
- Low levels of mechanization
- Low income
- Low productivity, recovery and inefficiency
- Largely unregulated, informal and illegal
- Often undertaken in conjunction with other economic activities, such as farming
- Often seasonal
- Usually undertaken in remote areas
- Health and safety concerns – extremely hazardous working conditions
- Large proportion of female mine workers (as compared to large-scale mining)
- Widespread use of child labour
- Environmentally damaging
- Public health concerns, especially in rush areas

Fig. 1. Child labour, with young girls transporting 25 kg headpans of raw gypsum, Gombe, Nigeria.

Fig. 2. Artisanal miners (cresseurs) screening copper oxide ore, Kipuhsi, Katanga, Democratic Republic of Congo.

Fig. 4. Artisanal quarrying for limestone aggregate, Abakalike, Nigeria.

Fig. 3. Environmental devastation caused by artisanal mining for tourmaline, Iseyin, Nigeria.

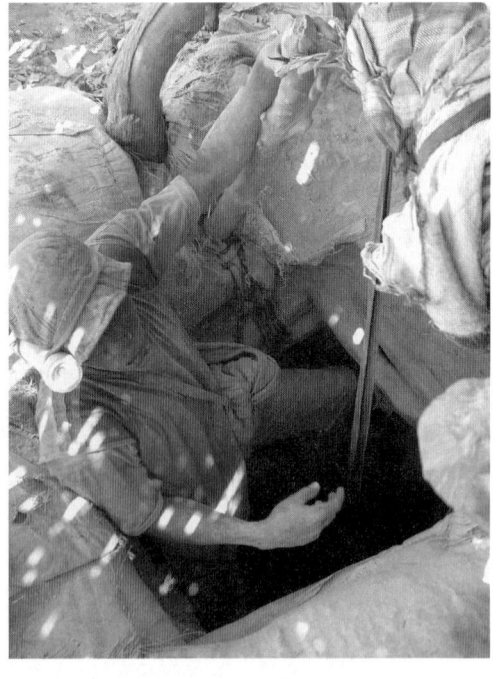

Fig. 5. Young artisanal gold miner (orpailleur) descending a 60 m vertical shaft, Komo Bagnou, Niger.

Fig. 6. Artisanal gold miners (*galemsey*) using crude sluice boxes, Tarkwa, Ghana.

and devastation, hazardous working conditions, poor health and safety practices, labour-intensive operations, and low levels of mechanization and technology, all of which are discussed within the following sections.

Economic impact

From a livelihoods perspective, ASM often provides the only means of obtaining income and can constitute the principal source of economic activity for many poor Africans in remote rural areas who have few livelihood alternatives. Nonetheless, since the 1970s, it has been recognized that the ASM sector does have the potential to economically empower disadvantaged and vulnerable groups and contribute to national poverty reduction efforts (Konopasek 1981; Hollaway 1991). On a local level it can provide a means of survival and 'decent' (as described

Fig. 7. Deforestation and river siltation caused by artisanal gold miners, Mgusu, Tanzania.

Fig. 8. Child labour with young boys amalgamating gold with mercury, Mgusu, Tanzania.

by the International Labour Office; ILO 1999) and productive work for the miners, generate new economic linkages, local investment and stimulate demand for locally produced goods and services, various types of infrastructure, and even some luxury goods. However, due to the transient and nomadic nature of many ASM operations, there are questions regarding legacy issues and the extent to which these new economic linkages will prove sustainable once ASM activities decrease or cease in an area.

On a national level, this can translate into foreign exchange earnings and tax revenue for national governments, provided that ASM is legalized and a mutually conducive environment is created (Mireku-Gyimah *et al.* 1996; D'Souza

Fig. 9. Women and children hand-washing barytes ore, Azara, Nigeria.

1998). In fact, it has been estimated (by the United Nations; ILO 1999) that gold and gemstones worth US$1 billion/year are produced by ASM in sub-Saharan Africa. But how much of this capital stays within the Continent and is reinvested?

Artisanal and small-scale mining can also be very disruptive, particularly when it takes the form of a sudden 'rush', such as the relatively recent rush for gemstone in the Ilakaka region in Madagascar, or coltan in the Kivu area of the Eastern République Démocratique du Congo, or even gold in the Lupa, Mpanda and Lake Victoria gold fields of Tanzania. When large numbers of migrants arrive, they may come into conflict with local communities. In addition, when a 'booming' economy develops around a 'rush' type mining activity, localised inflation brought about by the newly acquired high purchasing power of those involved in the mining, often poses extreme difficulties to those who are not involved in mining (D'Souza, 2000). This can sometimes provoke violence and introduces new social problems resulting in xenophobia and antagonism within some communities and sometimes between different ethnic groups. In addition, increased pressure on local services, such as water supply and health provision, which are in already scarce in many remote rural areas of the Africa, also poses difficulties and becomes yet another potential source of conflict between the miners and the local, indigenous populations. When the rush is over and mining activities have subsided, local people may conclude that they have seen few lasting benefits. Most of the profits will have disappeared while the legacy of social and environmental damage persists. However, this is not always the case, as smaller mining sites often comprise people from many different regions (and even countries) that have established themselves in remote areas away from nearby villages. In such communities there is often a sense of solidarity among these informal communities.

Poverty

As a sector, ASM has often been marginalized geographically and politically, and a key question today is whether this situation is changing. Both the ASM sector and governments are caught in what has been termed by the World Bank as a 'negative circle' of cause and effect (World Bank 1985). This is essentially a poverty trap, which results from a denial of choices and opportunities to miners who are forced to live in a marginal and increasingly vulnerable environment (Fig. 10).

The use of inadequate mining and processing techniques leads to very low productivity of operations and low recovery of valuable minerals (in some instances losses can be up to a staggering 75% for alluvial gold and 90% for primary lode gold deposits; D'Souza 1998), which in turn results in low revenues and the inability to accumulate funds for investment. In these circumstances, ASM therefore becomes unsustainable. The lack of funds to improve methods and acquire appropriate equipment traps these miners in crude and inefficient mining and processing, closing the negative circle and perpetuating a subsistence life for the miners and their families. It is important to remember that it is this vicious circle that must be broken if we are to really help this vulnerable sector in sub-Saharan Africa. These impoverished miners are currently more concerned about, and focused on, basic survival and mere sustenance than the luxury of 'sustainability', which is currently being discussed within the international mining arena.

ASM assistance schemes

Activities in ASM are often viewed negatively by many, especially those who are uninformed

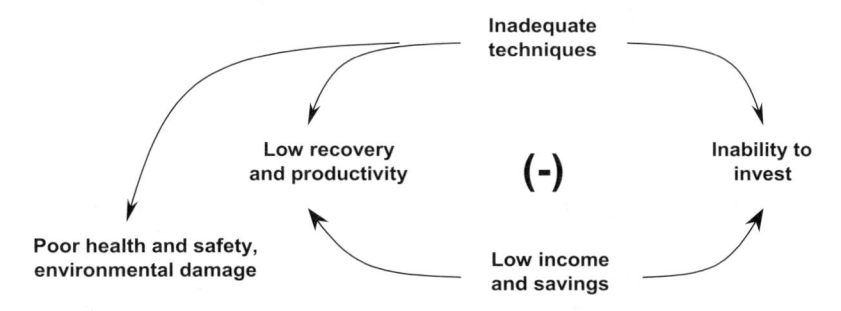

Fig. 10. Negative circle affecting ASM communities (World Bank).

about the sector. Concerns are wide-ranging, from the use of child labour and the potential for environmental damage, poor health and safety, smuggling and the use of ASM revenue to finance conflicts. Other problems involve the social disruption caused by 'rush' operations, migrant alien workers, the high incidence of prostitution, and the spread of sexually transmitted diseases where migrant workers are involved. The complexity of the sector makes it difficult to assist with any quick-fix solutions and often deters many from trying to understand, constructively intervene and help.

In the past, most *ad hoc* efforts by donor agencies (such as DFID, UNDP, UNIDO and the World Bank) have endeavoured to solve one aspect of the ASM sector alone and ignored others. While being well meaning, these one-dimensional projects have provided little more than superficial gains and little long-lasting impact. Indeed, many projects have suffered from a lack of resources, competent management, political will, result-oriented actions and monitoring. Over the last few decades various approaches have been employed in order to 'deal' with the ASM sector and these have been closely associated with the viewpoint and agenda of the implementing donor agency. These themes have changed from definitional issues in the 1970s (Barnea 1978) to a concentration on technical issues in the 1980s (United Nations 1978, 1989; Wels 1983; World Bank 1995). The early 1990s witnessed a move towards integration of technical, environmental, legal, social and economic issues, and by the late 1990s the focus had shifted towards legalization, the relationship between large mining companies, gender mainstreaming and child labour issues (Davidson 1993; Labonne 1993; UN 1994; Burke 1995; D'Souza 1998). The current decade has started to see community- related issues and sustainable livelihoods approaches being adopted as a potential means to provide assistance.

Many ASM development opponents are quick to draw attention to the fact that the ASM sector is littered with examples of unsustainable interventions, with projects having failed for a large number of reasons. Mostly, however, they have failed because of the approach taken in designing and implementing the project rather than the type of project *per se*. Many of the project interventions that have aimed at improving the technology available to ASM operators have failed because the needs and concerns of miners were not understood or the interests of ASM was secondary to those of the Government, donor or outside 'experts', who conceived and managed these interventions. Future interventions must ensure they are locally owned and driven, are informed on robust research data, are strategic and link to other key policy initiatives and sectors and also build on existing capacity.

In the past there was also a lack of communication between the numerous donors, organizations and institutions involved in providing funds and assistance to the sector (D'Souza 1998). This lack of systematic long-term donor engagement and cross-sectoral understanding has had an adverse impact on the sector, because the millions of dollars allocated by the donor agencies have not been effectively used. There is now an urgent need to take into account the existing socio-economic system and consider how ASM can best contribute to poverty reduction and sustainable development through integration into rural community development plans. Future assistance projects must find better ways of integrating miners into the rest of the economy and encouraging ASM communities to invest their revenues in other forms of economic activity as well as in communal services. There must be a clear objective to foster and encourage local stakeholder buy-in and ownership, and ensure we provide incentives for the continuation of any assistance scheme while appealing to the self-interest of the miners and their communities. Only by adopting such an approach will the sector be able to contribute to sustainable rural livelihoods and contribute to the fulfilment of the Millennium Development Goals (MDGs) by 2015.

Governance and institutional capacity

Despite the donor interest in this sector for over three decades and the numerous international fora, conferences and meetings, the truth, albeit unfortunate, is that many African governments have been unsure as to what their long-term goals are for the ASM sector (D'Souza 2002). They appeared to be caught between the shorter-term national economic benefits that can be gained from encouraging foreign large-scale mining and the idealized vision of having a formalized, mainly local, ASM sector. At the extreme, some national governments even considered the sector illegal and attempted to ban it through various unscrupulous means. In many cases they simply neglected it, thereby allowing the negative social and environmental impacts to be aggravated. Some may claim that this neglect is deliberate, as there are some influential and well-connected business and political 'players' who actually benefit from this status quo, which allows them to use the unregulated nature of the sector to camouflage more sinister criminal activities including smuggling and money laundering (D'Souza 1998). The indubitable truth is that not all African governments are convinced or even committed to assisting and formalizing their ASM sectors. To a large extent, ASM has always been viewed politically as a marginal sector because of its geographic remoteness and rudimentary nature. Some governments ignore this sector because they often perceive it to have little political influence and, in most instances, to provide no tax revenues. Other governments actually fear political reprisals if they attempt to police and regulate the sector and again simply resort to an apathetic stance.

Many claim that it is this unclear government stance and apathy, coupled with a lack of real motivation from the ASM miners themselves, that has stifled development and caused many assistance schemes to falter in Africa, while equivalent attempts have been reasonably successful in Latin America. Regardless of the ambivalent political will, the nations of Africa today face three stark choices (D'Souza 2002):

- to try to stop ASM and thus eliminate the associated problems;
- to maintain the status quo and attempt to undertake the occasional *ad hoc* assistance project;
- to undertake a radical reform of the sector by tackling *all* of the relevant issues.

Based on the recent discussions of the members of the new African Mining Partnership (a New Partnership for Africa's Development (NEPAD) initiative) and the objectives of the recently inaugurated (August 2005) Communities and Small-scale Mining (CASM) Africa network, one can assume that it is this final option that is now favoured. Therefore, there is an urgent need to educate and convince all African governments that this strategy is the best for all stakeholders including themselves. At the national level, ASM issues need to be aligned with the MDGs and mainstreamed in to the national development strategies to help improve the contribution of the ASM sector to development processes within sustainable communities. A very positive example of such prioritization is in Nigeria where the President himself has spearheaded ASM development.

One of the urgent and most critical steps will be to ensure that ASM issues are mainstreamed within the current National Poverty Reduction Strategy Papers (PRSPs) in order to lever donor aid and funds (including funds through the Highly Indebted Poor Countries (HIPC) initiative) towards ASM communities (D'Souza 2001). Likewise, poverty reduction must be mainstreamed into a formally endorsed and clearly enunciated national mineral and mining policy. There will be a need to focus on the means to allow the sector to align itself with the principles of sustainable development. The sector must also mitigate the threats to the social, economic and biophysical systems throughout Africa. This will necessitate national multi-stakeholder workshops and real capacity building to raise the profile of ASM and to build partnerships with governments and the large-scale mining sector to raise the general level of awareness of all the ASM issues and challenges.

Many governments are cognisant of the institutional shortfalls and have also acknowledged that such reforms will also rely on restructuring, strengthening the capacity of, and increasing the resources of the national ministry responsible for mining and other relevant institutions to ensure that they can effectively undertake their statutory functions, enforce appropriate legislation, properly monitor the ASM sector and provide much needed extension services (D'Souza, 1998; MMSD 2002; UNECA 2002). In addition to modern institutions of government, traditional institutions can often have equal or greater significance in terms of ASM development in many African countries. Such traditional institutions constitute key stakeholders and should be consulted to ensure sustainability of any ASM formalization and development process.

Decentralization, the process of devolving resources and power from central (national) government to more local structures, is a clearly defined developmental priority in many sub-Saharan African countries. Attempts to decentralize the management of ASM in Africa have tended to focus on regional mining bureau/ small-scale mining centres deigned to carry out a range of outreach type functions on behalf of central ministries. In countries like Ghana, Moçambique, Tanzania and Zambia, the tasks undertaken by provincial mines offices or regional/district mining centres include ensuring compliance with legislation, demarcation of mining rights, collection of revenues, simplification of concession application procedures, monitoring production, and provision of technical advice/support to miners. However, lack of capacity, resources, equipment, and transport has severely limited the effectiveness and efficiency of many such well-meaning extension services.

ASM legislative framework

It is evident that the isolation of ASM from the mainstream of economic development prevents it from becoming a recognized economic activity, contributing to government revenue or attracting investment (D'Souza 1998). In many African countries (such as Nigeria, Cameroun, Congo, Moçambique, Madagascar, Angola and the République Démocratique du Congo), more than 75% of miners operate illegally (ILO 1999). This situation must be changed, and ASM development encouraged, to generate rural employment and reduce poverty, while protecting the rights of indigenous people and in particular women and children. Legalization is the first step to formalizing the sector into a sustainable activity, but consideration must be given to a variety of issues including (D'Souza 2000; UNECA 2002):

- providing for the right to exploit a particular deposit by ASM and encouraging the development of specific minerals deposits that are more amenable to ASM;
- establishing an independent and fully decentralized licensing registry office that is transparent, non-discretionary and non-discriminatory;
- adopting the 'first come–first served' principle;
- providing full and transferable mining title and security of tenure to enhance liquidity;
- codifying the necessary elements for a modern mining cadastre;
- curbing the illegal trade in precious minerals;
- encouraging the formation of ASM associations and/or cooperatives;
- generating stable employment opportunities in rural areas;
- mitigating severe environmental and health and safety effects of uncontrolled ASM;
- encouraging the entry of nationals into ASM and eliminating alien workers (especially in conflict zones);
- protecting the rights of indigenous peoples;
- addressing misunderstandings and conflict over land and mineral rights;
- promoting co-existence with the private large-scale mining sector; and
- including incentives for compliance, formalization and local added value.

Unfortunately, the mining laws of many African countries still fail to identify the real needs of ASM operations and do not recognize the importance of the sector (D'Souza 1998). Some countries like Côte d'Ivoire, Ethiopia, Ghana, Tanzania, Zambia and Zimbabwe have enacted some very competent and specific ASM legislation; however, even these nations realize they have to go even further before they can really positively impact on the sector. Inaccessible and stringent legislation without empathy for the reality of ASM livelihoods can be counter-productive. Legislation and its 'practice' must be appropriate to the realities of the ASM sector and the capacity of staff to manage and enforce legislation. In addition, poverty eradication needs to be included in national mineral policies and more importantly within the National Poverty Reduction Strategy Papers (PRSPs) of individual countries. This may mean that many countries have to revisit their polices and current PRSP to provide a strategic framework for ASM formalization and development.

It is important not to be naïve: there should also be strong incentives for these miners to participate in the formal sector and comply with legislation and answer the question; 'What's in it for me ...?' For many, simply registering their business to appease the government is a tortuous and expensive process, costly in both time and money, and offers limited if any advantages. So why bother? This suggests that they see more disadvantages from working within the formal sector. One must convince them otherwise. Therefore there is also an urgent need to convince governments to devise tangible incentives and commit to assisting with other problems like access to water and affordable fuel, or facilitate linkages with mineral buyers and private sector mining companies in order to encourage the sector to legally comply and formalize.

Formation of ASM cooperatives

The general global consensus (including numerous DFID, UNECA and World Bank studies) is that the formation of miners' organizations, cooperatives or associations provides a single 'voice' that can help the miners in conducting pricing or workplace negotiations, mobilizing assistance programmes, conducting awareness campaigns among its members, and organizing security and other mine site related activities (D'Souza 1998). The formation of miners cooperatives would also aid the administration and enforcement of the legislation while facilitating dialogue and interaction between the government, the large-scale mining sector, the actual miners and other relevant stakeholders.

Although a conducive legislative and institutional structure is already in place in some countries to aid the formation of miners' cooperatives, the reality on the ground is somewhat different. Many local mining entrepreneurs are sceptical and somewhat reticent about the formation of such cooperatives and currently prefer to work alone and purposefully demean the formation of such associations. There are also many intermediaries, sponsors, vendors and mineral traders who will feel distinctly threatened by the formation and empowerment of such cooperatives, who could then simply bypass them and market their mineral produce directly to the end-use buyers. However, most of the mineral markets in sub-Saharan Africa could certainly benefit from a streamlining of the commodity chain to reduce the numbers of sponsors who all seek to maximize their share of the profits perpetuating the subsistence life of those actual artisanal miners at the bottom of the commodity chain.

Therefore, any encouragement or promotion to organize into cooperatives must fully take into account the peculiar cultural, traditional (it may also be more appropriate to advocate family of community based cooperatives in some regions) and situational circumstances in the various areas of each country and attempt to highlight the benefits of pooling resources. Wherever possible, it would be prudent to build on existing local and traditional organizations such as the peasants' associations in Ethiopia or the womens' mining group in Guinée or Malawi. However, such organizations need to be resourced and empowered or they risk becoming futile entities (like the Regional Miners Associations (REMAS) in Tanzania) that hardly serve their objectives due to low financial capacity and weak leadership. Any promotion process should also acknowledge issues of illiteracy, lack of unity, loyalty, conflict, administrative and leadership skills and available resources, while ensuring that the motives of potential members to join any cooperative are not fraudulent or self-seeking.

Gender mainstreaming (GM)

Over time, women's involvement in mining activities has tended to increase, particularly in sub-Saharan Africa. Women are engaged in most aspects of ASM and are also involved in ancillary activities resulting from family-based activity. According to the International Labour Organisation (ILO 1999) 45–50% of all ASM workers in Africa are women (varying from 5% in Gabon and RSA, 10% in Malawi, 26% in Tanzania, 30% in Zambia and Moçambique, 35% in Guinée, 45% in Ghana and Burkina Faso, 50% in Mali to more than 50% in Zimbabwe). In most instances the women tend to undertake the more ancillary operations including manual transport, ore crushing, washing, sorting and mineral dressing. In general, the women work long and arduous hours for far less pay than their male counterparts, while also being expected to undertake all their traditional domestic family duties. The fact that women are often limited to engaging in lower status and lower paid activities stems from cultural perceptions and traditional beliefs of appropriate work for men and women as well as from some outdated legislation that makes it illegal for women to work underground (Drechsler 2001). In some countries, and with specific minerals, ASM operations are almost entirely female run, including salt in Nigeria, trona in Niger, lime in Malawi and the handnapping of stone around Lusaka in Zambia. Frequently, these enterprises are better managed than those run by men, even though women find it more difficult to get financial, legal, or technical support (MMSD 2002; D'Souza 2004b). Women also provide a variety of services to ASM communities, such as cooking, petty trading and sexual services.

Although many African constitutions protect their position, many women still face traditional or cultural obstacles in asserting their formal rights. Women face problems such as illiteracy, insufficient technical knowledge, sexist/chauvinist attitudes, patriarchal views, social taboos (e.g. women are considered bad luck in some mining areas and in other areas are denied access during their menstruation period) and family/domestic responsibilities (ILO 1999). Future policy (Labonne 1996) in Africa must focus more sharply on removing gender-based constraints, and incorporate ways to give women more power in their communities and their households through the enactment of gender-neutral legislation and guidelines (in accordance with the ILO Convention on the Elimination of All Forms of Discrimination against Women). In addition, some of the occurrences of child labour in ASM can be attributed to the fact that women in some mining areas are forced to bring their children along to work. Such situations lead to arguments that it is the women's involvement in ASM activities that is to blame for child labour and that it is therefore not desirable to have women in ASM. However, it is the lack of child-care provision for the children, as well as the limited opportunities women have to make alternative arrangements, that is to blame for the presence of some children on ASM sites.

Although there are numerous women's mining associations (e.g. the newly formed, 2003, African Women in Mining Network) in Africa they need assistance as they have yet to physically implement any real strategies to materially change the conditions under which women work in the mines. More effort needs to be given to GM in the ASM sector, as experiences from other development work has unequivocally shown that empowering women with greater rights could make the single biggest positive difference to reducing poverty in the wider ASM communities by directly augmenting family incomes.

Child labour (CL) elimination

Even though most African countries have ratified the International Labour Organisation (ILO) 'Convention on the Rights of the Child' and the 'Worst Forms of Child Labour Convention' and signed the Organisation of African Unity (OAU) charter on the rights of children, child labour (CL) is still prevalent in many ASM operations in sub-Saharan Africa (ILO 1999; Jennings 1999; MMSD 2002). A combination of economic decline, poor education, poverty, conflict, rural remoteness, a large informal working sector, poor governance, disease and HIV/AIDS has created conditions that are rife for the exploitation of CL together with a depressing increase in orphanhood. Likewise, the ease of opportunity to exploit children, the growing proportion of the population under the age of fifteen and the fact that child work is often considered part of the socialisation process by many rural peoples has resulted in a high prevalence of CL in rural Africa (Mwami et al. 2001).

In most African nations, a general lack of law enforcement and investment in the social infrastructure has resulted in CL being common practice in the ASM sector. In general, children work in the mines to help their parents, and to supplement the family income in order to provide basic goods simply to survive, and their contribution is highly valued by their families. In addition, some children are forced to either drop out of school, due to a lack of funds to support their education, or have to work part time in the mines to meet their educational needs. Child labour in the ASM sector in Africa is a result of (Bhaltora 2003):

- worsening poverty in rural areas;
- a general lack of awareness;
- illegality and the informal nature of the ASM sector;
- past government apathy and lack of capacity and resources to police the sector and enforce legislation;
- inadequate governance, no coordinated effort to stop CL and a lack of CL law enforcement;
- lack of empowered miners' associations/ ASM unions, organized community-based organizations, (CBOs) or sector-specific non-governmental organizations, (NGOs);
- disintegration of the traditional extended family and significant gender inequality;
- growing proportion of children and rising number of orphans;
- remoteness and isolation of many mining areas;
- traditional and/or religious chauvinist attitudes toward girls and their educational needs;
- lack of opportunities or incentives to go to school or continue with education in mining areas;
- lack of post primary education, few job prospects, vocational training, regular employment or livelihood choices in rural areas;
- a lack of improved ASM performance and access to fair mineral markets;
- the plundering of high-value 'conflict minerals' in war zones.

CL is also exacerbated by the nomadic nature of ASM that is detrimental to the traditional family structure, particularly when the miners have to travel far from home and are away for indefinite periods of time. Sometimes, families accompany the men on their quests. In such circumstances the children really suffer as they lose out on an education and invariably become involved in the mining.

Although there are no detailed data on CL in he ASM sector of most African countries, preliminary research by DFID and the ILO (in countries like Guinée, Ghana, Ethiopia, Mali, Zambia, Madagascar, Tanzania, Niger and Burkina Faso) has shown that there is generally an equal incidence of boys and girls working directly in mines, with the children aged around 8–16 years. In those mines where CL is prevalent, the girls tend to be engaged in ore transport, washing and mineral dressing, while the boys assist with the actual mining (ILO 1999). Most of these children fall within the group who should have completed their primary education but now face a very limited chance of obtaining a secondary school education and/or vocational training. The depressing reality is that working in the ASM sector is often the only available livelihood option to the many thousands of impoverished children throughout the Continent.

These children are engaged in many aspects of the mining operation from rock-breaking and transport to washing, crushing and dressing, and usually work at least 6–8 hours per day. The children are often forced to undertake physically demanding work in hazardous conditions or in direct sunshine and dusty areas. Indeed, young children are particularly 'useful' in underground vein deposits (e.g. the 'snake-boys' in the tanzanite mines at Merelani in Tanzania). Although some children working in the ASM sector are paid in cash, some are not paid but work simply to survive and receive only basic sustenance (in-kind payment). Child labour exploiting young girls is also prevalent in the related service sectors (including domestic and sex work) in ASM communities, nearby mining villages and towns, and these children can often work more hours per day. At present there are insufficient data to determine whether these children remain in the ASM sector, or manage to find alternative employment. Whether or not the children manage to escape the ASM sector, there is unfortunately always a continuous influx of 'willing' new young labour. This is especially true for young girls in some areas where rural families often choose not to send the girls to school as they expect them to be married at a young age. Likewise, the means of recruitment is unclear, whether it be through 'voluntary' family or community relationships/understandings or 'forced', with the children working as conscripted or bonded labour.

Many past CL elimination projects in the ASM sector in Africa have only concentrated on this aspect in isolation and have been too academic, insufficiently researched and inadequately resourced. Likewise, many ASM-focused *ad hoc* donor projects in Africa have treated CL as a minor and low-priority issue. Future efforts and interventions must appreciate that there is certainly no quick-fix solution to the complex CL problem in ASM (c.f. Latin America where better governance has resulted in greater relative success in CL elimination interventions; (Astorga & Duran 1992, 1994) and reduction solutions must be integrated into a holistic approach to assist the ASM sector that considers all the constraints, issues and challenges. In the short term, complete eradication of CL in ASM is probably unrealistic in many African countries, although work in the most extreme and hazardous conditions such as underground mining must be eliminated. Experiences from the ILO work in countries like Tanzania, Madagascar and Burkina Faso may provide some valuable insights for future CL elimination schemes. Future interventions should fully adopt the current general consensus that CL elimination will only be achieved through a concerted effort by the government and other relevant ASM stakeholders and include:

- better governance of the rural economy by traditional, local and national Government.
- raising the living standards of ASM communities through financial empowerment of the miners and overall rural economic development (to hopefully lower the value and need of CL).
- identification of other sustainable and equivalent income-generating livelihoods in the remote mining areas, especially appropriate agriculture;
- efforts to strive to raise CL awareness in communities (and the distinction between child work and labour), especially of the worst forms of CL, with information for parents on the negative physical and mental effects of CL (e.g. the ILO programme at Merelani tanzanite mines in Tanzania);
- seeking advice on a realistic CL reduction policy to cater for rehabilitated child soldiers, orphans and female children in extreme poverty;
- encouraging more stringent law enforcement and monitoring of the ASM sector through institutional capacity building and adequate resourcing and adoption of a policy that places greater accountability for CL on the mining permit holders;
- seeking to raise mine productivity and assist with access to fair and equitable mineral markets;
- establishment of community-based family support services (e.g. child/baby crèche facilities such as those installed at the Alga site in Burkina Faso) and development of sustainable child protection 'social-security' schemes, especially for orphans;
- provision of improved, affordable and accessible education in mining areas and identification of the means to motivate children, especially girls, to remain in school;
- provision of vocational training to help school leavers acquire useful and appropriate trades/skills in mining areas;
- adoption of a gender-sensitive approach that gives particular emphasis to the role of women in the ASM sector.

Health and safety (H&S) management

For a number of reasons, the health and safety (H&S) risks to which artisanal and small-scale miners are exposed can be significantly greater than for large-scale mining (statistics on accidents and fatalities in the ASM sector are grossly inaccurate, as accidents in the ASM sector are under-reported or not reported at all;

(ILO 1999; MMSD 2002). Most obviously, the informal and unregulated nature of ASM means that it usually operates beyond the scope of legislation or enforcement of health and safety guidelines. The ASM miners are particularly vulnerable to exposure to mercury, dust and other chemicals, the effects of noise and vibration, poor ventilation and lighting, over-exertion, inadequate work space and inappropriate equipment. Such working conditions result in respiratory infections, asthenia, arthropathy and various dermatological, muscular and orthopaedic ailments. Often these individuals are not aware of the risks they are taking and even simple safety items including basic personnel protection equipment (PPE) represents a costly investment with no immediate return. Moreover, some miners have introduced maladapted mining methods, inappropriate or more mechanized equipment or techniques without the complementary safety measures, and often these individuals are not aware of the risks they are running. Official studies by the ILO (1999) have highlighted the varied reasons for accidents in ASM operations, from both the management and the operational perspective, as well as in relation to the workplace and equipment.

The activity of greatest concern for many is the use of mercury and the inhalation of methyl-mercury by ASM gold miners, at risk to their own health, that of others and the consequential bio-accumulation within their local ecosystem. Various studies have categorically shown that miners who use mercury have elevated levels of mercury in their blood, hair and urine. The inappropriate use of mercury often arises due to a lack of knowledge of the amalgamation and smelting process. A first step should be to alert people to the danger to themselves, their children, and the environment, followed by encouragement to adopt mercury pollution abatement techniques. These include the use of simple methods to capture mercury vapour through retorts or through the adoption of enhanced comminution, more efficient gravimetric concentration or improved amalgamation techniques. Another option is to introduce alternative forms of gold extraction that do not involve mercury, such as froth flotation, salt-electrolytic processing, coal–gold agglomeration, the IGoli® mercury-free gold-extraction process, cyanidation and the 'blow-and-tap' method of direct smelting.

Unfortunately, in spite of the best intentions, many governments have been largely unsuccessful at being able to raise standards immediately, simply through legislation and enforcement (D'Souza 1998). A more realistic approach that should be promoted needs to centre around increasing awareness of the risks and to demonstrate less dangerous alternatives that are appropriate to local circumstances, and that allow mining communities to make better-informed choices. Government must ensure that all regulations and guidelines (Walle & Jennings 2001) are formulated through a consultative approach, including discussions with the relevant stakeholders.

Specific legalisation addressing the issues on H&S for the ASM sector are rare, and most African mining jurisdictions have H&S covered under general labour regulations that address the entire mining sector (UNECA 2002). A few countries in Africa that have attempted to formulate ASM-specific H&S guidelines include the new Mineral & Mining Policy of South Africa and the 1971 Mining Regulations of Zambia. One of the most interesting examples of sector-specific guidelines are the localized regulations adopted at the Merelani tanzanite mine in Tanzania. The formulation of these guidelines dates back to the tragic flooding of the mines in 1998 that killed 70 miners. After this accident the Mines & Construction Workers Union and Arusha Mines Association, in collaboration with the Government Mines Zonal Officer, drafted regulations that would hopefully avoid a repeat accident. Such site-specific regulations may not be ideal (and they did not prevent another fatal accident in 2002), but they have ended many unsafe and hazardous working practices that were designed through a fully consultative process.

The dangerous environment also extends beyond the mines. This is something that most African governments can definitely do without given the current high infant mortality rates and the low life expectancies in many rural areas. Those engaged in ASM are already some of the poorest people in Africa and are therefore likely to have inadequate sanitation, with little access to clean water or basic health care (D'Souza 2002). These problems are likely to be even worse where migrant miners have converged around a freshly discovered 'rush' area or settled in unorganized camps. Such remote and temporary settlements are unlikely to have public health facilities. In addition to harbouring diseases related to poor sanitation, they are also breeding grounds for crime, alcoholism, narcotics and substance abuse, prostitution and sexually transmitted diseases (STDs) including HIV/AIDS. Other threats include malaria and yellow fever, spread by mosquitoes breeding in water-filled pits left by the miners, and also cholera, dysentery, diarrhoea, tuberculosis, bilharzia, and other parasitic and infectious diseases are common in such informal and remote mining camps.

Environmental management

The environmental destructiveness is perhaps the single most visible aspect of ASM throughout sub-Saharan Africa (UN 1994; Landner 1995, 1997; D'Souza 1998; MMSD 2002). However, a lack of awareness (e.g. in many Sahel countries, miners believe that gold nuggets can be found near tree roots, resulting in aggressive deforestation), particularly of the less visible or long-term environmental impacts of activities, combined with a lack of information about affordable methods to reduce impacts and a lack of obvious incentives to change, all contribute to significant environmental problems within the African ASM sector. The numerous environmental impacts of ASM include:

- destruction of natural habitat and arable land;
- deforestation, destruction of landforms and soil erosion;
- land degradation and ground subsidence;
- land instability;
- danger from failure of structures and dams;
- abandoned workings and equipment;
- adverse changes in river regime and ecology due to pollution, sedimentation and flow modification;
- alteration of the water table;
- soil contamination from treatment residues and chemical/diesel spillage;
- drainage from mining sites, including acid mine drainage and discharged mine water;
- direct dumping of mine waste;
- sediment/minewater runoff from ASM sites;
- effluents from mineral processing operations;
- sewage effluent from the ASM sites;
- leaching of pollutants from tailings residues, disposal areas and contaminated soils;
- air emissions from minerals processing diesel equipment and blasting activities; and
- dust emissions from sites close to villages and habitats.

Since most of the ASM operations are subsistence activities that struggle to survive from day to day, the miners are forced to focus more on immediate concerns rather than the long-term consequences of their activities. This is compounded by the fact that in the past most mining ministries, or environmental protection agencies, lacked the capacity to effectively monitor or control these informal activities that occur in remote and sometimes inaccessible locations. To add to this problem, in many African regions, poverty is intimately linked with environmental management. These links include the enormous burden of disease that afflicts the rural populace through pollution of water and air and their dependence on natural resources and the local ecosystem, which when degraded, can undermine their very livelihoods.

By their very nature, ASM workings tend to be informal and often illegal, and increased environmental legislation will be ineffective in their regulation. Better enforcement of existing legislation is one approach that could help limit or curtail operations. Therefore, governments need to develop appropriate, easily understood and enforceable legislation that will draw the ASM sector into national programmes for environmental management and protection (McMahon *et al.* 1999; Parsons 1999). They must ensure that simple environmental management guidelines are developed in order to encourage miners to adopt working methods that are appropriate to the sector in terms of efficiency, health and safety standards and the minimization of negative environmental impacts (Mutagwaba & Hangi 1995). This concept of sector-specific and appropriate environmental management guidelines that recognizes the special circumstance of the informal sector has also been endorsed by the UN Economic Commission for Africa (UNECA 2002) and the UN Department for Economic and Social Affairs (UNDESA) Yaoundé seminar on ASM in Africa and is embedded in the seminal 'Yaoundé Vision Statement' for ASM development in Africa.

One cannot ignore the fact that the ASM sector throughout sub-Saharan Africa needs to comply with the requirements for minimizing the numerous negative environmental impacts attributable to its unregulated activity. However, in jurisdictions where environmental (and H&S) regulations do not differentiate between ASM and large-scale mining needs, problems with compliance by the ASM sector quickly develop and deteriorate. The concept that the ASM sector should be regulated by exactly the same environmental legislation as the large-scale mining sector is rather naïve and will simply result in non-compliance and the continued 'illegality' of the ASM sector. Conducting Environment Impact Assessments (EIAs), for example, requires specialized knowledge and is not cheap. Unless specific mechanisms are put in place to enable the miners to get both the technical and financial assistance necessary for such an undertaking, it is ridiculous to expect that the impoverished artisanal miners will conduct EIAs to an equivalent standard as the large-scale mining sector. One potentially worthy approach is adopted in South Africa (by the National Steering Committee of Service Providers to the Small-Scale Mining Sector (NSC)) where they actually fund the requisite EIAs in all projects in their jurisdiction. However, critics will quickly cite that the problem with such an approach is that only a select few projects can comply with the regulations, while the rest simply revert to being clandestine and illegal operations. Probably a better concept would be to develop greatly simplified EIA requirements applicable to ASM. This has been attempted by the Department of Minerals & Energy in Kimberly, South Africa, and by the Ghanaian Minerals Commission in collaboration with the Environmental Protection Agency (EPA). In the latter case of Ghana, this EIA requirement through the EPA is fulfilled by the completion of a simple two-page standard document that is easy to understand and complete. This form is purchased from the EPA and on satisfactory completion it is certified by the EPA, thereby allowing the miner to obtain his mining permit. However, this EPA requirement and certification has unfortunately developed into a bottleneck in the mining licensing process, and miners are once again deterred form legalizing, many still choosing to remain as unregistered and operate illegally as *galemsey*.

Another key concern within the ASM sectors of many African countries that needs to be addressed is the legacy and required rehabilitation of past, present and future ASM activity. Again the reality is that the financial requirements for rehabilitation of the mined areas are usually beyond the capacity of the miners. Therefore, there is a growing argument for the need for a prescribed rehabilitation fee. This concept is in accordance with the accepted UN Economic Commission for Africa (UNECA 2002) 'best practice' for ASM mine rehabilitation providing the 'prescribed fee' does not substantially reduce the miners' earnings. In Ghana, for example, the Government (through the Precious Minerals Marketing Corporation (PMMC) and the Minerals Commission) deducts 3% of the value of sales made by miners to supposedly fund the rehabilitation of past and present mined areas. Another worthy example is from Zimbabwe, where the chromite miners, who have a tributor arrangement with the commercial private sector company ZimAlloys, are paid extra per tonne of ore sold providing they undertake rehabilitation of the mined areas.

Environmental protection is not simply a matter of legislature, governments must ensure that artisanal miners are offered awareness campaigns and then provided with quality education and training and appropriate equipment to avoid or mitigate many of the environmental problems. It must also be remembered that artisanal miners usually accept working practice changes reluctantly, especially when the benefits of change are of a long-term nature. Indeed, this attitude has stifled the adoption of mercury distillation retorts in gold mining areas in many countries despite their apparent health and economic benefits vigorously promoted by the UN Industrial Development Organisation (UNIDO). Experience has shown that compliance with environmental guidelines will only be enhanced through specially tailored programmes or appropriate 'tool-kits' that take full account of the cultural background, gender issues, geographical isolation, capabilities and the working environment of ASM communities (British Geological Survey 1999). Therefore, any advice will have to take cognizance of this issue and will need to be clearly worded in an appropriate local language (or illustrated considering issues of illiteracy) and be relevant to the local conditions and culture. All potential methods of communication, especially visual aids, should be examined and appropriate communication media adopted. A prudent approach would be to establish a set of positive inducements to encourage artisanal workers to adopt better and more environmentally sound practices. Positive inducements could include access to markets for sale of commodities, access to credit facilities, equipment supplies, education and the provision of communal mining equipment such as crushers and concentrators, and so on. Potential monetary benefits associated with the adoption of new environmentally friendly working techniques also need to be clearly demonstrated.

ASM technology and training

Throughout Africa, technological issues associated with mining and processing are among the main constraints that limit ASM from attaining its full potential (Abate 1991; Holloway 1993; D'Souza 1998; Peak *et al.* 1998). In most instances mining methods are crude, inefficient and hazardous and are frequently maladapted from the large-scale mining sector. Even rudimentary mine planning and grade control concepts are non-existent and the hand tools, equipment and techniques used by the miners are very simple and often inappropriate. Usually the techniques, tools and equipment adopted for routine unit operations in the mines (drilling, rock breaking/blasting, ore/waste handling, hoisting, ventilation, and drainage) are inadequate. Likewise, techniques and equipment used for processing (Hosford 1993) and mineral dressing (including comminution, sizing/classification, and concentration) are also crude, being passed unchanged from previous generations, and overall recoveries can be exceedingly low. The lack of knowledge and funds to improve methods and techniques or acquire appropriate equipment forces many in the African ASM sector to adopt inefficient mining and processing techniques (Taupitz & Malango 1993). Such conditions thereby perpetuate haphazard mining, high dilution and low recovery situations leading to high grading, the possible sterilization of highly economic deposits, and in the case of gemstones, the physical destruction of the sought after and prized crystals.

When discussing areas of concern, and what are their main desires, many African miners cite lack of equipment and technology as a key problem. It is because of this apparent demand, and because of the past orientation of many development interventions, that technological improvement and 'best practice' adoption has been, for some intervening agencies, viewed as the key for improving ASM for both environmental protection and increased productivity (Priester et al. 1993). Beside these two objectives, the selection of appropriate 'best practice' techniques and technology (UNECA 2002) should be based on a variety of other practical aspects and criteria such as the full acceptance by the miners, the local structural geology and orebody characteristics, the specific mineralogical and/or metallurgical characteristics, health and safety, financial gain, affordability, the ease of use, and the availability of necessary equipment and materials all within the specific country context. However, field experience has shown that the key benefits to the miners of any 'best practice' technology must be easy to demonstrate in terms of a more efficient and lucrative operation rather than rely on arguments based on long-term health and environmental implications. Therefore, in order to gain acceptance and be adopted, any new or alternative technology must be fully accessible, inexpensive to operate, easy to duplicate, use and maintain, utilize readily available local resources, and it must generate obvious financial benefits through more efficient and expedient recovery.

In some countries there are also many 'social' barriers that have to be overcome before acceptance and adoption of new technology, as there is a traditional and almost cultural affinity to the reliance on some preferred equipment and techniques (British Geological Survey 1999). In some established mining areas, miners can be very sceptical and loath to adopt any different methods or technologies regardless of who is advocating there use. Also, there are numerous iniquitous and superstitious cultural beliefs regarding mining activities, some of which have further immoral and dangerous connotations ranging from intercourse with young virgins and blood sacrifices (Niger and Burkina Faso), the presence of women (Zambia), to the wearing of shoes (Tanzania) in the mines and even the extreme belief that money derived from mining is cursed (Tanzania and Burkina Faso). This is especially true of the imposition of technology through solutions perceived to be developed behind closed doors in a 'Western' non-African environment. Any strategy for the introduction of new technology must not be arrogant or patronizing, but should always ensure due consideration to the cultural diversity, level of knowledge and expertise, and varied perceptions of individuals in the ASM communities. Therefore respect and understanding for the varying degrees of permanence, local knowledge, adaptability to new methods, and other issues (including ethnicity, culture, gender, age, poverty, motives, and so on) should always be given before selecting an optimum alternative technique or specifying different equipment. Indeed, experience from Latin America has shown that the widespread adoption of alternative technologies may take several years and, because miners are suspicious of outsiders they are a particularly difficult group to persuade to try new technology unless a mutual trust and understanding can be built and maintained. Future training and technology transfer interventions should build on experiences of the reasonably successful training schemes that have been conducted in Zimbabwe, Ghana, Tanzania, Namibia, Mali and Burkina Faso (UNECA 2002).

Given the fact that many in the ASM sector use inappropriate technology or have limited access to, or knowledge of, geology or mining, there is also a desperate need to provide accessible, free (or affordable) multi-disciplined advice, training services and products to miners (D'Souza 1998). Such training services also need to be delivered in appropriate locations (either with static or roving training centres) and at appropriate times of the day (when the miners have finished work). Such proposed training extension services need to act as a 'one-stop-shop' in terms of outreach to the miners and their communities (with access to rights and entitlements) and should have a duty to advise on best/good practice in accordance with the national environmental and health and safety policy and guidelines for ASM (UNECA 2002). Training should include mining and processing (appropriate locally fabricated technology), sensitization to sustainable environmental management and H&S guidelines, legislation, access to credit and profitable marketing, mineral pricing and evaluation, business, bookkeeping and management skills, and community health/issues. The linking of the training of different aspects of ASM and SME business management will be essential to allow the ASM sector to thrive and grow. In theory, these training centres could also act as a conduit for more general education including general healthcare, water management and HIV/AIDS awareness for the wider community.

All the training should be appropriately structured to ensure the level of detail and methods of instruction match the targeted educational level and technological capacities of the miners. All advice and assistance services should therefore take account of the cultural background, gender issues, geographical isolation, capabilities and the working environment of ASM communities in each locality. Advice should be clearly worded in an appropriate local language (or illustrated considering potential issues of illiteracy) and be relevant to the local conditions and culture. Indeeds it is important that the literacy standards of the miners are not assumed to ensure that information dissemination is aimed at the miners with the poorest literacy levels. For this reason, minimum written information and maximum use of 'pictograms' may be important in the dissemination and uptake of some critical messages. All potential methods of communication and appropriate media should be investigated and tested, such as leaflets, posters, videos, local radio broadcasts, theatrical plays, meetings with CBOs, group leaders, village elders/traditional rulers, and so on.

Co-existence with large-scale mining

The relationship between large-scale mining companies and the ASM sector in many African countries is poorly understood and has been troubled with mutual mistrust, antagonism, resentment, intimidation, threats and increasing conflict (D'Souza 2002; MMSD 2002). Both sides of the conflict have their own misguided preconceptions and strong feelings regarding each other and their alleged rights. For example the numerous companies have viewed the ASM sector as 'trespassers' on their legally endorsed concessions, while the miners may see the granting of such concessions to a large, and usually foreign, company as depriving them of their 'traditional' land and rightful livelihoods as indigenes. Such conflicts are often heightened by the contentious issue of land tenure caused by lack of appreciation and understanding by the ASM sector of the legal difference between land and mineral ownership rights. In many instances this is then further compounded by the fact the mining regulatory body only operates at the national level and is ineffective in dealing with these local conflicts. In most cases conflict arises simply because the two sectors are often competing for the same mineral resources although in many countries the ASM sector is completely overshadowed by the ever-growing large-scale sector. This competition is no coincidence as mining and exploration companies have often used the artisanal miners as unpaid 'geologists', as they are often very efficient prospectors, and subsequently concentrate exploration wherever there is ASM activity, historical or recent (Bills *et al.* 1991). A notable example was the 'discovery' of the 30 Moz gold Bulyanhulu deposit by Placer Dome Inc. in northern Tanzania back in 1976 that was actually founded on historic ASM workings. Conversely, ASM miners have often illegally congregated around virgin exploration sites and newly developed large-scale mine sites taking advantage of the better access and perhaps re-mining some of the mining company's tailings, waste or marginal ground. There is also a disparity between the objectives of the ASM sector scraping a livelihood by essentially high-grading shallow near surface deposits and the commercial large-scale sector seeking to bulk mine a larger but lower grade global deposit of which the high-grade areas constitutes only a minor but essential component. Many examples of this type of competition have occurred particularly in countries with growing large-scale sectors like Ghana, Guinée, Mali and Tanzania.

In much of Africa during the colonial (and early post-independence) period the usual colonial oppressive force was adopted to suppress and 'punish' any 'native' artisanal activity. Post-independence, and once more modern operations started, some companies, with little morality, have tried to keep the artisanal miners at bay through force and intimidation and built expensive systems of security. Other companies adopted a laissez-faire attitude that simply tolerated the presence of illegal ASM providing it did not encroach or impact on their operations. Today some companies have learnt that building constructive relationships works better than resorting to force and trying to shut the ASM miners down and hope that the 'problem' simply disappears (Davidson 1998). However, given the continued growth of the large-scale mining sector in many African countries, the irrefutable challenge remains of how to constructively engage and attempt to coexist with the increasingly vulnerable, and in some cases volatile, ASM sector.

One of the first areas of concern for many of these companies is exactly how to engage and gain the trust of the ASM sector. Some actually claim that this initial engagement issue has stalled and even prevented the private sector from working with the ASM sector. Also, on the issue of engagement, the lack of a coherent and single voice to represent the ASM sector in most African countries makes it difficult for the companies to know with whom to formulate a dialogue. Constructive dialogue with the ASM sector would undoubtedly be eased if the ASM sector was empowered with a single voice, through legitimate and democratic associations that could sit on the national Chamber of Mines and really represent the artisanal miners and communities. Engagement, dialogue and trust building may also be helped if a third party such as an NGO, international donor or international industry body (like the Global Mining Initiative (GMI) or the Global Mining Dialogue (GMD)) could intervene and mediate, thereby forming a tripartite relationship that allows both the mining companies and the ASM sector to understand each other and hopefully slowly develop mutual trust and ultimately even respect for each other.

It is highly unlikely that the ASM sector in sub-Saharan Africa will be truly empowered and formalized unless the private sector is engaged as a full partner from the very outset of any proposed assistance interventions. The fact remains that international donor funding and national government resources explicitly for the ASM sector are very limited. This may change if ASM can be formally embedded into the National Poverty Reduction Strategy Papers (PRSPs), but even this action does not guarantee the availability of future funds. Therefore, in many instances, the only realistic and sufficiently resourced partner available to assist the ASM sector is the private mining sector. This claim is not naïve and there is no false assertion that the desires of the private sector to engage with the ASM sector are completely altruistic. Likewise, there is little dispute over the fact that the private sector has its own agenda, including the need to improve its global image and fulfill some of its obligations of corporate social responsibility (CSR). Future interventions must ensure that any relationship or partnership is mutually beneficial and will not take advantage of the lack of knowledge or education of the ASM sector as in the case of many past one-sided 'negotiations'. Some past 'negotiations' and 'conflict settlements' have been little more than deceitful public relations scams resulting in intimidation and humiliation of the ASM communities and an indiscriminate abuse of basic human rights. Subordination, intimidation, treachery, the abuse of human rights or the imposition of unjust regulations on the ASM sector by the large-scale sector or corrupt government representatives should no longer be tolerated.

The private mining sector (especially through the International Council on Mining & Metals (ICMM)) is aware that the global public (including their shareholders) often view the mining sector as 'one industry' and do not differentiate between the activities of the ASM and the large-scale mechanized mining conducted by the big transnational mining companies (TMCs) some of which are floated on the major stock exchanges of the world. Therefore, issues of child labour, threats of environmental devastation, or mercury pollution caused by ASM activity in Africa are viewed by the global public as problems caused by the mining industry as a whole. This problem and the overall image of the mining sector is often exacerbated and politically sensitized by the local and international media and some NGOs. Unfortunately, in some countries like Ghana and Tanzania, the media and some dubious NGOs seem compelled to spread unfounded or misinformed propaganda about the mining sector and continue to misinform the public about the mining sector's contribution to the national economy, choose not to report on the numerous positive actions undertaken by the industry or grossly exaggerates any problems or negative impacts or incidents of conflict. Governments are also to blame, as such ignorance is a direct result of miscommunication on the part of the ministry responsible for mining, which often does not disseminate information pertaining to various mineral rights or the mining sector's real contributions. Correspondingly, many companies are keen to help formalize and 'sanitize' these ASM operations in their locality to hopefully alleviate some of these negative images of operating in Africa that unjustly tarnish their global reputation for environmental and social compliance.

Conversely, the mining private sector should not be seen as the alternative rural social service or a regional charity in the mining areas. The majority of mining companies operating in Africa fully accept a degree of CSR and many regularly go beyond mere basics to appease the government requirements and really do genuinely care for their local communities. However, one must question whether it is reasonable to expect these mining companies to also transform into philanthropic organizations, sacrifice their profits and undertake all the social welfare and infrastructure projects in their locality that government should be responsible for.

Many local Chambers of Mines rightly emphasize the fact that even though their members do pay a variety of reasonable taxes to the national government, these taxes are not repatriated into the mining communities. These local communities find it easier to accuse the companies of neglect rather than question the motives and accountability of their own government. The need for transparency in revenue and earnings from the mining sector is essential and the ministry responsible for mining should disseminate information regarding the mining sectors' financial, social and environmental contributions. The concept of encouraging governments to declare its full earnings from the mining sector and then ensure a proportion is repatriated into the regions where the minerals have been won is in line with the current Extractive Industries Transparency Initiative (EITI), already signed by Ghana and Nigeria (with Congo, Angola, Chad and Gabon currently in ongoing discussions).

Based on experiences in Africa, there are numerous examples (MMSD 2002; UNECA 2002) of partnering that could be investigated and possibly duplicated in other areas. For example, mining companies have helped ASM communities by finding alternative employment (SEMOS in Mali, Anglogold Ashanti in Tanzania), assisting with the development of basic social infrastructure, extending micro-credit (SEMOS in Mali), setting aside areas for them to mine (Ingwe Coal Mining Company in RSA, Gold Fields and Bogoso Gold in Ghana or ALMA/Benicon in Moçambique), providing training (SEMOS in Mali), giving H&S and mines rescue advice (Anglogold Ashanti in Tanzania) or buying their produce (Zimasco and ZimAlloys in Zimbabwe or Gold Fields in Ghana). Also worth noting is the current UK's Department for International Development (DFID) project that will work with Anglogold Ashanti in Tanzania and Gold Fields in Ghana to determine and test pragmatic strategies for long-term and sustainable co-existence with ASM communities in the vicinity of their mining concessions.

Mineral trading and markets

Informality and illegality of the ASM sector throughout sub-Saharan Africa also extends to the marketing of mineral products. In truth, the economic dynamics of the ASM commodity chain, from mine to end user, in Africa are not fully understood, especially for gold and gemstones (D'Souza 2002). However, the proliferation of parallel markets for such precious commodities is testimony that there are problems with the existing mineral marketing arrangements. The lack of formal protection by government for the informal ASM activities increases the risk that the artisanal miners are exploited by the lease owners, and a multitude of unscrupulous intermediaries, vendors or mineral traders, and they rarely get fair prices, especially when debt-bonded to these sponsors. In order to ensure a regular cash flow from small amounts of production mined in remote areas, mining communities certainly cannot afford to stockpile their mineral winnings and typically have to sell their products as quickly as they can. Combined with a lack of real 'voice', up-to-date price information, bargaining power, or the comprehension to fully understand their predicament, they are usually exploited and get very low prices. On the occasions when miners do get a windfall, with unexpected extra money from above average production, it is most often misspent and propagates the negative impacts of alcoholism, gambling, substance abuse and prostitution and they remain in poverty. Those working further along the commodity chain, such as sponsors, vendors, mineral traders, intermediaries and manufacturers, take all the profits with little regard for the impoverished miners and labourers. This inequitable distribution of the profits is one of the major obstacles to achieving poverty reduction in the mining communities throughout sub-Saharan Africa.

It is important to remember that illicit marketing (e.g. in 1998 the World Bank estimated that between US$100 and 200 million worth of minerals were illegal traded in Madagascar) in many African countries is primarily the result of inadequate government policies when 'official' prices are too low or when an overvalued local currency rate and high inflation depress the effective price, or due to the traditional strong ties (debt bondage) established by sponsors through the pre-financing of ASM operations. In addition, since the majority of mining sites in Africa are in remote areas, the lack of basic rural infrastructure is a major obstacle to the marketing of mineral produce, especially for low-value bulk industrial and construction minerals.

Unfortunately, one cannot ignore the illegal use of ASM produce to finance and perpetuate past rebel activities, as with the 'blood diamonds' in Angola, Sierra Leone (Campbell 2004), Liberia, République Centrafricaine and the République Démocratique du Congo (RDC) or the coltan boom that devastated eastern RDC recently (D'Souza 2003). In these cases, the links between informality and illegality tend to be self-reinforcing and, rather than being the engine for promoting sustainable development, ASM can in such context quickly become the fuel for further conflict. Hopefully, the recent peace agreements in these areas and the Kimberly Process will establish a means of monitoring the international diamond trade and ensure these 'blood diamonds' do not enter the market. The Kimberly Process was launched in 2001 and is a voluntary scheme that imposes requirements upon participants to certify that shipments of rough diamonds are disassociated with conflict. Although riddled with problems, including a lack of stringent policy and statistical inaccuracy, the initiative has been identified in many policy-making circles as a critical step forward in helping to eradicate the conflicts associated with ASM diamond production in many African countries.

There are a number of ways (many of which have already been attempted in several African countries and with some positive results from Tanzania and Ghana in particular) in which the profits retained by mining ASM communities could be increased through simple interventions including (D'Souza 2002; UNECA 2002):

- establishment of official buying centres in remote mining areas where miners receive a fair price (i.e. the margin does not exceed an appropriate percentage);
- a system of licensed buyers/vendors who visit mining areas and buy small amounts of mine output. If there were sufficient licensed agents, competition between them should prevent the purchase price from being so low that it leads to widespread black market transactions;
- the use of commercial medium/large mining companies or mineral trading companies who would act as assured buyers (tributor arrangement) for specific mineral commodities (providing this did not reduce ASM bargaining power and result in a quasi-slavery dependency or excessive debt bondage);
- links with the growing 'fair and community trade' movement in industrialized countries;
- increasing the awareness within the ASM sector regarding market prices (e.g. publishing prices in local newspapers or broadcasting on local radio stations) and methods of adding value to mineral commodities by establishing appropriate processing industries (including lapidary and jewellery manufacture).

The liberalization of the domestic mineral markets through simplification of licensing procedures for private mineral dealers should hopefully steer most dealers through legal channels. Mineral marketing arrangements are best fully opened to the private sector, with the government acting only as the regulator and monitor. Any proposed marketing scheme must also discourage monopoly organizations from controlling mineral markets and pricing of commodities, as well as foster competition among private sector buyers and sellers to ensure that ASM communities have a number of options for disposal of their output and are not trapped by predatory sponsors.

In some African countries, the absence of any independent and competent laboratories to undertake mineralogical studies and assay samples restricts the marketing capabilities of the ASM mineral producers. Without such independent services or their own valuation knowledge they are unable to verify buyer and vendor claims of mineral grade or quality. They are therefore highly susceptible to being cheated on price and are consistently underpaid for their produce. This results in the continued non-equitable distribution of mineral profits and results in the actual artisanal miners being poorly compensated for their efforts.

ASM mineral added value

Almost all sub-Saharan African countries with ASM production of semi-precious minerals have some degree of value adding, as witnessed in the numerous craft and tourist curio markets throughout the continent where traders sell a variety of products made from locally produced minerals. However, these goods are viewed as 'ethnic' souvenirs and trinkets that do not meet the standards of the lucrative international jewellery markets. Ghana, Tanzania and Madagascar have all attempted to overcome this problem by initiating capacity-building schemes that can add value to mineral products, which can compete on the international markets and satisfy the pernickety tastes of Western buyers.

Although added value to mineral products is something that many governments are keen to encourage, there does not appear to be any real incentives or guidelines for mineral producers to invest in such associated mineral industries. Added value would result in higher minerals-related profits being retained in the country, increased revenue for the national government and the creation of additional skilled jobs. For instance, the gemstone and gold sectors would benefit from the development of a parallel lapidary and jewellery sector that could focus on both the growing African demand for jewellery and the international market. Such African-made, and influenced, jewellery could then be promoted and sold through the major tourist hubs in Africa like Cape Town, Johannesburg, Nairobi, Mombasa or Arusha that already promote Africa, her people and culture. For example, in Ghana the Precious Minerals Marketing Corporation (PMMC) has established a jewellery business that refines gold and cuts diamonds to produce jewellery based on authentic Ghanaian (Asante Adrinka) and African designs. In Tanzania the planned Arusha Gemstone Carving Centre will aim to train Tanzanians in lapidary skills, whilst in Nigeria there is a planned Gemstone Village to add value to the numerous locally won precious and semi-precious stones that will build on the recent success of development of the added-value gemstone sector in Madagascar.

However, the other mineral sectors must not be ignored and, in particular, added value must be encouraged for the ASM industrial minerals sector where there is a great potential for import substitution and the development of a multitude of sizeable downstream mineral industries.

Finance and credit

Financial empowerment of the ASM sector is the fundamental prerequisite for the success of the overall development of the ASM sector in sub-Saharan Africa (D'Souza 2002). Most artisanal miners would like to step up to higher levels of productivity and output by mechanizing more of their activities and developing new reserves (ILO 1999). However, few are able to gain the necessary capital, as most banks are exceedingly wary of ASM producers as they lack acceptable forms of collateral (this can be as much as 20% of the requested equity). Although there is a very strong and competitive banking sector in some African countries, they are not conversant with the concept of realistic micro credit for the ASM sector or the specifics of mining project finance for the small- to medium-scale mine operators and are currently unwilling to provide soft loans for mining ventures. The lack of appropriate financing mechanisms has resulted in many miners throughout Africa having to resort to seeking credit from the multitude of predatory sponsors that plague the sector and hence become trapped to endlessly work for these sponsors in a debt-bondage situation (D'Souza 1998).

Many countries in Africa (including Zambia, Ghana, South Africa, Moçambique, Zimbabwe and Tanzania) have attempted to initiate some form of loan/equity-based financial scheme, equipment hiring or other form of credit with varying degrees of success, particularly with regard to loan repayment (UNECA 2002). The reasons why the establishment of a sustainable revolving loan scheme has proved elusive in many cases is still unknown, although inadequate debt servicing, unrealistic interest rates, selecting the wrong partners, and high administration and set-up costs are all major factors.

Any approach to financing should also be tailored to suit the mineral commodity because the 'turnover' and income pattern from precious commodities with high unit value differs fundamentally from industrial minerals produced for local markets or from bulk commodities for domestic use. Governments could investigate a variety of financing mechanisms in consultation with all the stakeholders, including some form of soft loan guarantees, grants, micro-credit schemes, small–medium enterprise (SME) development grants/loans, and/or equipment leasing and hire.

The overall objective must be to establish a long-term, transparent, fair and sustainable financing scheme, which ensures adequate debt servicing based on firm business terms, and results in the long-term sustainability of the scheme. The proposed schemes must also ensure they target the actual miners/labourers through empowering cooperatives that break the existing debt-bondage links common in many areas, rather than sustain the voracious sponsors and mineral traders. However, any future schemes must be simple to administer and monitor, and take into account the special needs of the ASM sector, as well as problems with exchange rate fluctuations, current exorbitant bank interest rates and possible local currency devaluation. Similarly, the ASM legislation must ensure that it provides full security of tenure for their mining titles, thereby allowing the miners to have a degree of credit worthiness such that they can transfer, pledge, transfer or trade their mineral rights in order to facilitate access to finance.

Pro-poor policy framework

What should be done to achieve these numerous objectives outlined above? Past studies undertaken by the UK's Department for International Development (DFID), the World Bank, the Mining, Minerals and Sustainable Development (MMSD) Project, and the UN Economic Commission for Africa (UNECA) have shown the following points essentially embody the most critical issues that must be addressed in order to formalize, develop and empower the ASM sector in sub-Saharan Africa (D'Souza 1998):

- delineation and definition of the mineral potential suitable for ASM exploitation;

- establishment and resourcing of governmental institution or units to manage the ASM sector;
- formulation of an enabling, appropriate and transparent legislative framework and licensing scheme specifically for ASM;
- formation of ASM associations, community organizations and/or cooperatives;
- provision of vocational outreach services – regional self-sustaining technical assistance and training schemes/centres (appropriate and best practice technology);
- provision of viable and sustainable financing/credit schemes for the ASM sector;
- provision of fair, transparent and effective marketing systems for ASM products;
- promotion of and assistance with added value schemes to ASM mineral products;
- definition and enforcement of relevant and consistent H&S safety standards and practices;
- definition and promotion of ASM best practice and appropriate technology transfer to increase productivity, reduce mercury use, and improve H&S and environmental management;
- promotion and implementation of child labour elimination programmes;
- testing of co-existence strategies of ASM with large-scale mining;
- promotion and implementation of gender mainstreaming and empowerment programmes;
- establishment and enforcement of appropriate and realistic environmental protection practices and principles.

The pro-poor and best-practice policy framework must be multi-dimensional: institutional, legal, technical and fiscal. A holistic approach must be adopted in order to increase human, financial, physical and social capital available to the sector. However, there must be an assurance that such macro-policy links with micro-reality of the ASM sector. In full consultation with all relevant and interested stakeholders, this growth-based and poverty reduction policy framework needs to be implemented to ensure a holistic programme of assistance that addresses all the issues. The policy also has to take into account the often divergent interests of the various ASM stakeholders.

It is hoped that such a 'generic' policy framework is adopted by the various countries of sub-Saharan Africa to formulate strategies that are pragmatic and that are fully supported by, and appropriate to, rural communities. In order to achieve this adoption, the 'generic' policy will have to be tailored to suit the political, cultural, geographical and economic peculiarities of each nation and all micro, meso and macro linkages must be identified and resourced. In addition, future assistance must incorporate some of the experiences of best practice from countries like Ghana, Tanzania, South Africa, Mali, Ethiopia, Zambia, Guinée, and Zimbabwe, which have made significant progress with the development and formalization of their ASM sectors.

Summary

It is clear that if this informal sector is to be turned into an economically viable industry that provides benefit to all concerned, including African governments, real reform is required. Broadly speaking, the short- and medium-term goals of assistance to the ASM sector in Africa should (D'Souza 2002)

- ensure poverty alleviation and wealth creation of the actual miners, labourers and rural communities is always the primary focus of any proposed intervention;
- promote the formation of family- or community-based mining cooperatives that are sensitive to the diversity of local cultures and traditions, build on existing or traditional CBOs and that help give 'voice' to and empower the miners and their communities;
- encourage local rural economic development by ensuring that mineral revenues are repatriated to the mining communities and invested in ways that bring sustained benefits including the development of basic social infrastructure;
- ensure ASM is adopted as part of a range of complimentary, alternative livelihoods that in combination lead to real rural community sustainable development;
- adopt a gender-sensitive approach that gives particular emphasis to the valued role of women while respecting the traditional values of local communities;
- reduce and eventually eliminate child labour in mining areas through the provision of viable alternatives (while differentiating from suitable child work that is part of normal rural socialization processes);
- avoid or mitigate negative environmental and social impacts as well as impacts on human health on the mining sites and within the mining communities;
- encourage fully equitable and liberalized markets for all mining products that eliminate oppressive mineral purchasing and marketing monopolies and that curb the illicit trading and smuggling or precious minerals;

- develop the collective capacity of artisanal miners to contribute to sustainable development and an upscaling of mining activity through the adoption of identified best/good practice, methods, technology and equipment;
- mainstream the ASM sector in national economic activity and ensure that other relevant government actors, cogent ministries and key stakeholders understand their obligations to ASM communities;
- ensure good relationships between the artisanal miners, communities, local and traditional leaders, regional and national government and all other relevant stakeholders.

There should be an emphasis on people rather than merely on technology and productivity increases and a realization that ASM activities will continue for at least as long as poverty drives them. Therefore the rights of individuals to choose and secure a livelihood, whether within or outside the ASM sector, must be respected.

The focus of any reform should be in accordance with the identified targets of the Millennium Development Goals (MDGs) to alleviate poverty by providing financial betterment and empowerment to ASM communities, assist in rural development, and should be part of an integrated Community Development Plan. Proper development of the ASM sector, especially industrial/construction minerals, may also reduce many African countries' importation demands and thereby improve self-sufficiency on national and regional levels, and will allow the efficient exploitation of small mineral deposits where only ASM methods are practical and economically feasible.

ASM mining exploits finite non-renewable resources so that the challenge for sub-Saharan Africa now is to capitalize on the livelihood opportunities while ensuring that ASM also contributes to other goals of sustainable livelihoods, rural development and national poverty reduction efforts for the long-term future of the Continent.

As a final note, it is worth remembering that the ASM sector in many African countries is already a victim of 'research fatigue', with countless academic studies, inappropriate data-gathering exercises, a plethora of dead paper policy and grandiose declarations that simply choke the ASM sector and that are yet to be implemented or even tested (D'Souza 2002; Pedro 2002). Future interventions and assistance projects should be wary of these past, well-meaning, but highly academic approaches that have failed to really impact on the physical lives of artisanal miners and their communities. Future projects should aspire to finally bridge the increasing gap between high-level policy and on-the-ground practice and guarantee that a realistic and pragmatic approach is adopted that ensures both immediate impact to the impoverished mining communities and long-term sustainability for the benefit of the entire Continent.

Future assistance needs to build on the momentum of work and recommendations of the UK's Department for International Development (DFID), the World Bank, the new African Mining Partnership (AMP), Mining, Minerals and Sustainable Development (MMSD) Project, the Extractive Industries Review (EIR), UN Economic Commission for Africa (UNECA), the International Labour Organisation (ILO), the UN Conference on Trade and Development (UNCTAD) – African Mining Network, the Global Mining Initiative (GMI), the Global Mining Dialogue (GMD), the International Council on Mining & Metals (ICMM), the Communities & Small-Scale Mining (CASM) initiative and even the new Equator Principles signed by many of the world's major financial institutions.

The current international donor and government interest in many African countries must be mobilized to resource and formulate real and sustained programmes of assistance that are properly managed and defined with results-oriented actions and that include a means of monitoring and assessment. Assistance to the sector desperately needs to move from 'what to do' rhetoric and repetitive academic research that has afflicted many African countries, to the more difficult and somewhat elusive 'how to do' pragmatic and tangible actions on the ground. The ASM sector is desperate for real demand-led actions rather than yet more 'developmental' self-perpetuating research and data-gathering exercises. Assistance to the sector should not be deterred, or be afraid of, the potential for some intervention failures and needs to prepare for the long haul to bring ASM in sub-Saharan Africa into the mainstream by addressing all the challenges that plague the sector in a holistic and planned manner through a concerted effort by all stakeholders.

There is now an urgent need to take into account the existing socio-economic system and consider how ASM can best contribute to poverty reduction and sustainable development through integration into rural community development plans. Development interventions must find better ways of integrating miners into the rest of the economy and encouraging ASM communities to invest their revenues in other forms of economic activity as well as in communal services. There must be a clear objective to foster and encourage local stakeholder buy-in and ownership, and ensure we provide incentives for the continuation of any assistance scheme while appealing to the self-interest of the miners and their communities. Only by adopting such an approach will the sector be able to contribute to sustainable rural livelihoods and contribute to the fulfilment of the Millennium Development Goals (MDGs) by 2015.

References

ABATE, R. 1991. Pre-designed modular plants for small-scale mining operations in Third World countries. *African Mining*, **91**.

AGID 1980. *Regional Workshop on Strategies for Small Scale Mining and Mineral Industries.* AGID, Mombassa, Kenya.

ALI, S. V. 1986. Small scale mining – towards a purposive definition? *Journal of Mines Metals Fuels*, **34**(8–9), 401–404.

ASTORGA, T. S. & DURAN, N. M. 1992. The young people in the rural artisanal mining sector. *Minerales*, **47**(198), 5–10.

ASTORGA, T. & DURAN, N. 1994. The rural youth artisan mining sector. *SMI Bulletin*, **7**, 12–15.

BARNEA, J. 1978. *Important for the Future.* Special Issue on the Conference on the Future of Small-Scale Mining. United Nations Institute for Training and Research (UNITAR), New York.

BHALOTORA, S. 2003. *Child Labour in Africa.* OECD DELSA/ELS/WD/SEM (2003–4).

BILLS, J. H., MARTINEAU, M. P. & PARK, J. G. 1991. Artisanal mining in the Lake Victoria goldfields, Tanzania. *Proceedings of African Mining*, **91**.

BORLA, G. 1996. Artisanal mining activities in developing countries. *GEAM Geoing. Ambient. Min.*, **33**(4), 171–177.

BRITISH GEOLOGICAL SURVEY 1999. *Technical Report WC/99/7 (1999): Socio-Economic Issues Relevant to the Uptake of New Technology in Small-Scale Gold Mining: Studies in Guyana and Zimbabwe.* British Geological Survey, UK.

BURKE, G. 1995. Policies for small-scale mining: the need for integration. Mining and mineral resource policy issues in Asia-Pacific: prospects for the 21st century. Proceedings of a conference held in Canberra, ACT.

BURKE, G. 1997. Policies for small-scale mining: the need for integration. *Raw Material Report*, **12**(3), 11–14.

CAMPBELL, G. 2004. *Blood Diamonds.* Westview Press, Colorado, USA.

CARMAN, J. S. 1985. The contribution of small-scale mining to world mineral production. *Natural Resources Forum*, **9**(2), 119–124.

DAVIDSON, J. 1993. *The Successful Development of Small-Scale Mining Enterprises in Developing Countries: An Overview.* Interregional Seminar, Guidelines for Development of Small/Medium Scale Mining, Harare, Zimbabwe.

DAVIDSON, J. 1998. Building partnerships with artisinal miners. *Mining Environmental Management*, 20–23.

DRECHSLER, B. 2001 Tanzania. *In: Small Scale Mining in Southern Africa*, MMSD, Research Topic 1. ITDG, London.

D'SOUZA, K. 1998. *Design and Implementation of a Model Scheme of Assistance to the ASM Sector.* Department of International Development (project) KaR 7181, unpublished.

D'SOUZA, K. 2000. Small-scale mining. *Proceedings of the Mining 2000 – Issues for the New Millennium Conference*, Dundee, UK, June 2000.

D'SOUZA, K. 2002. *Artisanal and Small-Scale Mining in Africa: A Reality Check.* Unpublished: Keynote presentation at the Seminar on Artisanal and Small-Scale Mining in Africa, Yaounde, Cameroon, November 2002.

D'SOUZA, K. 2003. *Coltan Mining in the DRC.* Proceedings of the Durban Process (Dian Fossey Gorilla Fund), Durban, South Africa, July 2003.

D'SOUZA, K. 2004a. *Reprofiling the Artisanal Mining Sector of Katanga.* Proceedings of the World Bank Mining Conference (Contribution of the Mining Sector to the Economic Development of the Province of Kantana), Lubumbashi, DR Congo, June 2004.

D'SOUZA, K. 2004b. *Changing Government and International Perspectives on Artisanal and Small-Scale Mining.* Proceedings of the World Bank National Mining Policy Dialogue Conference, Abuja, Nigeria, July 2004.

GHOSE, A. K. 1994. *Small Scale Mining – A Global Overview.* A. A. Balkema, Rotterdam, Netherlands.

HOLLAWAY, J. 1991. Role of small-scale mining in Africa: building on the informal sector. Proceeding of the African Mining 91 Conference, Johannesburg.

HOLLAWAY, J. 1993. Review of technology for the successful development of small scale mining. *Chamber of Mines Journal Zimbabwe*, **35**(3), 19–25.

HOSFORD, P. A. J. 1993. Innovations in mineral processing technology for small scale mining. *Chamber of Mines Journal Zimbabwe*, **35**(3), 26–27, 29, 35, 41.

ILO 1999. *Social and Labour Issues in Small-Scale Mines: Report for Discussion at the Tripartite Meeting on Social and Labour Issues in Small-Scale Mines.* Sectoral Activities Programme, TMSSM/1999. ILO, Geneva.

JENNINGS, N. S. 1999. *Child Labour in Small-Scale Mining: Examples from Niger, Peru and Philippines.* International Labour Office.

KONOPASEK, R. 1981. The significance of small scale mining in developing countries. *BHM Berg-u. Huttenm. Mh.*, **126**(12), 521–523, 2 refs.

LABONNE, B. 1993. *Interregional Seminar, Guidelines for Development of Small/Medium Scale Mining.* Harare, Zimbabwe.

LABONNE, B. 1996. Artisanal mining: an economic stepping-stone for women. *Nat. Resour. Forum*, **20**(2), 117–122.

LANDNER, L. (ed) *Environment and Mining in Eastern and Southern Africa.* Selected papers from an international conference, 23–27 October 1995, Mwanza, Tanzania.

LANDNER, L. (ed) *Small-Scale Mining in African Countries: Prospects, Policy and Environmental Impacts.* Proceedings of an international conference 29 September – 1 October 1997, Dar es Salaam, Tanzania.

MCMAHON, G., EVIA, J., PASCÓ-FONT, A., SANCHEZ, J. 1999. *An Environmental Study of Artisanal, Small and Medium Mining in Bolivia, Chile and Peru.* World Bank Technical Paper No. 429, World Bank, Washington DC, USA.

MIREKU-GYIMAH, D., OPARE-BAIDOO, S. & COBBLAH, A. 1996. *Small-scale gold mining, and its impact on the economy of Ghana.* Proceedings of a conference, Surface Mining 1996. Johannesburg.

MINING, MINERALS & SUSTAINABLE DEVELOPMENT (MMSD) 2002. *Breaking New Ground*, Chapter 13. Earthscan Publications Ltd, UK.

MUTAGWABA, W. & HANGI, A. 1995. Environmentally sustainable artisanal gold mining in the Lake Victoria regions, Tanzania. Papers presented at the Third African Mining Conference, African Mining 1995. Windhoek, Namibia.

MWAMI, J. A., SANGA, A. J. & NYONI, J. 2001. *Child Labour in Mining: A Rapid Assessment. Investigating the Worst Forms of Child Labour.* No. 15. ILO, Geneva, Switzerland.

PARSONS, A. 1999. *Small-Scale Mining and the Environment.* UNEP, Paris, France.

PEAKE, V., JOHNSON, R. A. & SVOTWA, R. 1998. A perspective on the provision of technical expertise to the small scale mining industry in southern Africa. *Mining in Africa*, **98**, Johannesburg, South Africa, September 1998.

PEDRO, A. 2002. *Poverty Eradication & Sustainable Livelihoods in ASM Communities in Africa: Food for Thought: From Paper Policy to Results-Oriented Actions.* Proceedings of the Seminar on Artisanal and Small-Scale Mining in Africa, Yaounde, Cameroon, November 2002.

PRIESTER, M., HENTSCHEL, T. & BENTHIN, B. 1993. *Tools for Mining – Techniques and Processes for Small Scale Mining.* Deutsche Gesellschaft fur Technische Zusammenarbeit (GTZ), Germany.

SECRETARY-GENERAL OF THE UNITED NATIONS 1996. Recent developments in small-scale mining. *Natural Resources Forum*, **20**(3), 215–225.

TAUPITZ, K. C. & MALANGO, V. 1993. *Making the Transition from Unmechanized Manual Mining to Industrial Small-scale Mining.* United Nations Interregional Seminar on Guidelines for the Development of Small/Medium Scale Mining, Harare, Zimbabwe.

TRAORE, P. A. 1994. *Constraints on Small-Scale Mining in Africa.* United Nations, Natural Resources Forum.

UNITED NATIONS 1978. The First International Conference, Mexico. *The Future of Small Scale Mining.* United Nations.

UNITED NATIONS DEPARTMENT OF TECHNICAL CO-OPERATION FOR DEVELOPMENT 1989. *Mining Policies and Planning in Developing Countries.* United Nations, New York, USA.

UNITED NATIONS 1994. *Natural Resources Forum, Small and Medium Scale Mining.* The Harare Seminar and Guidelines. United Nations, New York, USA.

UNITED NATIONS ECONOMIC COMMISSION FOR AFRICA 1992. Status of small-scale mining in Africa and strategy for its development. UNECA, Addis Ababa, Ethiopia.

UNITED NATIONS ECONOMIC COMMISSION FOR AFRICA (UNECA) 2002. *Compendium of Best Practices in Small-Scale Mining in Africa.* United Nations, Addis Ababa, Ethiopia.

WALLE, M. & JENNINGS, N. 2001. *Safety & Health in Small-Scale Surface Mines: A Handbook.* Sectoral Activities Programme, International Labour Organisation Office, Geneva, Switzerland.

WEBER-FAHR, M. *ET AL.* 2002. Mining. *In: Volume 2 – Macroeconomic and Sectoral Approaches.* World Bank, Washington DC, USA.

WORLD BANK 1995. *A Comprehensive Strategy Toward Artisanal Mining.* The World Bank, Industry and Mining Division, Industry and Energy Department, Washington DC, USA.

WELS, T. A. 1983. Small-scale mining – the forgotten partner. *Transactions of the Institute of Mining & Metallurgy* A, **92**.

FarmLime: low-cost lime for small-scale farming

CLIVE J. MITCHELL

British Geological Survey, Keyworth, Nottingham NG12 5GG, UK (e-mail:cjmi@bgs.ac.uk)

Abstract: In the less-developed parts of southern Africa, where agriculture is crucial in daily survival, agricultural lime is often difficult to obtain. This is due to the scarcity of production sites, high transport costs and inadequate support for farmers from government extension services. In Zambia there are upwards of 700 000 small-scale farmers who struggle to farm on acid soils and as a result have poor crop yields. They cannot afford to use agricultural lime, which would solve the problem, and as a result are trapped in a cycle of poverty. In an attempt to address this problem, the UK research project 'FarmLime', based in Zambia, investigated a means of producing affordable agricultural lime using simple, locally available technology. Dolomite suitable for agricultural lime occurs throughout Zambia, including those farming districts with acidic soils. It is estimated that small-scale production using partly manual methods could produce agricultural lime for US$25–30 per tonne. Demonstration crop trials were successful in demonstrating the benefits of using agricultural lime to small-scale farmers. Where the price of maize is high and the cost of lime is low, the economic benefits of its use are high. However, even if there is a demonstrable economic benefit, the use of agricultural lime will be constrained by the lack of cash in the rural economy; one potential solution to this could be bartering of crops for agricultural lime.

Agricultural lime is a commodity taken for granted in most developed countries. It is used in farming as a soil conditioner to prevent acidification, provide a source of plant nutrients (calcium and magnesium) and improve the physical properties of the soil. In the UK it is produced in modern processing plants alongside aggregate and other products such as mineral fillers. The project 'FarmLime: low-cost lime for small-scale farming' was funded by the UK Government Department for International Development (DfID). The aim of DfID research is to 'alleviate poverty' through improved access to knowledge and technology, specifically by 'enhancing the productive capacity (of less-developed countries) in an environmentally sensitive manner'. The findings of a previous DfID research project 'Local development of affordable lime in southern Africa' concluded that agricultural lime could be produced on a small scale from local dolomite within farming districts. The aim of the FarmLime project was to test these conclusions in the farming districts of Northern Zambia, which are broadly representative of farming in southern Africa.

Agricultural lime production and consumption in Zambia

Agricultural lime is available from five companies in Zambia, located in the main 'economic corridor' extending from Lusaka in the south to the Copperbelt in the north. Lilyvale Farm (Kabwe, Central Province) is a coffee plantation that also produces agricultural lime and represents the old style of operation, which was developed to serve the needs of the large agricultural estates. Minedeco Small Mines in Lusaka is a former parastatal company that produces mineral fillers for local industry as well as a small volume of agricultural lime. Ndola Lime Ltd (Ndola, Copperbelt Province) is a large quick lime and hydrated lime plant that produces agricultural lime as a byproduct 'dust' from the crushing plant. These operations represent the old nationalized industry established in the 1960s. However, they now have to exist without state subsidies alongside more competitive rivals and as a result all are struggling. Uniturtle Industries (Z) Ltd, Lusaka, is a company that mainly produces decorative stone and has in the last few years expanded into production of mineral fillers, agricultural lime and stock feed. Hi-Qwalime Mining Ltd (Mkushi, Central Province) is an operation dedicated to producing agricultural lime. These operations have been encouraged since denationalization of the state mining sector and have been funded by private investment with strong South African connections.

The current consumption of agricultural lime in Zambia is about 40 000 tonnes per year, mostly by large-scale commercial farmers. This is a relatively low volume, as the latent demand for agricultural lime is thought to be about 150 000 tonnes per year (Shitumbanuma & Simukanga 1995). To achieve this level of consumption would require a dramatic increase in the use of agricultural lime, especially by small-scale farmers; however, they do not have a tradition of using this agricultural input.

Socio-economic survey

Small-scale farmers who use agricultural lime can improve their crop yields, produce more food for their families and (potentially) generate some income. Surveys carried out by questionnaire returns and interviewing of small-scale farmers representative of farming districts in Zambia (Mitchell *et al.* 1997, 2003) found that they do not use lime for the following reasons.

- Farmers are either unaware or unconvinced of the need for agricultural lime. Farmers receive little advice on the use of agricultural lime from agricultural extension workers (field workers providing advice and assistance to farmers), there is a lack of information regarding its quality and use, and poor packaging and labelling.
- The Ministry of Agriculture, Food & Fisheries and the University of Zambia both carry out soil testing for farmers. However, little soil sampling and testing is carried out for small-scale farmers, because it is expensive and time consuming. Farmers do not know their soil pH and the amount of lime to apply.
- Agricultural lime is expensive and often the farmers do not have the cash to buy it. An alternative form of payment, such as bartering, could partly replace cash payment. Most of the producers are reluctant to supply the small volumes (50 kg) required by small-scale farmers as they prefer to supply in large volumes to commercial farmers. Also, transportation is expensive, typically adding US$1 per tonne per 10 km travelled.

Government and non-governmental organization (NGO) food security programmes have attempted to stimulate the demand for agricultural lime by small-scale farmers without much success (Mengo 2002). However, this is probably the key to future development of small-scale farming in southern Africa.

Carbonate resource assessment

Most of the provinces in Zambia have carbonate resources (limestone and dolomite). However, not all are suitable for the production of agricultural lime. The key parameters used to determine the suitability of these rocks as a source of agricultural lime are as follows:

- *Neutralization ability*, expressed as the weight percentage calcium carbonate equivalent (CCE) and often referred to as the neutralization value (NV). This is the amount of calcite and/or equivalent dolomite and is the key property in reducing soil acidity. A CCE value of 80% is considered to be a minimum for agricultural lime.
- *Plant nutrient content*, expressed as the weight percentage calcium oxide (CaO) and magnesium oxide (MgO). An MgO content of 6% is considered to be a minimum for agricultural lime (Tether & Money 1989).
- *Ease of pulverization*, otherwise known as the grindability index, is the weight percentage of fine particles generated in a milling trial. This indicates the hardness of the carbonate and, indirectly, the amount of energy required to produce agricultural lime of the required fineness.
- *Agronomic effectiveness*, otherwise known as the reactivity, is a measure of the rate at which agricultural lime neutralizes soil acidity.

Dolomite is preferred for the production of agricultural lime as it contains both of the important plant nutrients (calcium and magnesium) and has a high neutralizing ability (ARC Southern 1996). Dolomite samples from across Zambia were evaluated and many were found to have comparable technical properties with commercially available agricultural lime (Table 1).

Lime production research

Research was carried out to determine if agricultural lime could be produced on a small scale, using appropriate technology and manual labour where appropriate. Commercial operations quarry, crush and grind carbonate rock to produce agricultural lime, typically with a particle-size distribution of 100% finer than 2 mm, 60% finer than 400 μm and up to 50% finer than 150 μm.

Small-scale production of agricultural lime would require quarrying of carbonate rock and would involve manual bush clearing and loose soil removal, drilling and blasting, and manual loading and hauling of the extracted rock, with an estimated cost of US$1.45 per tonne of rock extracted (Table 2). Field trials were carried

Table 1. *Comparison with commercially available agricultural lime*

Sample site	Neutralization ability/CCE (wt%)	Plant nutrient content CaO (wt%)	MgO (wt%)	Ease of pulverization grindability (wt% <75 μm)	Agronomic effectiveness/ reactivity (%)
Solwezi dolomite	104.3	32.2	19.3	72.5	80.4
Mkushi dolomite	103	30.5	19.1	71.5	77.1
Lilyvale Farm dolomite	106	30.7	21.1	97.2	84.2
Ndola Lime Ltd limestone	98.7	53.7	1.9	81.4	90.42

out to produce agricultural lime. Crushing of the dolomite to an appropriate size (10 mm) for the milling stage was carried out manually using sledgehammers. This was found to be an expensive process as it was slow and labour-intensive, costing up to US$10 per tonne. As an alternative, a manually-operated jaw crusher could be used to speed up the crushing process; it is estimated that this could reduce the cost of the crushing to less than US$1 per tonne. The milling trials were based on the TD Hammer Mill (originally designed for maize milling), which was modified for milling of dolomite. The mill is powered by a single-piston petrol engine, which is used to drive the mill hammers directly via the engine shaft. The rock is fed into the milling chamber via a launder and the ground material falls through a product sieve. The milling cost was estimated to be approximately US$20 per tonne of (bagged) agricultural lime produced.

The estimated costs of producing agricultural lime using a small-scale operation are summarized in Table 2; these were based on Zambian labour and equipment costs. The overall cost is about US$32 per tonne; use of a manual jaw crusher could reduce the production costs to US$25 per tonne.

Table 2. *Summary of unit operation costs for agricultural lime production*

Unit operation	Cost per tonne (US$)
Bush clearing and soil removal	0.20
Drilling and blasting	0.70
Loading and haulage	0.55
Manual crushing	10
Milling (and bagging)	20.10
Total production cost	31.55

Diversification and spontaneous uptake

The Zambian Poverty Reduction Strategy Paper (PRSP), produced by the Zambian Ministry of Finance and Economic Development (MOFED), focuses on measures to achieve strong sustained economic growth (5–8% pa) with enhanced agricultural productivity given the highest priority. Another key goal of the PRSP is diversification of the small-scale mining sector, currently dominated by gemstone mining, into industrial minerals. Industrial minerals form an integral part of the sustainable development of an economy as they are driven by genuine industrial needs and are more likely to be the basis of long-term economic activity. This compares with gemstone mining, which often has problems associated with migrant labour and 'boom town' scenarios, cross-border smuggling and illegal mining.

Diversification of an existing mineral producer, FRECA Mining & Manufacturing Ltd in Solwezi, has occurred through 'spontaneous uptake' of the agricultural lime production methods advocated by the FarmLime project following field trials of the modified TD Hammer Mill. FRECA employ small-scale miners to produce construction aggregate using traditional mining and crushing methods. Fires are set on the dolomite outcrops, which when hot are doused with water to promote cracking. The dolomite is then removed using picks and crowbars. The rock is crushed by hand using hammers and sieved to remove the dust, which is sold as agricultural lime (in 50 kg bags) to local farmers. FRECA have fabricated a hammer mill, which they will use to produce agricultural lime from the crushed dolomite.

Crop trials

Demonstration crop trials were carried out to show farmers the benefits of using agricultural lime. The sites chosen for the demonstration trials were located in Mkushi district, which has acid soils and a source of dolomite suitable for agricultural lime production. The agricultural lime used was that produced using the modified TD Hammer Mill. The amount required for the demonstration sites was based on the exchangeable aluminium content of the soil and the neutralizing ability of the agricultural lime to be used. This was determined to be between 200 and 450 kg per hectare. Each site was divided into ten plots – five for maize and five for groundnuts – including one with no lime, two with lime, and two with double the amount of lime. The lime was added as either a 'spot application' (i.e. applied to the immediate vicinity of the plant only) or 'broadcast application' (i.e. spread over the entire plot).

The crops were planted according to the Zambian MAFF (Ministry of Agriculture, Food & Fisheries) recommendations for small-scale farmers. The farmer, his family and neighbours, tended the plots, which were revisited to monitor the progress the following year. The crops were harvested and data on grain weight are given in Table 3. The maize and groundnuts from the limed plots generally gave higher yields than the unlimed plots.

During the agricultural lime demonstration, farmers neighbouring the trial sites started to use agricultural lime; understanding this aspect of the crop trials will be important for future promotion of agricultural lime.

Cost benefit analysis

The benefits of using agricultural lime will largely depend on economic factors, such as the cost of the lime and the increase in crop yields to the small-scale farmer. A cost benefit analysis was carried out to quantify the economic benefits for small-scale farmers in using agricultural lime. Farmers will directly benefit if the value of the additional crops produced exceeds the cost of the agricultural lime used. This can be quantified as a value cost ratio (VCR) as follows:

$$VCR = \frac{\text{(Weight of additional crop produced)} \times \text{(Unit value of the crop)}}{\text{(Cost of using agricultural lime)}}$$

The UN Food & Agriculture Organisation (FAO) believes that for small-scale farmers a VCR above 2 is required for uptake of new inputs (i.e. the value of the additional crops produced is at least double the cost of using agricultural lime). Value cost ratios were determined for several scenarios based on varying maize yields, maize price and input costs (Table 4). It is estimated that small-scale farmers should be able to increase their maize yields from their current average of about 1.5 tonnes per hectare per tonne of lime as follows:

- Extremely acid soil (pH < 4.5): increase maize yield by 5 tonnes per hectare
- Very acid soils (pH < 5): increase maize yield by 3 tonnes per hectare
- Acid soils (pH 5–6): increase maize yield by 1.5 tonnes per hectare

The value of maize produced can vary greatly, from US$200 per tonne (realized by commercial farmers close to maize mills and selling at the optimum time) to less than US$50 per tonne (realized by small-scale farmers in remote locations selling as the need arises). The cost of agricultural lime is typically US$20 ex-works (i.e. collected from the producer) and a further US$1 is added per tonne for every 10 kilometres the lime has to be transported (this can multiply the cost several fold for remote farmers). Two maize values (US$50 and US$110 per tonne) and two agricultural lime costs (US$30 and US$60 per tonne) were used in the calculations.

The use of agricultural lime was found to be economically beneficial in most cases. The results show that farmers who live close to the source of their inputs, farm highly acid soils,

Table 3. *Response of crops to lime application at trial sites in Mkushi, Zambia*

Agricultural lime treatment	Maize (grain yield kg/ha)	Groundnuts (grain yield kg/ha)
Control (unlimed)	2490–5200	180
Lime (spot)	3050–5490	200–320
Lime (broadcast)	3570–6760	210–240
Double lime (spot)	3430–4880	150–200
Double (broadcast)	3110–5610	90–320

Table 4. *Value cost ratio (VCR) scenarios for maize*

	Value cost ratio (VCR)			
	High lime cost per tonne (US$60)		Low lime cost per tonne (US$30)	
	High maize price per tonne (US$110)	Low maize price per tonne (US$50)	High maize price per tonne (US$110)	Low maize price per tonne (US$50)
Extremely acid soil	9.1	4.2	18	8.3
Very acid soils	5.5	2.5	11	5
Acid soil	2.7	1.3	5.5	2.5

and can realize a high price for their maize will see the greatest improvements in their incomes.

Conclusions

Agricultural lime is a widely used commodity in the developed world. However, in less-developed countries such as many in southern Africa, this is not the case. Small-scale farmers are unaware or unconvinced of the need for agricultural lime and hence their use of lime is minimal. Demonstrating the benefits of using lime is an important way to persuade farmers to use lime. Once persuaded, they need a simple and cheap soil test to determine agricultural lime demand. Agricultural lime should be made available to all small-scale farmers; for those who cannot afford to buy lime an alternative form of payment should be allowed, such as bartering.

There are ample resources of dolomite rock available throughout Zambia for agricultural lime production. The neutralizing value (NV) and the plant nutrient content are the most important properties to determine in an evaluation of their suitability for use as agricultural lime. Small-scale production of agricultural lime would require quarrying (involving drilling and blasting), manual crushing and milling. The modified TD Hammer mill is capable of producing 1–2 tonnes of agricultural lime per day. The production cost was found to be relatively high (US$32 per tonne); this cost could be reduced if a cheaper crushing method were employed.

Enhanced agricultural productivity and diversification of the small-scale mining sector are two key objectives of the Zambian Poverty Reduction Strategy. Spontaneous uptake of the agricultural lime production method by an existing small-scale operator in Solwezi met these objectives.

Demonstration plots (maize and groundnut) were established in an attempt to convince small-scale farmers of the benefits of using agricultural lime. The highly acidic soil of the trial plots required lime addition of up to 500 kg per hectare. The maize yield increased nearly threefold to 6.7 tonnes per hectare and the groundnut yield increased nearly sevenfold to 320 kg per hectare.

In most cases the use of agricultural lime by small-scale farmers will have economic benefits. However, its use will depend on the small-scale farmer having the cash to purchase it and any significant increase in its use will probably require government intervention. The key to uptake is demonstration of the benefits of using agricultural lime. The FarmLime project summary report is available as a download from http://www.mineralsuk.com/britmin/farmline.pdf.

I would like to thank the following for their enthusiastic contribution to this work: Stephen Simukanga, Victor Shitumbanuma, Diana Banda, Briton Walker, Ellie Steadman, Boniface Muibeya, Moffat Mwanza, Mathew Mtonga, David Kapindula and the small-scale farmers of the Mkushi and Solwezi farming districts in Zambia.

References

ARC SOUTHERN 1996. *Arclime – Agricultural Magnesian Limestone*. Special products factsheet SP6. Amey Roadstone Corporation Southern.

MENGO, V. 2002. Kapiri woman sees hope in Heifer Project International. *Zambia Daily Mail*, March 11, p. 7.

MITCHELL, C. J., INGLETHORPE, S. D. J., TAWODZERA, P., BRADWELL, S. & EVANS, E. J. 1997. *Local Development of Affordable Lime in Southern Africa*. British Geological Survey Technical Report WC/97/20.

MITCHELL, C. J., SIMUKANGA, S. ET AL. 2003 *FarmLime Summary Report*. British Geological

Commissioned Technical Report CR/03/066 N, ISBN 0 85272 4624.

SHITUMBANUMA, V. & SIMUKANGA, S. 1995. Potential use of carbonate rocks for agriculture in Zambia. In: MAMBWE, S. H., SIMUKANGA, S., SIKAZWE, O. N. & KAMONA, F. (eds) *Proceedings of International Conference on Industrial Minerals: Investment Opportunuities in Southern Africa*, 7–9 June 1995, Pamodzi Hotel, Lusaka, Zambia, Council of Geosciences, South Africa, 161–173.

TETHER, J. & MONEY, N. J. 1989. A review of agricultural minerals in Zambia. *Proceedings of the East and Southeast African Geology Workshop on Fertiliser Minerals*, Lusaka, December 8–10, 1987.

Mineral resources and their economic significance in national development: Bangladesh perspective

AFIA AKHTAR

Geological Survey of Bangladesh, Segunbagicha, Dhaka, Bangladesh
(e-mail: afia@agni.com)

Abstract: Modern urbanization, industrialization, transportation and communication systems are the achievements of worldwide sustainable mineral resource development and their proper utilization in various sectors. Sustainable mineral resources have played, and are still playing, a vital role in shaping the modern civilized industrial world. This means that the sustainable socio-economic infrastructure of any country is an indication of its richness in natural resources, its technological know how, its ability to explore and exploit mineral resources, and, finally, its wisdom in utilizing those resources properly in the development activities of the nation. In development activities, countries of the developing world are generally far behind compared with countries in the developed world. This is mainly due to a lack of adequate natural resources, properly educated human resources and good socio-economic conditions. Although Bangladesh is a small country, it has a number of mineral resources such as natural gas, oil, coal, hard rock, limestone, white clay, glass sand and mineral sand. At present, natural gas is the only mineral commodity significantly contributing to the national economy. More than 90% of the country's energy needs are met by gas, total reserves of which are 21.35 trillion cubic feet (TCF) and 12.43 TCF, respectively. Huge reserves of hard rock (granodiorite, quartzdiorite, gneiss) and coal in northwest Bangladesh will help, in the near future, to meet the growing demand for construction materials and energy for the ever-growing population. Total coal reserves are 1753 million tons (MT), the market value of which is more than US$110 billion. Hard rock reserves are 115 million tons, valued at over US$3 billion. Fully fledged extraction of these resources would help to alleviate the country's poverty through industrialization. It is expected that coal will soon be extracted on a commercial basis, of which 70 to 80% will be used in power generation. The mineral resources so far found in Bangladesh are meagre in comparison to its high population. To meet the growing demand of the population, more mineral resources need to be discovered and developed, otherwise sustainable development cannot be achieved. However, it is difficult for developing countries like Bangladesh to carry out the necessary activities for exploration and exploitation of hidden mineral resources without foreign assistance. This is a major drawback for Bangladesh. To progress towards an endurable sustainable society, a nation such as Bangladesh must give priority to the development of its existing mineral resources, which can play a major role in helping to reshape the country's socio-economic infrastructure.

The term sustainable development is increasingly used by the concerned people of the world. Development means the general improvement of life from grassroots level in all spheres; maintaining that development is the goal of sustainable development. The achievement of sustainability depends on various factors, including (a) availability of natural resources (especially mineral resources); (b) exploration and exploitation of natural resources and their proper development; (c) development of mineral-based industries; (d) preservation, conservation and proper utilization of natural resources; and (e) dependence on renewable resources as much as possible; these are the key factors for sustainable mineral development in the world whether it is in developed or the developing countries. Sustainable mineral development means the effective management and utilization of mineral resources in various sectors by developing mineral-based industries to meet growing demand.

The sustainable development of a nation's infrastructure is dependent on:

- relative richness in mineral resources or availability of sufficient mineral resources;
- technological know how;

From: MARKER, B. R., PETTERSON, M. G., MCEVOY, F. & STEPHENSON, M. H. (eds) 2005. *Sustainable Minerals Operations in the Developing World*. Geological Society, London, Special Publications, **250**, 127–134.
0305-8719/05/$15.00 © The Geological Society of London 2005.

- capability to explore and exploit new mineral resources;
- management capability to develop these resources properly; and
- wisdom in utilizing these resources properly in development activities.

Countries of the developing world are far behind in achieving sustainable mineral development in comparison to the countries of the developed world, mainly because of the following factors:

- lack of good socio-economic infrastructure;
- dependency on foreign expertise, which costs foreign currency;
- lack of modern technological knowledge;
- lack of sufficiently educated and skilled human resources;
- lack of self-confidence; and
- misuse of natural resources due to unawareness/negligence.

Bangladesh and its mineral resources

Bangladesh is a relatively small developing country with an area of 150 000 square kilometre located in South East Asia. Although it is a small country, it has a number of mineral resources, including natural gas, oil, coal, hard rock, limestone, white clay, glass sand and mineral sand (Fig. 1).

Natural gas

All natural gas fields are located on the northeastern and southeastern parts of the country (Fig. 1). There are 22 gas fields and, of these, 16 are in the production stage while the rest require development. The bulk of the gas supply comes mainly from four major gas fields — Bakhrabad, Titas, Habigonj and Kailastila. More than 90% of the country's energy need is met by natural gas, with the remaining 10% met by gas condensate and hydroelectricity. At present, natural gas is the only mineral commodity contributing significantly to the national economy. Gas is being used in various sectors, such as power generation, fertilizer production, industrial uses, commercial and domestic purposes. From 64 exploratory boreholes, 22 gas fields have been discovered, a better ratio than that of any other country (Shamsuddin 2002). According to Petrobangla's sources, the total and recoverable reserves are 26.2 TCF and 16.1 TCF, respectively. So far, 4 TCF gas has been exploited, making the present recoverable reserve about 12 TCF. A few years ago, Bangladesh was self-sufficient in gas production, but, at present, production is much less than the total demand, which is now 1288 MCF (million cubic feet). For the last two years, production has been increasing by 53 MCF per year, whereas demand has been increasing by 432 MCF per year (*The Daily Inqilab* 2 September 2003). If demand for gas continues to increase at the same rate, then it will be very difficult for Bangladesh to cope in the future. So, attention has focused on finding more gas fields on the mainland as well as offshore. It is economically positive that Bangladesh is now trying to discover new gas fields on its own initiative, rather than depending on foreign expertise.

Oil

The oil field in Sylhet, in northeastern Bangladesh, was the only significant deposit discovered in 1986, probable and recoverable reserves of which were 27 and 6 MMBBL (million barrels), respectively (Petrobangla 1992). Oil was extracted from a single hole in the Sylhet field from 1987 to 1993, and then production ceased. During 1990, 118 449 barrels of oil were extracted, which at that time was economically viable for Bangladesh (Akhtar & Hasan 1994). At present, BAPEX (Bangladesh Petroleum Exploration Company Ltd.) is endeavouring to discover new oil plays. Otherwise, in the absence of oil, it will be possible to extract gas from this field, which will help to increase the gas production of the country. Other possible places also need detailed investigation.

Coal

Good quality coal deposits of Permian age have been recorded from five areas in northwestern Bangladesh. The coal is a high-quality bituminous grade with low sulphur content and is suitable for power generation. For this purpose a power plant of 250 MW capacity is now under construction. Mining development has been going on to extract coal only from the Barapukuria coal field of Dinajpur District. It is expected that production on a commercial basis will start by 2006. Previously, it was expected that fully fledged extraction of coal would be started in 2001, but because of sudden underground water flow, mining work was postponed. However, mining work has restarted. Target production is 3300 metric tons, giving an annual target of 1.2 million tons at a market value of US$90 million (Akhtar 2000). As a result, US$45 million will be saved in foreign currency.

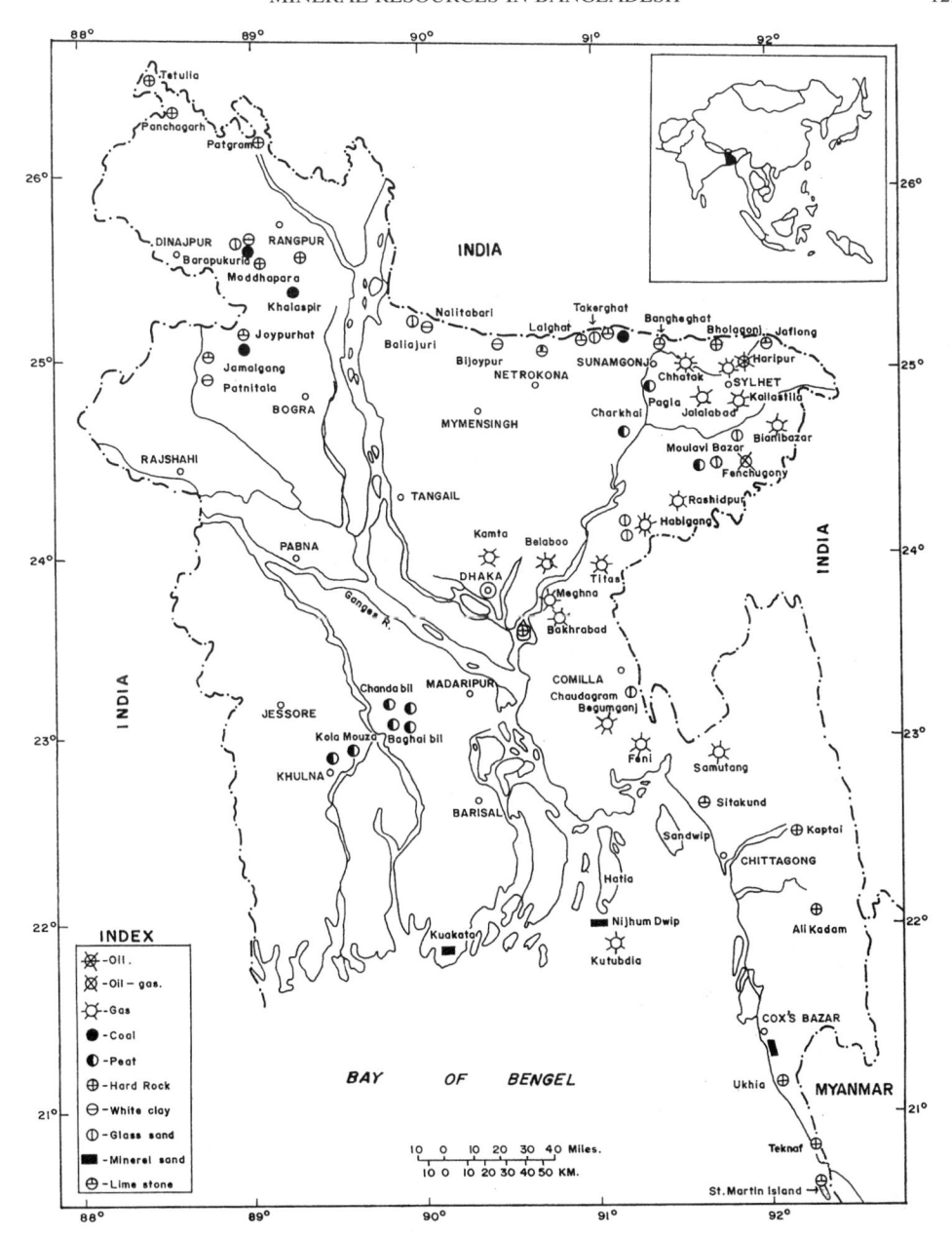

Fig. 1. Mineral deposits of Bangladesh.

Eighty percent of the mined coal used for power generation will help to solve the electrification problems of northern Bangladesh. Industrial and agricultural sectors will also benefit from electrification. The remaining 20% will be used for other purposes such as railway transportation, fuel for cooking, various industrial uses for producing heat and raising steam, for brick fields to produce heat and for carbonization to produce coke, tar, and so on. The development of Khalaspir coal in the Rangpur District is currently under consideration, but extraction of Jamalganoj coal from the Bogra District is not feasible because of its depth. It is possible that methane gas may be extracted from the Jamalganoj coalfield, and for this purpose a pilot project has been undertaken. (Hussain 2000). In the meantime, a number of organizations have shown interest in extracting methane from this field. Dighipara coalfield in Dinajpur District is another potential coalfield, with the coal reported to be at a depth of 327 m, for which the extraction is considered to be economically viable. Continued exploration will further define coal reserves.

Peat

Peat deposits of Quaternary age have been recorded from the southwestern and northeastern side of the country, the total reserves of which are more than 170 MT. Peat is currently being extracted and used mainly in brick fields, as well as for cooking and other purposes by local people. This usage is slowing down deforestation and the burning of vegetation for domestic use.

Hard rock

Construction materials are in relative short supply in a country formed mainly through deltaic aggradation. Hard rock of Pre-Cambrian age consisting of granodiorite, quartzdiorite, gneiss and pegmatite has been recorded from Dinajpur, in northwest Bangladesh. These are very good quality construction materials, suitable for housing apartments, commercial buildings, roads and highways, bridges, dams, river training, embankment construction, flood control, railway ballast and also for decoration purposes. These rocks typically contain manganese, zinc and vanadium, which are suitable for the manufacture of batteries, gas-holders and electronics equipment. Mining development is ongoing and it is hoped that by 2006 commercial extraction will be possible. It is also expected that fully fledged hard rock extraction will meet the country's demand. There is also the potential for exporting stone and stone products. In the first phase, 115 MT will be extracted. Estimated annual production will be 1.65 MT, the market value of which is more than US$47 million. Present annual hard rock imports are around 1.8 MT. It is hoped to extract annually 2.0 MT of hard rock after the first few years. Other construction materials of Recent age have also been recorded from other parts of Bangladesh, mainly from southeastern and extreme northwestern parts of the country. These materials are being extracted and used in the construction of buildings, bridges, roads and highways.

White clay or china clay

White clay or china clay deposits of Plio-Pleistocene and Cretaceous ages are located on the northern and northwestern sides of the country. Estimated reserves are around 43 MT. Clay is currently being extracted and it is suitable for ceramics, sanitaryware, insulators and for cement manufacture. Other uses of white clay include the manufacturer of paper, rubber and plastic.

Glass sands

Glass sands of the Cretaceous to Holocene age are available in northeastern, northern and northwestern parts of Bangladesh, the total reserves of which are around 117 MT, valued at US$13950 million. Glass sand is of medium grade in quality, but some are comparatively free from impurities and suitable for making colourless glassware. Sand is being extracted from some places, and is used for making coloured glass sheet, bottles, window glass and so on.

Limestone

Limestone deposits of Eocene and Pleistocene age are recorded from northwestern, northern and southeastern parts of the country, the total reserves of which are around 300 MT, valued at US$5000 million. A big deposit of limestone was found in 1965 at Joypurhat, in northwestern Bangladesh, with a total reserve of 270 MT at a depth of 518 m and over an area of 2.5 square miles. The limestone of Joypurhat is of good quality and suitable for cement and lime production. In 1978, the mining of limestone was being carried out there, with an extraction target of 1 MT per year. Later on, mining was postponed on the basis that the extraction of the limestone and the use of that limestone to prepare clinker and then cement would not be economically feasible. It was estimated that it might be possible to extract 2 MT of limestone instead of 1 MT (Hussain 2000). In 1978, the approximate cement demand of the country was about 4 MT per year, and the extraction of 2 MT of limestone per year would bring substantial economic benefit to the country. Contracts have been signed with a number of foreign countries (Germany, Saudi Arabia, Kuwait and others) at different times since the discovery of the Joypurhat limestones, but because of financial problems the projects could not be imple- mented. At present, negotiations and the necessary steps for the development of limestone from Joypurhat are ongoing with Poland under the 'Joypurhat Limestone Mine and Cement Project' (Haque, unpublished paper). Identified limestone resources from different parts of the country could reduce the importing of limestone if the necessary steps could be taken to develop and utilize them properly.

Mineral sands

Beach sand around Cox's Bazar, in the southern part of Bangladesh, contains several tens of million tons of heavy mineral sands, which are economically very important and valuable in the international market. Twenty million tons of beach sand can contain up to 4 MT of mineral sand (BAEC 1994), with zircon, rutile, ilmenite, leucoxene, kyanite, monazite, magnetite and garnet also present. The estimated reserves of mineral sands are: zircon 158 117 tons, rutile 70 274 tons, ilmenite 1 025 558 tons, leucoxene 96 709 tons, kyanite 90 746 tons, garnet 222 761 tons, magnetite 80 599 tons and monazite 17 352 tons. The total value of these minerals is approximately US$50 million. The average mineral content is 23% in Cox's Bazar beach sand. The mineral content of the beach sand of Bangladesh is much higher than that of many other countries. For example, beach sand in Australia contains 5% minerals, whereas beach sand in Bangladesh contains 15–23% minerals. However, effective steps have yet to be taken to commercially

Table 1. *Mineral resources of Bangladesh along with their reserves and market price*

Mineral resources	Reserves	Value in US$ (million)	Remarks
Coal	1753 MT (proved)	89 760	Mining development of Barapukuria coalfield is ongoing and it is hoped that commercial extraction will be possible by 2006
Peat	170.55 MT	5122	Extraction of peat is ongoing from Kola Mouja and is being used for various purposes
Limestone	270 + 29 MT	5000	Extracted mainly from Takerghat and is being used in Chhatak Cement Factory. A project has been undertaken in cooperation with Poland to extract limestone from Joypurhat, the biggest deposit in Bangladesh
China clay	42.52 MT	6375	Extracted mainly from Bijoypur area and is being used in making ceramics, tiles, and so on
Mineral sand	Average presence in beach sand is 23%		An Australian company is associated for exploration and extraction of mineral sand from the beach sand of Cox's Bazar
Glass sand	117 MT	13 950	Glass sand is being extracted mainly from Chaddagram and is used in the glass-making industries to prepare glasswares and glass sheets
Hard rock	Huge reserves	2849	Mining development is ongoing at Maddhyapara and it is expected that by 2006, hard rock will be extracted commercially (in the first phase 115 million tons will be extracted)
Construction materials	10.32 MT	95	Construction materials are being used from these places, especially from the Bholaganj area
Natural gas	12 TCF (proved reserve)		The only mineral commodity contributing significantly to the national economy; 95% of the country's energy need is met by natural gas
Oil	6 MMBBL to 27 MMBBL		A remarkable amount of oil was extracted during 1987–1993 from the Sylhet Gas Field

Sources: MEMR & GSB.

extract these valuable resources. At present, the authorities are considering developing these resources. It is expected that in the next 10 to 20 years, it will be possible to extract heavy minerals, at an estimated value of 6 to 10 billion dollars (*The Daily Inqilab* 30 October 2003). Australia (International Titanium Private Ltd.) has taken out a mining lease over 30 500 hectares of beach for the extraction of mineral sands. According to the agreement, the Bangladesh Government would get a 12% royalty from the extracted minerals. Zircon, rutile, garnet and monazite have also been reported in the Brahmaputra valley, the percentages of which are more than that of beach sands, but these deposits require further exploration.

The mineral deposits of Bangladesh and their estimated reserves and market values are summarized in Table 1 (MEMR 1991–1995; Akhtar 2001; GSB 2002). Fully fledged commercial extraction and proper utilization of mineral resources within Bangladesh will bring significant benefit to the national economy and will help reshape the country's socio-economic infrastructure, thus contributing to the development of the country. Proper management and development of mineral resources can bring sustainable development for a nation; otherwise mismanagement will be responsible for retardation of development (Fig. 2).

Mineral-based industries

Occurrences of metallic deposits generally enhance the development of mineral-based industries in any country. Unfortunately, Bangladesh lacks such metallic mineral deposits and that is one of the major constraints for the growth of mineral-based industries in Bangladesh. In spite of this drawback, on the basis of imported commodities and whatever mineral resources Bangladesh has, there is some minerals-based industry. Key players are outlined in Table 2.

Ceramics produced using local clay or of imported china clay mainly from India, China and Norway are of export quality, and foreign currencies are saved by exporting them. Most companies use about 95% local clay for their production of export quality ceramics. Titas Gas Transmission & Distribution Company Ltd., Bakhrabad Gas System Ltd. and Jalalabad Gas Transmission & Distribution System Ltd under Petrobangla are responsible for distributing and marketing of

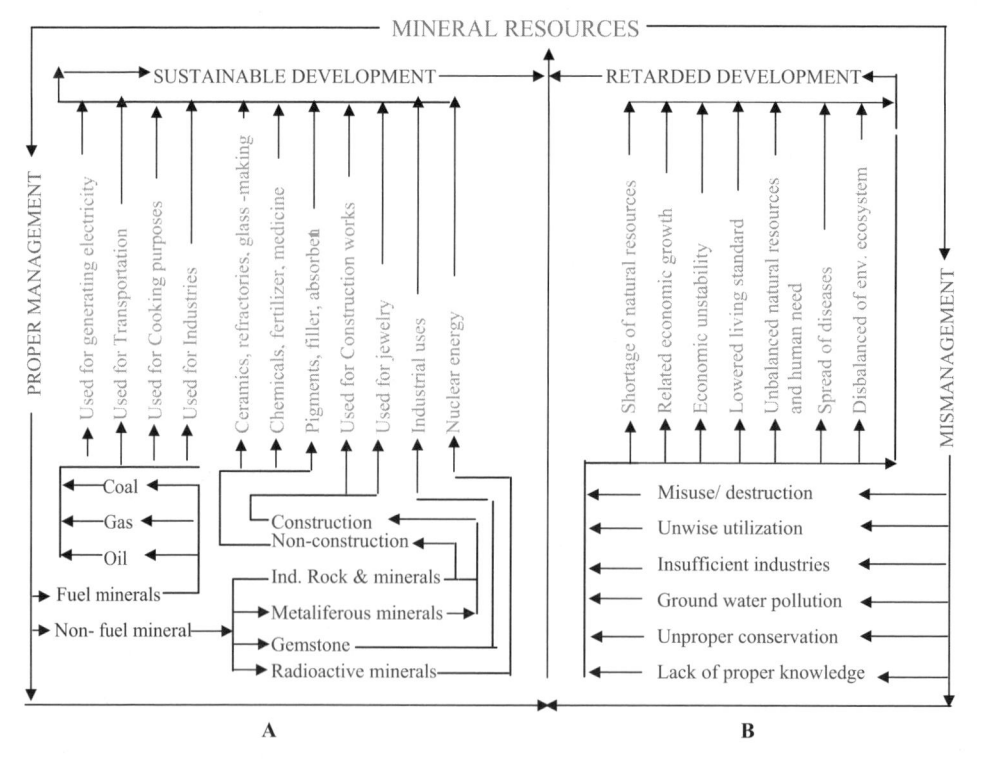

Fig. 2. (A) Proper management of natural resources for sustainable development; (B) Mismanagement of natural resources causing retarded development.

Table 2. *Mineral-based industries in Bangladesh*

Commodity	Major operating companies	Some major industries utilizing mineral resource
Fertilizer	Bangladesh Chemical Fertilizer Corporation	Ashuganj Fertilizer Factory near Titas gasfield Ghorasal Fertilizer Factory, north of Dhaka Other factories in Dhaka, Chittagong and Jamalpur
Cement		Chhatak Cement Factory, Sylhet Clinker Grinding Factory, Chattagong
Gas, natural	Bangladesh Oil, Gas and Mineral Corporation (Pterobangla)	Bakhrabad Gas System Ltd., Comilla Titas Gas Transmission & Distribution Co. Ltd. Jalalabad Gas T & D System Ltd. Bangladesh Gas Field Com. Ltd. Brahmannbaria Sylhet Gas Field Company Ltd., Haripur Bangladesh Petroleum Exploration Company Ltd., Dhaka Paschimanchal Gas Company Ltd., Dhaka
Petroleum, refined	Bangladesh Petroleum Corporation (BPC)	Eastern Refining Ltd., Chittagong
Steel	Bangladesh Steel and Engineering Corporation	Chittagong Steel Mill Industry
White clay or china clay		Monno Ceramic industries Ltd. R.A.K. Ceramics Shinepukur Ceramics Tajma Ceramic, Bogra Bengal Fine Ceramic Ltd., Savar Fuang Ceramics People Ceramic Industries Standard Ceramics Tajmahal Ceramics China–Bangla Ceramics Bangladesh Insulator & Sanitaryware Factory Ltd.
Glass sand		More than 15 glass making factories and ceramic industries using local glass sand for their production

Source: Akhtar & Hasan 1995 (including additions/information).

natural gas. The yearly capacity of Eastern oil refining is about 1.5 million barrels and that of steel plant is about 0.25 MT. To meet the demand of the country, a few more fertilizer plants will be established very soon. A plan has already been made to expand the Chhatak Cement plant and to set up a new plant at Surma.

Conclusions

Bangladesh has a number of industrial non-metallic minerals and mineral fuel deposits. Mining and quarrying of these minerals, except gas, plays a minor part in the national economy. It is expected that the fully fledged extraction of coal and hard rock in the near future will create commercial opportunities that will bring substantial benefits to the economy of Bangladesh. At present, natural gas is the only mineral commodity that is contributing a great deal to the growth of the national economy. The living standard of any nation is, in part, a measure of the productiveness of its hidden mineral resources and the development of mineral-based industries. However, for the successful exploration and exploitation of hidden mineral resources, one of the principle factors is a sound economic condition, and that is the major drawback for developing countries like Bangladesh. The economic significance of minerals to a country's national development depends not only on mineral development, but also on the proper utilization of those mineral resources. Such utilization will help to

- bring economic stability;
- develop mineral-based industries;
- increase job facilities/employment opportunities;
- increase per capita income;
- raise overall living standards;
- alleviate poverty;
- reduce crime and corruption; and thus
- help to achieve sustainable development.

All these achievements depend strongly on active participation of the country's population, from top to bottom, in developing activities in all sectors of society. To get mass participation, the following steps should be taken:

- develop human resources by giving proper education emphasizing the geosciences;
- develop local expertise in order to reduce dependency on foreign expertise;
- encourage the population to actively participate in development activities;
- make people understand about the contribution of mineral resources to their society;
- motivate the population not to misuse or destroy natural resources;
- convince people to conserve and protect natural resources for the nation's benefit.

Therefore, to achieve an endurable and sustainable society, a nation must give importance to the development of mineral resources and mineral-based industries, and also to the development of human resources. Careful development and management of mineral and human resources will undoubtedly help to restructure and reshape the country's socio-economic infrastructure.

References

AKHTAR, A. 2001. The role of a national geological survey in the development of a sustainable society, with special reference to Bangladesh. *Episodes*, **24**(2), 135–138.

AKHTAR, A. 2000. Coal and hard rock resources in Bangladesh. *Episodes*, **23**(1), 25–28.

AKHTAR, A. & HASAN, M. N. 1994. Fossil fuel in Bangladesh. *Proc. 2nd SEGMITE International Conference*, Karachi, Pakistan, 79–84.

AKHTAR, A. & HASAN, M. N. 1995. Mineral deposits and mineral based industries in Bangladesh. *Proc. Int. Symp. and Field Workshop on Phosphorites and Other Industrial Minerals*, Abbottabad, Pakistan.

BAEC 1994. *Brochure – Beach Sand Exploration Centre, Cox's Bazar.* BAEC, Dhaka, Bangladesh.

HAQUE, E. *Joypurhat Limestone Mine and Cement Project* (handout).

HUSSAIN, K. M. 2000. Mining development in Bangladesh: problem and possibility. *The Daily Inqilab*, Dhaka, 13 March 2000.

GSB 2002. *Discovery of Mineral Resources and - Geoscientific Activities by Geological Survey of Bangladesh* (file report) GSB, Dhaka, Bangladesh.

MEMR 1991–1995. Geological Survey of Bangladesh (GSB). Unnyan Probabah, Ministry of Energy and Mineral Resources, Government of the People's Republic of Bangladesh, 75–84.

PETROBANGLA 1992. *Brochure – Exploration Opportunities in Bangladesh.* Petrocentre, Dhaka, Bangladesh.

PETROBANGLA 2000. *Brochure – Exploration Opportunities in Bangladesh.* Petrocentre, Dhaka, Bangladesh.

SHAMSUDDIN 2002. Present reserve and prospect of gas in Bangladesh. *Orthakatha*, 16–31 July 2002.

The Daily Inqilab 2003. A remarkable decision to drill gas well by own initiatives, 2 September 2003.

The Daily Inqilab 2003. Economic prospect of mineral sand, 30 October 2003.

Obstacles in the sustainable development of artisanal and small-scale mines in Pakistan and remedial measures

VIQAR HUSAIN

Department of Geology, University of Karachi, Karachi, Pakistan
(e-mail: viqarpk@yahoo.com)

Abstract: Pakistan is a large country with diverse geology and geography. It possesses many industrial rocks and minerals, including precious stones, marble and granite. Some metallic mineral deposits, and large reserves of coal/lignite, oil and natural gas also occur. Pakistan's mining industry is dominated by thousands of artisanal and small-scale mines, which lack capital, technical know how, modern equipment and trained manpower. Further, local mining practices cause much damage to mineral deposits and are very hazardous to the health of mine workers and the environment. The mining sector is backward due to lack of political will and pragmatic mining laws, and absence of technical and financial support by the government agencies to the small mining units. Moreover, the narrow base of the domestic mineral industry leads to poor demand in the local market and low mineral production. Additionally, the lack of infrastructure and the poor law and order situation adversely affect the mineral industry. Further, mineral-rich districts of the country are socio-economically backward due to hilly terrain or arid climate and, hence, less suitable for agriculture. Sustainable development of artisanal and small-scale mines is possible if the necessary legal, technological, financial, commercial, social and environmental support is provided on a long-term basis.

Pakistan has an area of 880 000 km^2, and a population of over 140 million. It possesses huge reserves of very good quality industrial rocks and minerals (Table 1). Pakistan's mineral-rich districts comprise parts of the Northern Areas, North West Frontier Province, Balochistan, Sindh, and central and southern Punjab, where all mineral resources including important industrial rocks and minerals, precious metals and gemstones occur. Mineral resources include clays of various kinds, gypsum, rock salt, rock phosphate, nepheline syenite, dolomite, limestone, barite, magnesite, soapstone, marble, granite and silica sand (Kazmi & Abbas 1991; Husain *et al.* 1993).

Important metallic minerals occurring in Pakistan are antimony, chromite, copper, iron and lead–zinc. Its important gemstone deposits include emerald, ruby, topaz, aquamarine, tourmaline, moonstone, quartz and red garnet (Kazmi & Donoghue 1990). Despite Pakistan's rich mineral potential, its mining and mineral industries are backward due to low technical and financial inputs, poor infrastructure and absence of a developed domestic consumer market (Griffiths 1987). The mining sector in Pakistan has also remained backward as all the minerals including coal, chromite and gemstones are mined by thousands of unorganized small and artisanal mining units owned and operated by family groups and other small entrepreneurs. These small mines are not provided with any financial, commercial and technological support from provincial or federal government (Husain & Bilqees 1991).

Further, the lack of clearcut mineral concession rules at the provincial level, the unsatisfactory law and order situation in mineral-bearing areas, coupled with outdated mining methods and complete absence of mineral processing equipment make the mining industry unproductive and unprofitable. Moreover, the narrow base of domestic mineral industry also leads to poor demand for minerals in the market and, hence, low mineral production (Table 2).

In fact, an analysis of the import data of many developing countries demonstrates that many domestically available industrial minerals are imported at considerable cost, utilizing valuable foreign exchange reserves (Walde 1985). Pakistan is no exception; a large number of indigenous raw materials and their products that are currently being imported occur in huge quantities within the country itself (Husain *et al.* 1993). Because of the risk involved in the development of mineral resources due to the uncertain demand and due to the lack of proper chemical and geological data on local minerals, the mining industry of Pakistan

Table 1. *Pakistan's mineral potential*

Minerals/rocks	Reserves (tons)
Abrasives	Very large deposits (unestimated)
Barite	30 million
Bauxitic clay/fire clay	283 million
Bentonite/fullers earth	56 million
China clay	6 million
Chromite	Fairly large deposits (unestimated)
Coal	19.7 billion
Copper	Not estimated
Dolomite	Very large deposits (unestimated)
Feldspar	Over 10 million
Fluorite/celestite	Small deposits (unestimated)
Gemstone	Significant deposits (unestimated)
Granite/other building stones	Huge and extensive deposits (unestimated)
Gypsum	Several billion
Lead/zinc	Over 100 million
Limestone	Huge and extensive deposits (unestimated)
Magnesite	Over 25 million
Marble	Huge and extensive deposits (unestimated)
Mica	Not estimated
Nephline syenite	6 billion
Ochre	Over 100 million
Quartz/quartzite	Very large deposits (unestimated)
Rock phosphate	About 30 million
Rock salt	Several billion
Silica sand	Over 1 billion
Soapstone	Fairly large deposits (unestimated)

Source: Kazmi & Abbas 1991.

Table 2. *Output of principal Pakistani minerals (1000s of tons)*

Mineral	2000	2001	2002
Aragonite/marble	582.0	468.00	401.70
Agr. clay	1275.00	1130.00	1128.20
Barites	21.23	27.15	26.10
Bauxite	8.67	5.88	8.76
Bentonite	13.85	10.65	11.22
Chalk	7.71	7.70	7.54
China clay	49.57	55.57	54.36
Chromite	26.84	9.92	22.32
Coal	3117.00	3262.00	3487.00
Dolomite	287.96	256.80	240.39
Feldspar	43.19	30.07	56.92
Fire clay	143.64	152.92	153.55
Fuller's Earth	15.29	13.37	14.12
Gypsum	377.00	624.00	417.00
Limestone	9884.00	9607.00	1481.00
Magnesite	3.61	3.03	4.43
Rock salt	1313.00	1393.00	1387.00
Silica sand	162.00	145.00	152.00
Soapstone	54.36	30.79	57.37
Sulphur	20.19	17.18	22.84
Urea	3887.63	4162.03	3747.60

Source: *Central Bulletin*, **51**(3): figures for calendar year, January to December 2002.

has not developed to its full potential. The huge cost of transportation of minerals from mines to markets, and uncertain operating conditions also increase the risk for investors (Awais 1991).

The mineral sector contributes only 1% to GDP and this is mainly in the area of industrial and construction minerals such as coal, lignite, limestone and gypsum (Dalton 1994). In addition, dolomite, rock salt, silica sand, marble, gypsum, feldspar, chromite, china clay, fire clay, fullers earth, soapstone, barite, magnesite and a number of semi-precious and precious stones are also mined in small quantities (Griffiths 1987).

Obstacles and remedial measures

Major constraints to effective development of small-scale mineral projects are the prohibitive cost of imported mining and processing machinery, delays in the acquisition of mining leases, and the lack of awareness of investment opportunities in the mining and mineral industry sectors (Arbisala & Adegbesan 1994). The aim of this paper is to identify obstacles to the sustainable development of artisanal and small-scale mines in Pakistan and suggest short- and long-term remedial measures.

Old mining laws

Mining legislation provides the framework that enables the responsible authorities to ensure that mineral development contributes to attainment of national goals and objectives. To achieve this end, there are some principles that are fundamental to any responsible mining legislation (Drolet 1975). Many factors – geological, economic, legal and political – influence the exploitation of minerals. Mineral resources, as distinct from a minerals industry, do not in themselves endow a nation with tangible wealth (Highley 1994).

In Pakistan, the first mineral policy was promulgated in 1995, but the provinces have not yet revised their mining concession rules, with the result that the national mineral policy has not yet been implemented. The mineral sector in Pakistan is dealt with at provincial level, unlike the petroleum and radioactive mineral sectors, which are controlled by the federal Government. At present, the mineral concession rules of the provinces of Pakistan are cumbersome, bureaucratic, non-transparent and not in line with modern mineral development practices that facilitate quick and efficient exploration (Dalton 1994). In contrast, the Pakistan Government has been very consistent in formulating and implementing petroleum regulations that have made the country's oil and gas sector modern and productive. Regular updating of petroleum concession rules by the Government has attracted about 20 foreign companies to explore oil and gas in the country and resulted in a number of new discoveries in recent years.

The growth of a mining industry directly depends greatly on the government and its agencies promulgating and implementing stable sets of legal, fiscal, commercial, technical, social and environmental laws. Mining laws should be simple, ensuring easy allotment and transfer of mining leases. In Pakistan, the ministry of solid minerals or mines should be separated from the existing Ministry of Petroleum and Natural Resources to focus solely on hitherto neglected mineral development and the mining sector (Husain 1994).

Low profile of solid minerals

According to Hill (1994), greater priority should be given to projects that will lead to local processing and consumption, as these are more likely to bring sustained benefits to the developing countries themselves and the local people in particular. However, ministries of mines and natural resources in developing countries, their institutions and universities, mostly continue to operate as if mining is synonymous with the exploitation (for export) of the country's reserves of gold, other metals, precious stones and phosphates. The importance of industrial and other minerals to national economic development is not clearly understood by the government and policymakers. Perhaps this is one of the main reasons for the lack of investment and funding opportunities in the mineral sector (Morgan 1994).

At present, the Pakistan Government's attention is focused on the oil and gas sector and a small number of large base metal and coal projects. Despite the fact that large quantities of mineral resources are readily available in Pakistan, imports are on the rise. Tax generated from oil and gas revenues is used for financing Pakistan's budget deficits and meeting other state expenditure needs. Hence, the Ministry of Petroleum and Natural Resource pays more attention to the development of petroleum and gas than solid minerals. As a result, development of mineral resources and their exploitation is not a priority. This is exacerbated by the fact that minerals are dealt with at a provincial level and subsequently do not receive sufficient attention from the federal government.

Poor mining conditions

Although wages, safety and welfare of mine workers and their families are regulated through the Mineral Concession Rules and the Mines Act 1923, in practice, mine safety equipment, medical facilities and other services provided to the miners and their families are far from satisfactory. The working conditions in mines are very unsafe, especially for underground miners, who are continuously exposed to extreme noise, vibrations, harmful chemicals and lethal gases. Mine workers commonly suffer from hearing impairment, physical and psychological stresses, tuberculosis and lung cancer (Jan 1994). Several mine workers die every year in coal and marble mines due to poor mining practices. Those found guilty of causing harm to the health and lives of mine workers remain unpunished and compensation paid by mining companies to victims is also very small. Primarily due to low investment, mining and processing equipment is outdated, and when operated by non-technical personnel leads to low mineral production compared to the size and reserves of the mineral deposits. Further, the ore wastage ratio is as high as 60% due to blasting and hand picking of high-grade ore and other profit-motivated mining practices. The minerals are also mined from subsurface pits and tunnels (up to a depth of 100–2000 feet) by unskilled workers endangering their lives. Subsequently, the mines are abandoned without rehabilitation of the affected land. Mining conditions in the solid mineral sector can be improved if mining laws are implemented that promote safe working conditions and protection of the environment. In summary, the Government should pay attention to mineral development.

Lack of marketing facilities

Most mines in the country are located in remote and mountainous areas, cut off from the road and rail networks. Lack of electricity and water supply makes mineral processing in close proximity to the majority of mines very difficult. As a result, raw material is transported to the cities in a low-value and unprocessed state. In addition, there is no concession in tariffs for transporting mineral commodities by railways to ports or big city markets (Husain & Bilqees 1991). The Government of Pakistan and its agencies do not provide marketing facilities or assistance to miners. There is no dissemination of marketing-related information to the miners, nor are they trained in marketing skills. Hence, it is common for small miners to sell their precious stones and other minerals to knowledgeable middle men at below market value, who then sell the products onwards at considerable profit in local and international markets (Bilqees 1997). The Government does intervene in other sectors, however. For example, agricultural products like wheat, rice and cotton are procured by Government marketing agencies, who set prices each year to ensure that farmers are not exploited by traders.

Artisanal and small mines should be provided with assistance in marketing their produce to Government marketing agencies. The miners should be trained in marketing skills and provided with loans for processing and marketing their output. The infrastructure from mines to markets should be developed by the Government. Local exhibitions of minerals and their products should be organized to promote marketing (Husain 1997). The Government should also assist in purchasing, refining and marketing the output, like other developing countries (Walde 1985).

Lack of financial assistance

Financial support by the Government and its financial institutions to the small-scale mines is totally lacking in Pakistan, whereas small and cottage industrial units in agriculture and other industrial sectors are supported by development financial institutions, public and private banks (Awais 1991). There is no mineral development bank to provide easy access to loans for small-scale and artisanal mines (Husain & Bilqees 1991). In most other sectors, the State Bank of Pakistan gives special purpose credit lines at concession rates to banks and other financial institutions for distribution. Mining so far has not been added to this category (Chaudhry 1994). Further, there are no mining trust funds for these mining units to provide risk capital, required for exploration and development of mines. The Government support for establishment and development of cooperatives to provide services to the small mining sector including soft-term loans is also lacking.

The Government should treat small mines like units in textile and other sectors. The leased mineral property should be accepted as collateral and credit banks should give loans to mining companies to cover financial risks in exploration, mining and to set up gem and other mineral processing units. Mineral taxation should be profit-based and remain limited to only royalties on production. The whole mining sector income should be given to the provincial Government for infrastructure development.

A revolving fund should be created by the Government to provide soft term loans to mine owners to purchase mining equipment and reduce their dependence on picks and shovels. Joint venture operations should be encouraged to strengthen financial standings of the small-scale mining operators, as has been successfully done in many developing countries like Chile, Botswana, Mexico and Zambia (Walde 1985). Government agencies may provide mining machinery and equipment to private mine owners on easy installments or for rent. Leasing companies can play a very useful role in such financing (Chaudhry 1994).

Lack of technical assistance

In Pakistan, technical extension services are not provided to the artisanal and small mines in the same way that they are given to farmers by Government agricultural departments throughout the country. However, geological mapping, mineral resource survey reports, research publications and mineral testing services provided by the federal and provincial mineral development agencies have all been useful to the small-scale mine owners to a limited extent. Periodic training to mine workers in safe mining practices, provided by provincial inspectorates of mines, has also been helpful. Overall exploration has not been carried out on a sustained basis and large areas of Pakistan, especially in North West Frontier Province and Federally Administered Tribal Areas, are totally unexplored. Basic geological information pertaining to large areas with rich mineral potential is not available (Dalton 1994).

Further identification and promotion of appropriate technologies and dissemination of related information is also lacking. Poor coordination among local scientific and mineral business communities (mineral producers, consumers and traders) has also hampered the development of the mineral sector in Pakistan (Husain 1994).

In order to boost the production of minerals and their exports, the Government should support technical organizations and development agencies to assist and motivate the mine owners for doing scientific mining. Measures should also be taken to provide training for miners and encourage manufacture of equipment suitable for small-scale mines.

The Government of Pakistan should pay attention to collecting and disseminating basic information on the mining industry in local languages. Established educational institutes dealing with mining should provide short courses to small mine operators on all aspects of small-scale mining. The Government should also set up institutes to train people in processing of minerals and coloured gemstones. The mining labour, supervisors, inspectors and other provincial Government officers supervising the mining operations need to be given extensive training so that the operation of mines is efficient (Chaudhry 1994). Technical cooperation with the mining industry should be strengthened by providing annual reports, newsletters, market survey reports, conference and symposia proceedings and annual mineral production statistics (Mcardle 1994).

Teaching and research in the fields of resource geology and the mining industry should be strengthened. There is much scope for increasing the pairing of university departments in developing and industrialized nations, as an extension of the bilateral cooperation that has been taking place with many geological surveys (Hill 1980). There is a need to constantly generate and disseminate data based on comprehensive characterization of raw materials using a range of mineralogical and chemical techniques, followed by development of methods for concentrating economically useful minerals, eliminating impurities or modifying properties (Lorenz 1985; Morgan 1997). Federal and provincial mineral development organizations and research institutions should provide a wide-ranging information service in the form of geological maps, geochemical and geophysical surveys, technical reports, mineral locality databases and a mines database that contains the mine plans of all abandoned mines (Mcardle 1994; Scott & Jackson 1994).

Environmental impact

In Pakistan, like other developing countries, the majority of mines are small quarries and pits of non-metallic minerals, which are typically abandoned once the technical and financial requirements of the operations go beyond the capacity of the small mine operators. Quarries are developed without observance of proper health and safety procedures, making them unsafe for quarry operators and miners (Hill 1993). The lack of mine rehabilitation and improper disposal of mine waste also greatly damages the environment (Ahmad *et al.* 1995). Quarrying of rocks and minerals leads to the degradation of the land, water and soil of the region.

There should be environmental guidelines to the small mining sector, including general clauses requiring minimizing of damage and provision for waste disposal (Walde 1985). There is a need to enforce rules related to environmental impacts, which should be continuously monitored during mine development and the post-mining stages. The development or use of environmentally friendly technologies should also be encouraged.

Social

The socio-economic condition of mine workers in Pakistan is poor, as the mines are located in remote areas of the country. Poor infrastructure, lack of housing facilities and absence of proper health clinics and schools near the mines make the lives of miners very tough. Further, poor wages, unsafe working conditions and employment badly affect the living standards of mine workers and their families.

Research on the technical and social aspects of small-scale mining should be identified and undertaken (Dahlberg 1994). The Government should enforce laws to protect employment and working conditions of miners. Adequate medical, educational and other services should also be provided to the workers' families by mine owners and provincial Governments by setting up hospitals and schools in mining districts.

Conclusions

Sustainable development of the mining sector in Pakistan is possible, if small mining units are provided with investment incentives and if regulatory institutions such as the provincial directorates of mineral development and inspectorates of mines are strengthened. Pakistan needs to formulate and implement simple and pragmatic legal, fiscal, commercial and environmental policies to ensure sustainable development of mineral resources.

Small-scale mining can play an effective role in Pakistan's socio-economic development, provided that national policies encourage local investors. The Government should also establish agencies to provide technical and financial support for small-scale mineral development.

I thank AGID's financial support in enabling me to attend the 'Sustainable Minerals in Developing World Conference' held in London on 24–25 November 2003. I am especially thankful to Dr. Mike G. Petterson for his support before and during the conference, and for encouragement to complete this paper. I am also grateful to Dr. S. A. Bilgrami for a helpful review of this paper.

References

AHMAD, A., DANRHWAR, S., AHMAD, I. & AHMED, A. 1995. Environmental impact of coal mining in Jhangh valley, eastern Salt Range, Pakistan. *In*: REHMAN, S. S., ABBASI, I. A. & MAJID, M. (eds) *Proceedings of National Symposium on Environmental Geology, Geological Hazards, Prediction, Initiation & Control*, Department of Geology, University of Peshawar, 173–182.

ARBISALA, A. O. & ADEGBESAN, A. B. 1994. Exploitation and export prospects of Nigerian industrial minerals. *In*: MATHERS, S. J. & NOTHOLT, A. J. G. (eds) *Industrial Minerals in Developing Countries*. AGID Report Series, Geoscience International Development, **18**, 107–110.

AWAIS, M. 1991. Small scale mining and mineral based industries in North West Frontier Province. *In*: BILQEES, R., HUSAIN, V., KHAN, A. M. & SIDDIQUI, F. (eds) *Proceedings of the First Society of Economic Geologists & Mineral Technologists Symposium on Export Promotion of Minerals and Mineral Products*, Peshawar, Pakistan, 27–33.

BILLQEES, R. 1997. Economic importance of Pakistan's industrial mineral resources. *In*: KAIFI, F. M. Z., HUSAIN, V. & BILQEES, R. (eds) *Proceedings Industrial Minerals for Development Workshop*, Peshawar, Pakistan, 25–27.

CHAUDHRY, M. B. 1994. Role of national and international financial institutions in development of mineral sector in Pakistan. *In*: KHAN, M. N. (ed) *Proceedings of the International Round Table Conference on Foreign Investment in Exploration and Mining in Pakistan*, Islamabad, Pakistan, 21–30.

DAHLBERG, H. 1994. Profile Small Mining International (SMI). *AGID News*, **76**, 23.

DALTON, D. L. 1994. Mineral investment in Pakistan. *In*: KHAN, M. N. (ed) *Proceedings of the International Round Table Conference on Foreign Investment in Exploration and Mining in Pakistan*, Islamabad, Pakistan, 173–188.

DROLET, J. P. 1975. Mining legislation and role of responsible authorities. In: *International Symposium on Technical Research in Mineralogy and Management of Mineral Patrimony*, Orleans-La source, France.

GRIFFITHS, J. 1987. Pakistan's mineral potential. Prince or pauper. *Industrial Minerals*, London, July, 20–43.

HIGHLEY, D. E. 1994. The role of industrial minerals in the economies of developing countries. *In*: MATHERS, S. J. & NOTHOLT, A. J. G. (eds) *Industrial Minerals in Developing Countries*, AGID report series, Geoscience International Development, **18**, 1–12.

HILL, N. R. 1980. Developing countries and the third world of the minerals industry. *4th Industrial Minerals International Congress*, Atlanta, USA.

HILL, N. R. 1993. Industrial minerals deserve more attention than they currently receive particularly as many countries move towards industrial development. *In*: MCDIVIT, J. F. (ed) *International Mineral Development Source Book*. Herrington Geoscience, UK, 11–15.

HILL, N. R. 1994. Process, consume and live or export. In: MATHERS, S. J. & NOTHOLT, A. J. G. (eds) *Industrial Minerals in Developing Countries*, AGID report series, Geoscience International Development, **18**, 23–26.

HUSAIN, V. 1994. Conference report and recommendations. *In*: KHAN, A. A., HUDA, Q. U. & HUSAIN, V. (eds) *Proceedings Second Society of Economic Geologists & Mineral Technologists Conference on Export Oriented Development of Mineral Resources & Mineral Based Industries*, Karachi, Pakistan, 5–8.

HUSAIN, V. 1997. Economic importance of Pakistan's industrial mineral resources. *In*: KAIFI, F. M. Z., HUSAIN, V. & BILQEES, R. (eds) *Proceedings Industrial Minerals for Development Workshop*, Peshawar, Pakistan, 30–31.

HUSAIN, V. & BILQEES, R. 1991. Developing Pakistan's mineral resources and minerals industry. *In*: BILQEES, R., HUSAIN, V., KHAN, A. M. & SIDDIQUI, F. (eds) *Proceeding of the First Society of Economic Geologists & Mineral Technologists Symposium on Export Promotion of Minerals and Mineral Products*, Peshawar, Pakistan, 17–26.

HUSAIN, V., BILQEES, R. & ABBAS, G. 1993. Pakistan's industrial mineral resources for 21st century. *Resource Geology Special Issue*, **16**, 263–270.

JAN, M. S. 1994. Mining labour laws in Pakistan. *In*: KHAN, M. N. (ed) *Proceedings of the International Round Table Conference on Foreign Investment in Exploration and Mining in Pakistan*, Islamabad, Pakistan, 289–292.

KAZMI, A. H. & ABBAS, S. 1991. Pakistan's mineral potential. *In*: *A Brief Review of the Mineral Wealth of Pakistan*. Geological Survey of Pakistan, Quetta, 43.

KAZMI, H. & DONOGHUE, M. 1990. Gemstone deposits of Pakistan *In*: *Gemstones of Pakistan*. Gemstone Corporation of Pakistan, Peshawar.

LORENZ, W. 1985. A model for estimating the economic and geological potential of non-metallic raw materials. *Natural Resources and Development*, **21**, 7–16.

MCARDLE, P. 1994. Ireland experience: GSI serves and promotes the mining industry. *In*: KHAN, M. N. (ed) *Proceedings of the International Round Table Conference on Foreign Investment in Exploration and Mining in Pakistan*, Islamabad, Pakistan, 71–76.

MORGAN, D. J. 1994. Industrial minerals assessment in developing countries. A decade of BGS experience. *In*: MATHERS, S. J. & NOTHOLT, A. J. G. (eds) *Industrial Minerals in Developing Countries*, AGID report series, Geoscience International Development, **18**, 39–54.

MORGAN, D. J. 1997. Economic importance of Pakistan's industrial mineral resources. *In*: KAIFI, F. M. Z., HUSAIN, V. & BILQEES, R. (eds) *Proceedings Industrial Minerals for Development Workshop*, Peshawar, Pakistan, 1–13.

SCOTT, P. W. & JACKSON, T. 1994. Industrial rocks and minerals in the Caribbean; geology, exploitation and attitudes. *In*: MATHERS, S. J. & NOTHOLT, A. J. G. (eds) *Industrial Minerals in Developing Countries*, AGID report series, Geoscience International Development, **18**, 95–106.

WALDE, T. W. 1985. Permanent sovereignty over natural resources. *In*: *Report of the United Nations Secretary General*. Natural Resources and Energy Division, Department of Technical Cooperation for Development, United Nations, New York, USA.

Mining and environmental problems in the Ib valley coalfield of Orissa, India

P. PARAMITA MISHRA

Centre for Economic and Social Studies (CESS), Begumpet, Hyderabad-16, India
(e-mail: prajnap@rediffmail.com)

Abstract: The exploitation of mineral resources through surface and underground mining has in the past caused a wide range of environmental problems such as health degradation, air, water and noise pollution, decline in agricultural production, deforestation, displacement and other socio-economic impacts. However, over the past number of years, stakeholders in the industry have been striving to avoid and mitigate the potential detrimental effects of mining on fragile ecosystems and local communities. Governments are increasingly formulating and adopting policies to ensure the sustainable development of their country's mining industry and mining companies are striving to be better environmental citizens. Environmental groups have become increasingly involved in mining disputes. However, a lot has to be achieved to ensure mining in carried out in a sustainable way. This paper concentrates on the environmental effects of coal mining in the Ib valley coalfield of Orissa, India. Background to the increasing awareness of the environmental issues associated with mining is provided in the first section of this paper. The second section discusses the general problems associated with mining, with particular reference to the Ib valley coalfield. In addition, measures undertaken by the companies operating the Ib valley coalfield to deal with environmental problems are presented.

During the closing decades of the 20th century, environmental issues emerged as a major concern for the survival and welfare of mankind throughout the world. Modern civilization, with rapidly advancing technology and fast growing economies, came under increasing threat from its own activities causing pollution of air, water and soil. This endangers people and the environment.

The United Nations (UN) Conference on Environment at Stockholm, in June 1972, heralded the first serious effort to take note of the environmental issues at the global level. Since then, concepts such as environmental sustainability and the carrying capacity of the landscape have become central themes for policy making around the world. Environmental problems remain acute in many countries and continue to increase in many parts of the world. However, there has been a quantum step in awareness of the scale of the problem and of the need for action in recent years. The UN Conference on Environment and Development in 1992, in Rio, known as the Earth Summit, provided further momentum for change.

Although the desirability of development is universally recognized, recent years have witnessed rising concern about whether environmental constraints will limit development and whether development will always cause serious environmental damage, in turn impairing the quality of life of this and future generations. A number of environmental problems require urgent attention. The protection of the environment is an essential part of development. Without adequate environmental protection, development is undermined; without development, resources will be inadequate for needed investment, and environmental protection will fail (World Bank 1992).

The quality of the environment and ecosystem health in India is declining at an alarming rate as India is trying to bridge the gap between 'developing' and 'developed' status. Since independence in 1947, India has implemented ten Five-year plans and three Annual plans; as a result, India is among the few developing countries that are now on the threshold of achieving the status of a developed country. With improvement in science and technology, India has earned international prestige among the community of nations. However, in the process, with limited effort to conserve natural capital, India is rapidly becoming a vast wasteland.

Environmental pollution in India is multi-dimensional. It affects land, water, forests, people, wildlife and habitats, energy and the atmosphere. Mining, as conducted at present, is one of the worst polluting activities. At the same time, mining is an important economic activity. Despite its economic importance, it is, in general, a poorly managed sector, causing considerable environmental damage, which has a high risk to human and ecosystem health and economic infrastructure.

Coal is the most important and abundant mineral in India. At present, fossil fuels (mainly coal, oil and natural gas) supply more than 90% of world energy demand. As oil's share of the energy market diminishes it will have to be replaced by another energy source that is abundant, flexible in use and easily extracted. Coal can satisfy all these requirements. Coal has been and will continue to be the prime source of energy in India due to the abundance of coal reserves, foreign exchange difficulties and the inadequate exploitation of hydropower potential. However, there are environmental issues to be addressed. This paper concentrates on the effects of coal mines on the environment.

Coal mining commenced in India in 1845 in Raniganj coalfields in Bihar. The growth of the coal industry was slow, unsystematic and inefficient and was based purely on consideration of maximizing short-term profit by private mine owners until 1971–1972, when the coal industry was nationalized. Coal India Limited (CIL) operates all coalfields in India, with the coalfields of Orissa coming under Mahanadi Coalfield Limited (MCL), a subsidiary of CIL. Following nationalization, exploration for coal and coal mining was undertaken in a more systematic way to meet the long-term requirements of the country.

The present study has the following main objectives.

(1) To review environmental problems of mining in general and those of the Ib valley coal field in particular.
(2) To explain measures that have been taken by the company to improve and protect the environment.
(3) To suggest some further actions to be taken to ameliorate the adverse impacts.

Ib valley coalfield, Orissa

Orissa is situated in the east coastal region of India. Its geographical area is almost 4.7% of India's landmass and its population is 36.7 million (2001 census), about 3.6% of India's population. The state is endowed with rich mineral resources and occupies an important position in India's mineral map. The mineral belt is spread over an area more than 6000 km^2. Orissa has 20% of India's total mineral resources, including chromite, nickel, bauxite, iron ore and coal.

The Geological Survey of India estimated a total of 246 billion tonnes of coal reserves as on 1 January 2004. Jharkhand State has the largest reserves on 72 billion tonnes, placing Orissa second, with 61 billion tonnes. Orissa has two coalfields: Talcher and Ib valley. The Ib valley coalfield is situated in Jharsuguda and Sundargarh districts of Orissa between latitudes 21° 32' to 22° 06' N and longitudes 83° 32' to 84° 10' E, covering an area of 1375 km^2. The coalfield is named after the river Ib, a tributary of the river Mahanadi. It was discovered during the construction of a bridge over the river Ib for the Bombay–Nagpur railway. The first mine to start was Himgir Rampur Colliery in 1909, followed by Orient Mine No. 1 in 1940, both underground. The coalfield has since been divided into five areas comprising five underground and seven opencast mines.

Coal reserves of this coalfield are about 22.3 billion tonnes (as on 1 January 2004) of which about 14.4 billion tonnes are at a depth of less than 300 m. The quality of coal varies from grade C to G (largely F), suitable for power generation. Details of coal quality and gradation are given in Table 1. This coalfield is well located to supply coal to power stations in western and southern India. Many private entrepreneurs are also interested in building and operating power plants within the coalfield in the neighbourhood of Hirakud water reservoir because of the ready availability of coal and water. Thus, the coalfield has gained importance in recent times.

Environmental impacts of coal mining in the Ib valley coalfield

Coal mining has been undertaken in the Ib valley area for nearly 100 years. An aim of MCL is to become the largest producer of coal in the world, while the state government wishes to gain royalties and taxes from this sector. Therefore, limited attention has been paid to the local environment and the indigenous people. The following are some of the major environmental problems that local people are facing.

Health. A visible expression of the effects of mining activities is provided in the effects on health. Occupational diseases like silicosis, flurosis, asbestosis, manganism, plumbirm, and so on, have been associated with mining societies all over the world. Prolonged exposure to coal dust

Table 1. *Gradation of coal**

Grade	Useful heat value (UHV) (kcal/kg) UHV = 8900-138 (A + M)	Corresponding ash% + moisture% (at 60% RH & 40°C)	Gross calorific value (kcal/kg) (at 5% moisture level)
A	>6200	Not >19.5	>6454
B	>5600 but not >6200	19.6–23.8	>6049 but not >6454
C	>4940 but not >5600	23.9–28.6	>5597 but not >6049
D	>4200 but not >4940	28.7–34.0	>5089 but not >5597
E	>3360 but not >4200	34.1–40.0	>4324 but not >5089
F	>2400 but not >3360	40.1–47.0	>3865 but not >4324
G	>1300 but not >2400	47.1–55.0	>3113 but not >3865

*The gradation of coal is based on useful heat value (UHV), and corresponding ash and moisture content.

is the cause of coal miners' pneumoconiosis (CMP) in which the lungs lose their natural flexibility and it becomes increasingly difficult to breath. Simple tasks like walking up stairs become impossible. It leads to death from suffocation.

In the region around the Talcher coalfield, where there are high levels of air pollutants (both coal dust and associated minerals), the occurrences of cancer, tuberculosis, bronchitis and other lung and skin diseases are soaring. The disease rates are high among tribal populations because they have no option but to live in the land closest to mines. Similar problems are developing elsewhere in the Ib valley coalfield. People and animals are inhaling fine dust particles up to 5 microns (1000 micron = 1 mm) in size. As a result, asthma and bronchitis have become major problems of this area. As a result, both the human and animal mortality rates have increased (SEEN 1996).

Air, water and noise pollution. Exploitation of mineral resources through surface and underground mining can cause wide-ranging environmental problem such as air, water and noise pollution. All methods of surface and underground mining produce wastes that can, if not managed appropriately, affect surface water quality. Since coal mining commenced in the Ib valley area, surface water pollution has become a major problem. Ground water levels in wells of adjoining villages have fallen drastically. People are facing an acute scarcity of water for their fields and daily needs. Blasting for coal, drilling and other related activities are causing noise pollution, with consequent stress and adverse effects on the mental health of workers. Blast vibration is damaging houses and other surface structures. Because of mining and other allied activities, such as operation of coal handling plant and transportation of coal, air pollution in this area is increasing.

In the Mahanadi river basin, several industrial locations, with many polluting industries and 10 coalmines, discharge huge quantities of water containing heavy metal and sulphur compounds during the monsoon, posing a serious environmental threat. The river Brahmani also has major industrial areas and a number of coal and chromite mines in its basin. High levels of industrial pollutants and drainage from coal mines are discharged into the rivers (Das 2001).

Tribal people, fishing villages, cattle and several species of birds and other wildlife previously existed with little difficulty on the river. Since the advent of coal mining, the river has become so polluted that it can no longer support aquatic life. Several species of sea turtles have become endangered. The Nandira river is dead. Coal mining, aluminium smelting, fertilizer and chemical production have turned the river into an industrial drain. The black water is poisoning and slowly killing people, animals, fish and plants as far away as 50 miles downstream. Agricultural productivity has dropped where farmers are dependent on this polluted water. Fishing communities have been wiped out.

Agricultural production. Pollution of streams that originate in the mining area affect agriculture. Paddy yield has declined because of the accumulation of silt and waste tailings in the fields brought by monsoon water. These harden in the summer and become solid, so that nothing grows on the field. The tailings make sand unfit even for construction purposes (Sharma 2001).

Most villagers near the Ib valley coalfield have abandoned agricultural production because MCL has taken over much of the area. In those villages where people are still cultivating, polluted water from the mines drains onto nearby fields, resulting in deterioration of soil quality and declining agricultural productivity. Grazing land is also affected and flora and fauna are becoming depleted.

Displacement. Exploitation of mineral resources has resulted in large-scale displacement of the population. Since Independence in 1947, mining has displaced 10 million people in India. Three-quarters of these are yet to be compensated. Land records show that nearly 150 million acres of once arable or homestead land in Orissa are now being mined, with the consequence that there are now an estimated 50 000 environmental refugees in the state, most of them tribal and poor people. The construction of Hirakud Dam uprooted a large number of tribal people in the 1950s and many of them were displaced for the second time when their compensated lands were found to be coal rich.

Similarly, exploitation of coal in the Ib valley has resulted in large-scale displacement. The displaced people have not been properly resettled or rehabilitated. More than 1000 people remain unemployed, although in some areas they have been employed as daily wage labourers. Although resettlement colonies have been established, there are insufficient infrastructural facilities. Many people do not have their written records of land ownership, so they are not being compensated accordingly.

According to the World Bank, expansion of the Belpahar coalmine in the Ib valley currently threatens 3941 people, including 1110 tribal people. At least 7421 Orissans will be forced to move due to coal mining expansion funded, in part, by the World Bank. Thus, Orissa's share of coal mining related displacement accounts for more than half of the total related to coal mining in the whole of India (SEEN 1996).

Biodiversity and ecosystem. Mining and associated activities result in loss of habitat and fragmentation of the contiguous block of tropical moist forest and grasslands. In the mining area biodiversity is low. Although several species of grass and herbs have sprung up in areas abandoned after mining, they are insufficient to stabilize the disturbed soil. Mining, especially opencast mining, is associated with slope stability problems such as landslides and rapid soil erosion.

According to a fact sheet produced by the U.S. Network for Global Economic Justice, Orissa's contribution to climate change is significant. Orissa's industries and coal-fired power plants will be emitting 164 million tonnes of carbon dioxide equivalent annually by the year 2005, or the equivalent of about 3% of the projected growth in man-made greenhouse gases anticipated globally over the coming decades.

Vibration. There have been many complaints regarding excessive vibrations from people living in the vicinity of the mining industry. Areeparampil (1996) reports window rattling and breakage and animal disturbance as a result of these vibrations. In hilly areas, there is a perceived risk of triggering landslides. These problems are the result of poorly designed blasting operations and have, in some instances, caused damage to houses and other surface structures.

Deforestation. Before mining, the Ib valley was densely forested. The state government gave nearly 10 231 hectares of lands to MCL for four coalmines. MCL gave Rs 1 095 088 188 to Sambalpur District Forest Officer (DFO) for an afforestation programme between 1991 to 1993, but no trees have yet been planted (Anon 2000). Afforestation of the over burden dump is being undertaken by MCL, but that is insufficient to maintain an ecological balance in that area. Problems associated with deforestation are well documented. In the Ib valley, threats to wildlife habitats and significant changes in rainfall are being recorded.

Socio-economic impact. In response to the creation of additional jobs in the area, the population has suddenly increased, but most jobs available in mines require skilled people. Therefore, there is large-scale immigration of people. This leads to higher demand for goods and the prices rise, creating problems for unskilled local people. Children are encouraged to leave school early to earn money.

From the above points it is clear that large-scale mining of coal from opencast and underground mines undertaken in recent years has had a remarkable impact on the environment. CIL, as a matter of policy, is committed to ameliorate the ill effects of mining in a systematic way. This has necessitated the evolving of a corporate environmental policy.

Measures taken by MCL to protect and improve the environment

The objective of CIL's corporate environmental policies is to lay down guidelines for its subsidiary companies, enabling them to develop methodologies and work plans to match the guiding principles of sustainable development. The policy requires that they should comply with environmental statutes, guidelines, and instructions of legal agencies and implementation of the provisions of the Environmental Management Plan (EMP).

The Environmental Policy aims to build up organizational capacity at all levels. It also integrates Environmental Impact Assessment (EIA) and EMP with mine planning, design, operation and mitigation of adverse environmental impact to the best possible extent, training environmental personnel and raising awareness at all levels.

The CIL strategy envisages that all mines must have an EMP to be environmentally acceptable; CIL thus plans to continuously strive to increase its technical capability for environmental protection and development of organizational capability to address the impact of mining. It also proposes to give annual awards for achieving excellence in environmental programmes to its subsidiary companies.

CIL intends to place in position environment units, termed 'cells', at all levels. It plans to have environmental cells at Unit, Colliery, Area, Company Headquarters and CIL corporate levels. The environment cell will be responsible for ensuring all mines have environmental clearance from the Ministry of Environment and Forest (MOEF). It will also deal with mines having an EMP cleared by CIL or Company Boards.

According to the Annual Report 2001–2002 of MCL, the following measures have been taken to protect and improve the environment.

Afforestation and land reclamation. The afforestation drive in the mines of MCL continued vigorously and 260 000 saplings of mixed species were planted during the period 2001–2002. Special attention has been given to vegetating slopes. An Over burden Dump Reclamation Plan, approved by an expert from the Indian Institute of Technology (IIT), Kanpur, appointed to CIL with World Bank assistance, has been implemented for 23 old dumps within six World Bank Projects. An eco-park project has already been started in Samaleswari open cast project. Spice, dry fruit and fruit-bearing trees were planted on dumps together with other multipurpose trees.

Air pollution control measure. Fixed point, auto start, fine nozzle mounted water sprinklers have been provided on roads, railway sidings and coal handling plants (CHPs). Dust control systems and dust extraction systems have been installed in CHPs and at transfer points elsewhere. Green belts have been created between residential areas and mine infrastructure for dust control. Tenders were invited for the supply of six multipurpose hydroseeder/firewater tenders to be used for seeding and watering plants on overburden dumps and also to be used as water sprinklers.

Water pollution control measures. Effluent from mines, workshops, settling ponds and spoil dumps are made to conform to standards of the Ministry of Environment and Forest (MOEF) before being discharged into the natural drains and streams/nullahs. Mine seepage water is made to settle in large sedimentation lagoons created in the quarry floor. Mine discharge treatment plants (MDTPs) have been provided in each mine before the water discharge point to the surface or nullahs. Garland drains have been provided near the toes of overburden dumps as well as at quarry margins. Water is monitored at all discharge points and care is taken to ensure that all water quality parameters remain within prescribed limits.

Noise and ground vibration control measures. Green belts have been created between residential areas and the mines and infrastructural facilities such as CHPs, railway sidings, and workshops to attenuate noise levels and to arrest the movement of dust. Workers exposed to noise above permissible limits have been provided with earmuffs and earplugs. No electric detonators are used for blasting. Controlled blasting practices are adopted whenever it is necessary.

Environmental monitoring and environmental audit. Regular monitoring of air, water, noise and soils is carried out together with micrometeorological studies using Government agencies whose laboratories are duly recognized by MOEF in this regard. A senior officer at headquarters scrutinizes the results of monitoring; any upward trend is immediately brought to the notice of the project authorities so that corrective measures can be undertaken. Results of monitoring are submitted to the State Pollution Control Board (SPCB) on a monthly basis and to MOEF on a half-yearly basis. An environmental audit has been conducted on each project by a multidisciplinary audit team and the resulting environmental statements were submitted to SPCB and MOEF.

Environmental awareness. Environment Week was celebrated in June 2001, starting with World Environment Day on 5 June in all the mines of MCL. An environmental awareness training programme is arranged at regular intervals for employees on the subject of environmental management in mining areas. Programmes undertaken during Environment Week include planting of fruit trees, painting, an essay competition among schoolchildren, wide circulation of environmental booklets, training manuals, screening of videocassettes on environment themes and organizing a cycle rally, street plays, and so on.

Environmental award and recognition. The Indo-German Environment Excellence Award for 2000–2001, organized by Greentech Foundation, New Delhi, and the Centre for the International Transfer of Environmental Technologies, Germany, was won by MCL for outstanding achievements in the field of environmental management. The Tata Energy Research Institute (TERI) has extended a special invitation for MCL to join the Corporate Roundtable Development of Strategies for the Environment (CORE), an exclusive group of companies, which included National Thermal Power Corporation (NTPC), Tata Iron and Steel Company (TISCO) and Oil and Natural Gas Commission (ONGC), and so on. MCL was short-listed by the award committee for the 'TERI Corporate Environment Award' 2001–2002.

Although the coal companies are sincere in implementing the various measures contained in environmental management plans, they are also facing some problems. The coalfield has other industries in the vicinity, which also contribute to the pollution of land, air, water and the immediate surroundings.

M/S Envirocare, Rourkela, Orissa, has published a report on environmental monitoring in this area. The suspended particulate matter (SPM) concentration at eight-hourly intervals for 24 hours for continuously 8 days in a season gives a clear status of pollution caused by SPM in the atmosphere. SPM concentration in some locations varies from 102 $\mu g/m^3$ to 425 $\mu g/m^3$. Surface water was compared with drinking water quality, IS 10500, and mine discharge was compared with IS 2490. It is observed that the drinking water quality sometimes does not meet the prescribed standard for suspended solids.

The State Pollution Control Board and state government needs to act as an overall coordinator to maintain a healthy environment by pinpointing which industries are responsible for adverse effects (Patnaik 1990).

Conclusions

Although the company has indicated that much work is on going for the protection of the environment, much remains to be done. The State Pollution Control Board has an important part to play. The media need to improve wider environmental awareness. Social work organizations and trade unions also have a significant role. The concept of carrying capacity will have to taken into consideration in decision making for mining projects. It is an important factor in identifying the most sustainable approaches to development.

Carrying capacity is defined as the physical limit for economic development or maximum rate of resource utilization and waste discharge that can be supported or assimilated by the regional environment indefinitely without affecting ecological integrity. The prediction of impacts of mining can be assessed using mathematical models or expert opinion. Better management of mining activity would enhance both economic and ecological value. However, costs and benefits of management practices need to be assessed in order to arrive at a realistic balance between the need for urgent economic development and that for protection and preservation of the environment.

It is thus necessary to start a comprehensive programme of data compilation and analysis in the mining industry for this purpose. Suitable guidelines need to be developed for methods of environmental impact analysis and assessment in the mining industry and steps to be taken from the planning and implementation stages through to closure, for the protection of the environment (Mahapatra 1990).

Mining industry professionals must find ways of mining with the minimum possible impacts on the environment, and where damage is inevitable steps should be taken to reclaim the land and afforest the area as soon as possible after mining is completed. Industrial activity creates environmental hazards, but these can be controlled very effectively if society cooperates with the industry. If society and the industry plan together and work for the improvement of the country as well as of the area, matters will be much improved. We should strive for development alongside protection of the environment instead of development at the cost of environment.

The author is thankful to Prof. V. Ratna Reddy, Dr. Brian Marker, Dr. Tony Reedman and Tim Colman for their valuable comments. Thanks to the Commonwealth Commission in the UK for funding me split-site doctoral scholarship.

References

AREEPARAMPIL, M. 1996. Displacement due to mining in Jharkhand. *Economic and Political Weekly*, **24**, 1524–1528.

ANONYMOUS 2000. Compensation money on deforestation is not utilized properly. *The Bitarka*, **2**(5), 13–16.

DAS, P. 2001. Orissa government blamed for declining quality of river water. *The Hindu*, 6th August.

MAHAPATRA, R. K. 1990. *Study of Environmental Pollution in IB Valley Coal Fields*. Unpublished Project Report submitted to Regional Engineering College, Rourkela, Orissa, India.

PATNAIK, L. N. 1990. *Environmental Impacts of Industrial and Mining Activities*. Ashish Publishing House, New Delhi.

SHARMA, R. 2001. Kudremukh concerns. *Frontline*, 14 September 2001, 64–71.

SUSTAINABLE ENERGY AND ECONOMY NETWORK (SEEN). 1996. The *World Bank's Juggernaut, The Coal-Fired Industrial Colonization of India's State of Orissa*. A report in association with District Action Group, Talcher-Angul, Orissa, India, and the International Trade Information Service, Washington, DC.

WORLD BANK. 1992. *World Development Report: Development and Environment*. Oxford University Press, New York.

The Gold Ridge Mine, Guadalcanal, Solomon Islands' first gold mine: a case study in stakeholder consultation

DONN H. TOLIA[1] & M. G. PETTERSON[2]

[1]*Geological Survey Division, Department of Mines and Energy, PO Box G37, Honiara, Solomon Islands (e-mail: donn@mines.gov.sb)*
[2]*Economic Minerals and Geochemical Baseline, British Geological Survey, Keyworth, Nottingham NG12 5GG, UK (e-mail: mgp@bgs.ac.uk)*

Abstract: The Pacific small island state of Solomon Islands gained independence from Britain in 1978. Solomon Islands has a population of around 400 000 mainly Melanesian people distributed across six moderately sized islands and hundreds of smaller islands. The traditional economy has been based on hunter-gatherer and small-scale farming activities, with the bulk of the population residing in self-sufficient rainforest and coastal villages. Melanesians have a particularly strong cultural attachment to land, which is considered to be within the custodianship of the community at large. Individual land ownership in the Western economic sense is largely unknown. Melanesian society and culture is strong and complex. Colonial and post-independence Solomon Islands has had to face the challenges of a transition from a traditional society to a partially urbanized society and a rapidly increasing population and changing economic drivers and dynamics. Mining and mineral development is one area of economic activity that holds the promise of generating hard currency quickly to develop the country, but that needs to be achieved in a sustainable manner. Gold Ridge is situated in Central Guadalcanal, some 22 km southeast of the country's capital town, Honiara. Gold Ridge hosts around 1.4 million ounces of epithermal volcanic-hosted gold. Ross Mining NL began the construction of Solomon Islands' first gold mine in 1997 and operated a highly successful gold mine between 1997 and 2000, when ethnic tensions (unrelated to the mine) closed the operations. This paper documents the painstaking negotiations and planning that took place from 1993 and particularly from 1995–1996 which paved the way forward for the development of a gold mine within a fragile tropical rainforest environment among traditional Melanesian people who had little prior knowledge of modern mining activities.

This paper describes the process through which Solomon Islands' first medium-scale commercial mine was opened. The initial discovery of gold on Guadalcanal was made by the first European explorer to 'discover' the Solomon Islands in 1568. It was during the 20th century that serious science-based exploration took place and during the 1990s that the gold prospect was raised in status from a prospect to a mine with economically proven reserves. Once the mining application was accepted by the Government of Solomon Islands, the key players (government, community and company) had to develop a consultation process through which mining could begin within a very traditional Melanesian culture and a pristine rainforest environment.

The planning of the Gold Ridge Mine took several years. The local community had little previous experience of commercial medium-scale mining operations, although they were familiar with artisanal mining and exploration (Figs 6a & b). Community consultation and participation was perhaps the key factor in moving the Gold Ridge project forwards from discovery to exploration and finally mine development (Figs 5a & b). Fiscal security was a prime concern to the mining company because of the initial capital outlay of around US$70 million (the largest single investment to date in Solomon Islands since independence in 1978) (Gold Gazette Australia 1998; Hughes 1997). The extraction of minerals from the ground through mining is a non-renewable activity insofar as replenishment of mineral resource in the ground is concerned. However, mining should be viewed in terms of soft sustainable development (see Shields & Solar 2005), through which capital assets realized from mining are reinvested in the local economy with the objective of providing longer-term livelihoods that last well beyond mine closure. These and other issues were addressed through a series of consultation exercises.

From: MARKER, B. R., PETTERSON, M. G., MCEVOY, F. & STEPHENSON, M. H. (eds) 2005. *Sustainable Minerals Operations in the Developing World.* Geological Society, London, Special Publications, **250**, 149–160.
0305-8719/05/$15.00 © The Geological Society of London 2005.

Background

Solomon Islands is situated east of Papua New Guinea (Fig. 1). It became an independent nation in 1978 after having been a Protectorate Nation of the UK for over a century. The population of Solomon Islands is around 400 000, with an annual birth rate of 2.8% (Solomon Islands Human Resources Development Report 2002). The Gold Ridge Mine (GRM) is situated in Central Guadalcanal, one of the larger islands of the Solomon Islands chain in the Pacific Ocean. Honiara, the capital of Solomon

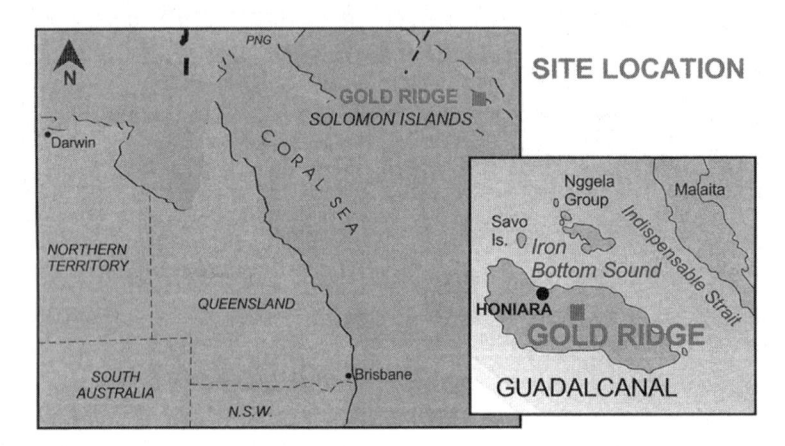

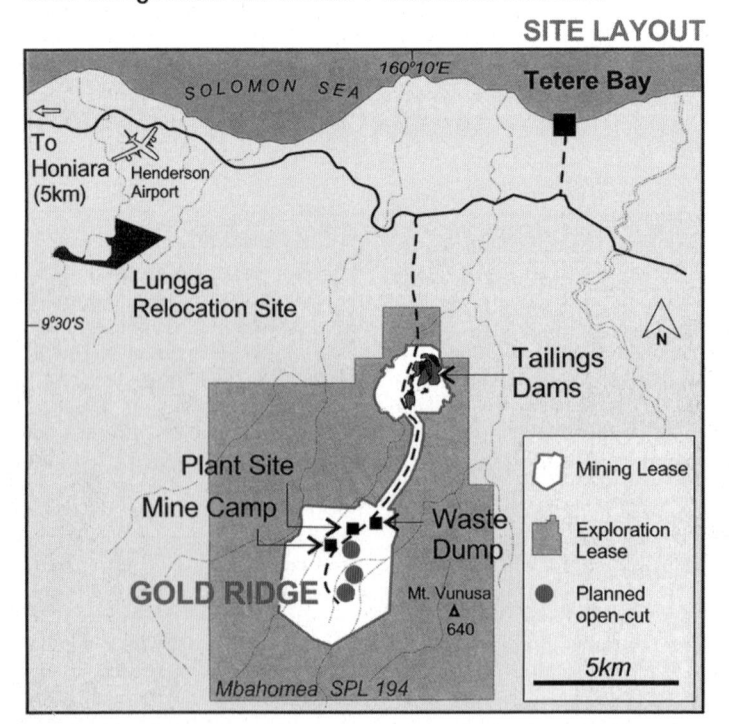

Fig. 1. Maps showing position of Guadalcanal, and position of mine and tailings lagoons (*source*: Gold Ridge Project, Gold Ridge Mining Limited, Solomon Islands, used with permission).

Fig. 2. Pre-mine Mines and Mineral Department visit to Gold Ridge, taking samples for the establishment of a Geochemical Baseline. Photograph includes images of typical local Mbahomea children returning home from school.

Islands, is also on Guadalcanal, some 22 km northwest of the GRM. Honiara hosts most of the services and infrastructure needed to service the mine operation. The main economy of the archipelago nation of Solomon Islands is based on copra, cocoa, palm oil, timber and fisheries. The development of the GRM in 1998 was welcomed as another revenue-generating economic activity, which contributed an estimated 30% to the Gross Domestic Product in 1998–1999 (Pyper 1996).

Gold was first discovered in 1568 in the Matepono river mouth area, along the northern coast of Guadalcanal. The Matepono river drains the Gold Ridge area. These observations were made by a Spanish explorer named Alvaro de Mendana, who subsequently named the islands after the famous biblical king, Solomon. It was not until 1936 that the gold discovery was traced back to the Gold Ridge hard-rock goldfield source, resulting in a number of attempts to work the goldfield through alluvial

Fig. 3. Company tours of the mine for local people during construction stage.

Fig. 4. Access road construction to the mine site, 1997.

mining in 1939. Small-scale mining was hampered by the Second World War, which reached the island in August 1942 (Grover 1955).

The establishment of the Solomon Islands Geological Survey in 1950 saw the start of systematic surveys in the area, which continued into the 1980s. It was not until 1983 that modern exploration commenced. Potential large tonnage low-grade deposits were identified by a number of international groups. In 1994–1996, Saracen Minerals very significantly improved the geological knowledge of the Gold Ridge deposit to the point where it finally looked like a serious mining proposition. Ross Mining NL of Brisbane then took the prospect over in 1996 and pursued its commercial interest with vigour and commitment. Ross Mining achieved their goal of mine construction in early 1997 and gold production by late 1998, thus giving birth to a modern mining industry in Solomon Islands (see Figs 2–6).

The GRM is situated within Mining Lease 01/97 (ML 01/97) having a total land area of 30 km^2. The Mining Lease was the right to mine granted to Gold Ridge Mining Limited, the locally incorporated subsidiary of Ross Mining NL by the Solomon Islands Government on 13 March 1997 for a period of 25 years.

The GRM is an open cut gold mining operation, which adopts modern mining concepts and methodologies, and is committed to acceptable 'best practice' standards in its operational conduct, with special attention to the environment and the community.

The Gold Ridge gold deposit is a volcanoclastic-hosted low sulphidation epithermal style gold deposit within the 500 m thick shallowly dipping Lower Pliocene Gold Ridge Volcanic Group (Walshaw 1974; Lum et al. 1991). The mine is a low-grade, high tonnage project with reserves of 26.17 million tonnes at 1.61 g/t gold (approx. 1 356 166 million oz. gold). The reserves are within four separate ore deposits, namely Valehaichichi (6.9 mt), Namachamata (6.7 mt), Kupers (1.1 mt), and Dawsons (11.6 mt), all situated in close proximity to one other. Continued mine-based exploration has resulted in increasing the reserves to 2 000 000 million ounces ((>60 t gold), which is moving the deposit forwards to world class status $\geq$100 t gold) (Ross Mining Annual Report 1998). The original mine lifetime expectancy was 10 years, but this will probably be extended in view of subsequent exploration discoveries and the recent interruption in production.

The Solomon Islands Mines and Minerals Act (1990) (Tolia 2000) and its subsequent Regulations of 1996 apply to the GRM, with specific reference and interdependence on a number of related specific laws.

The GRM area in central Guadalcanal enjoys a tropical climate with average daytime temperatures of 28–32°C. Year round seasonal variation is predominantly marked by wet and dry seasons. The annual rainfall on site is approximately 3000–5000 mm, which falls daily during the wet season. The tropical climate of Solomon Islands and its position at around nine degrees

Fig. 5. Pre-mine stakeholder consultations. (**a**) Inaugural meeting of the Gold Ridge Landowners' and Community Association of Ngalakasia School, 5 February 1996; (**b**) Gold Ridge Landowners' and Community Association Executive with Eugene Iliescu (Ross Mining) and Nicholas Biliki from the S.I. Government (or Eugene's right).

south of the equator ensures it receives its share of tropical cyclones. All these geographical and climatic factors were incorporated into the design of the mine, as well as taking other factors into account (e.g. relatively high seismicity and steep topography). The tailings disposal site is situated in a dry blind valley in the northern foothills of the Guadalcanal Highlands, some 8 km from the plant site, with a 300 m negative head-height difference (Fig. 1). The tailings dam wall is a conventional rock and clay fill structure, designed and constructed to meet industry best-practice standards and to be maintained progressively with increased tailing deposition. At the time of its premature closure in 2000, the GRM employed a total of 583 staff, of which 102 were expatriates and 481 Solomon Islanders (Delta Gold 2000).

Gold Ridge people, society and communal structure

Solomon Islands is inhabited predominantly by Melanesians (the dominant ethnic group), who have inhabited the islands for at least 50 000 years. The Gold Ridge area is occupied by the Malango linguistic-ethnic group. Local people belong to the Mbahomea tribe and the language they speak is one of the 19 indigenous language groups present on Guadalcanal (Naitoro 1995). Over 1000 people lived within the area affected by mining and its immediate vicinity (some 100 square kilometres) prior to the recent mine development. These people place a very high value on land in recognition of the special dependence of human society and land, which is key to environmental management and survival within

Fig. 6. (a) and (b) Local alluvial gold mining prior to the opening of Gold Ridge.

traditional socio-economic systems. This high cultural value placed on land by Melanesians results in the prevalent strongly held view that the land must be honoured, respected and cared for responsibly to ensure that future generations can likewise enjoy and be sustained by the land. Land is a highly prized capital asset with deep social significance. At times, land acts as a religious 'store' or 'bank'. This view is, perhaps, a real example of the essence of the 'sustainable development' paradigm enunciated by Brundtland (1987), whereby every generation ensures that the most fundamental economic and social asset society possesses (the land) is cared for and passed on in an enhanced or equivalent state to the next generation. It is no accident that the most important word in the *Pijin* language is 'blong' (meaning belong, to be part of), which emphasizes strong land, language and tribal roots. Melanesians are mainly subsistence crop and livestock (pigs and chickens) farmers.

Land ownership is a complex concept in Solomon Islands. Property rights over land are communally owned by kinship or clans. Social and cultural values of land (values that may exceed 'commercial value' in a Western context), affecting controlling customary land ownership attitudes, are flexible and change over time. Ownership patterns are complex and undocumented except through oral history. Land boundaries are poorly defined (McGavin 1993).

The Gold Ridge area Mbahomea peoples' traditional social structure is matrilineal, with land entitlement inherited from the female line. The legendary origins of the people are from two genealogical descents and subsequent clans. There are 17 clans with customary land title land rights within the Mining Lease area. The Gold Ridge Community and Landowners Association (GRCLA) was formed, consisting of representatives of 16 clans to negotiate with the Government and Mining Company. The landowners of the tailings dam area were not represented by this association.

The identification of clan land ownership was fundamental to mining negotiations. Land ownership had to be defined through a range of criteria that included the detailed clan knowledge of sites of archaeological, historical, religious and cultural importance. These sites hold particular significance to each clan or individual group (Naitoro 1995). A detailed knowledge of the location and significance of such sites builds up the credibility of clan claims over land ownership.

The settlement pattern in the Gold Ridge area is semi-nomadic (based on transitory village development patterns). The availability of fertile soil to cultivate yams and taro is a determining factor of continual patterns of movement (Naitoro 1995).

Prior to the discovery of gold, the Gold Ridge population was low but grew quickly, leading to the development of permanent villages, as an economy based around alluvial gold mining developed during the early 1980s (Figs 5 and 6). As with other island interior regions in Solomon Islands, Gold Ridge had depopulated during the 20th century as people moved to the coast to take advantage of developing economic, educational, health and transport infrastructure. To this day, interiors of most of the islands are undeveloped, mainly comprising dense, pristine primary or secondary rainforest. The population of Gold Ridge grew to over 1000 people when the current large mine was licensed to proceed to the full mining stage in 1996. This repopulation largely comprised people returning to the area who were part of the indigenous tribal groups but had relocated to Honiara and other coastal settlements. The issue of population relocation was one of the key challenges that had to be addressed before mining could take place.

Stakeholder consultation process

Bearing in mind the very high cultural value the Mbahomea people place on land, it was imperative to identify the genuine 'stakeholders' in the GRM. This paper defines 'stakeholders' as those people who are directly affected by the mining project, or have legitimate governance, commercial or employment interests in the mine. Stakeholders included a wide range of people, including National Government (represented by the Mines and Minerals Department), the Company (Ross Mining NL and its Shareholders), the Gold Ridge Community and Landowners Association, the Provincial Government of Guadalcanal, Gold Ridge Mining Limited (representing the employees and management), downstream and locally impacted communities, the artisanal mining industry, members of the general public, insurance and financial companies, and support service industries.

It was considered to be of paramount importance for the long-term security and success of the project that serious and protracted consultations occurred between relevant stakeholders at all stages of project development (Figs 5a & b). The aim was to achieve consensual or near-consensual decision making (McGavin 1993).

Pre-mine stakeholder consultation process

Before acquiring the Gold Ridge Project (GRP), Ross Mining decided to place a high priority on environmental and community baseline studies (Orr 1999) (Fig. 2). This meant that the company had to undertake a range of impact assessment studies (environmental, social, archaeological), which were independently reviewed by experts working for the Commonwealth Fund for Technical Co-operation (Orr 1999). This approach was encouraged by the Government and Solomon Island law. Good community relationships were considered by Ross to be of high importance and the company looked to the Department of Mines and Minerals for assistance in developing and maintaining healthy relations (Fig. 3).

Prior to pursuing the consultation processes for the GRM, all stakeholders noted the need for a coordinated and structured approach. This led to the establishment landowners of the Gold Ridge Community & Landowners Association (GRCLA) the Gold Ridge Project Office within the Department of Mines and Minerals, various committees (The Gold Ridge Co-ordination, Environment and Social Impact Assessment, Advisory Committees) and the Mining Agreement Negotiating Team by the Government.

The various committees and bodies formed by the Government had different roles and were composed of representatives from almost all sectors of the Government system. Non-governmental organizations (e.g. environmental and social NGOs), landowners and the immediate concerned downstream communities were also involved. This inclusive approach attempted to ensure the participation of all key stakeholder groups.

Initial consultations focused on landowner identification and demarcation of land ownership boundaries in so far as this was possible within the cultural constraints of the area. Landowners had to be identified and individual parcels of land clearly defined before legal land acquisition could take place. There is no written testimony to land ownership in Melanesia. Land is communally owned through local ethnic historical claims. This knowledge has been passed down from generation to generation through oral traditions. Identifying key people or groups of people who have a genuine land ownership right in a modern Western legal sense is a difficult process. Inevitably, the prospect of generous compensation results in some false claims being made. Each claim over land ownership had to be scrutinised and tested using pre-agreed methods. The Government played a vital and pivotal role with regard to land ownership, which involved consultation, land surveying and field-verification of claims and land knowledge, and negotiations for the initial surface access right to the land. When access rights were granted for the reevaluation of the Gold Ridge gold deposit, Ross Mining were able to employ local people to assist with detailed exploration and mine development.

Ross Mining played an active and cooperative role with the local community, Government and other stakeholders and attempted to operate transparent managerial and planning practices. All the technical aspects of the project and any written reports were presented to the Government for review and approval. The Government used Rio Tinto Consulting Services of London, UK, for an independent review of the Feasibility Study Report and other subsequent technical outputs of the pre-mine and mine construction phase of development. Further assistance was sought from the Economic, Legal and Advisory Service of the Commonwealth Secretariat Funds for Technical Cooperation, and the GRCLA were provided with independent legal and mining expert advisers from Papua New Guinea (a Melanesian country that has several decades of mining experience) for their negotiations. The success of this consultative and review process was confirmed to some degree by the smooth running of the mine during the early mining period (1997–2000).

The stakeholder consultation process was the largest ever conducted in Solomon Islands. Proof of its success was the establishment of very important consensual agreements between the three major stakeholders (Government, Company and Community), which took almost three onerous years to complete.

Several subsidiary understandings were also made, including the Memorandum of Understanding between the company and the Guadalcanal Provincial Government and the Downstream Communities.

Mining stage consultation processes

As the GRM went into the operation stage, the consultation process between the three main stakeholders continued. This was seen to be necessary in the sense that, continuing consultation, interaction and communication plays an important role in ensuring the maintenance of cooperation and positive relationships, transparency, implementation of agreement obligations, and communications regarding changes in project direction and project progress (e.g. McGavin 1993).

The company established, within its organizational structure, a community and employment relationship department whose role was to deal with socio-economic impacts (responding to concerns raised by community) of the dislocated populace, downstream communities and the general population of Solomon Islands. This department continuously monitored social impacts and developed a range of mitigation measures to address specific problems that arose from time to time within the affected communities. Perhaps the clearest example of social impact involved the resettled bush communities of Gold Ridge. These people were rainforest people living in a relatively remote area cut off from the rest of Solomon Islands for much of the year. They are also 'Highland' people who lived at an altitude of 500–1000 m. These people were resettled in the coastal plain of Guadalcanal some 20 km east of Honiara. The people were closely consulted over village and house design and the amenities they wanted within the new village. A church, school, medical facility and meeting house were built as an integral part of the new village. The people had access to a very significantly increased amount of personal finance, possessions, and range of facilities than they had ever experienced in the interior of Guadalcanal. As with any change, it took time for the people to adjust to a very different environment including a hotter, dustier and drier climate. There have been no serious studies to date documenting the experience of this resettled community, but such a study would prove useful to future mineral developers in Solomon Islands and elsewhere.

Under the requirements of the Special Conditions of the ML, an Environment and Social Impact Assessment Committee was formed, the role of which was to provide a forum for respective stakeholders to meet with company representatives on a quarterly basis to discuss concerns and issues of interest. The Committee consisted of Government, non-government organizations (NGOs), landowners and the downstream communities.

Owing to concerns held by some sections of the community regarding the environmental performance of the GRM after commencement of operation, an environmental audit was considered necessary and was conducted in 1999 (Orr 1999). This was a reflection of the lack of an indigenous mining culture in the country and unfamiliarity with mining infrastructure and management strategies. The environmental audit was undertaken independently under the auspices of the Department of Mines and Minerals with the objective of providing an independent professional and impartial assessment of the sites and environmental performance of the Gold Ridge Mine. The audit had to identify any particular problems and recommend mitigation and future monitoring/checking strategies. All audit results had to be communicated to stakeholders in an appropriate format. The audit was overseen by an independent audit management committee comprising, the National Government, Provincial Government, NGO representatives, and Company representatives. Following this audit, the Gold Ridge Mine was recommended to undertake environmental auditing every two years, adopt even closer and wider stakeholder consultation exercises, and contribute to greater public awareness raising education programmes. Communication and consultation exercises are deemed to be matters of importance in Solomon Islands as poor communications between company and stakeholders can become a source of risk for the GRM.

The following communication and consultation protocols were adopted for the GRM during the mining stage: training and re-training programmes established, monthly reports drafted, continued awareness programmes devised, radio-based message and reporting to be expanded (the radio is the most important form of communication in Solomon Islands), an accident record database to be kept up to date, regular personal contacts between appointed company staff and stakeholders to be undertaken, periodic auditing of a range of activities to occur in a pre-agreed manner, and community participation to be encouraged where appropriate.

The sustainability of benefits from the Gold Ridge Mine

The contractual arrangement established between the three key stakeholders governs the kind and level of direct benefits accrued from the mining operation to local communities. Social, environment and economical impacts of the GRM to the host community (Government, GRCLA, and the public) are to be constantly monitored. Positive impacts must outweigh the negative impacts: this is the key performance indicator to be used for the decision making with respect to establishing and further developing the GRM.

Benefits to the Landowners from the Gold Ridge Mine are varied and not necessarily always measurable in cash terms. To compensate for the loss of immediate land-use benefits, negative environmental impacts, loss of access to heritage and cultural sites, and disruption of lifestyle and dislocation of the community population, a package of benefits was established that included a compensation premium (lump sum upfront payment; annual land rental payments to landowners; 1.3% gross royalty payment − 1.0% to Landowners and 0.3% to Guadalcanal Province Government); 250 000 fully paid shares for Landowners in Ross Mining (worth US$1 million in 1997); 500 ha of freehold land from the Government for the relocation village; a newly built relocation village equipped with a five classroom school, a clinic, a church, a cultural and woman's centre, soccer and netball pitches, a police post, and a fully reticulated water supply; commercial land in the centre of Honiara provided free; secondary and tertiary educational scholarships; preferential employment and training priority in the mine; a Regional Tax Credit Scheme available (US$50 000 per annum); small business opportunities (nurtured in conjunction with the internationally recognized NGO Solomon Islands Development Trust and Australian Expert Services Overseas Program, AESOP); a joint venture (20% equity) in subcontracts for catering and security services; and improved roads within the surrounding communities; and improved site amenities, and infrastructure.

These benefits, if properly utilized and managed, are deemed to constitute a fair benefits package that reaches a wide range of stakeholders and could provide sufficient economic stimulus and investment for the future sustenance of economic activities after mine closure.

The Government (which, in theory at least, represents the entire Solomon Islands) benefits from the GRM as follows: by attracting a 1.5% gross royalty (export tax); by attracting revenue from a range of other taxation obligations; increased employment, educational and training opportunities; improved road, power and water supply infrastructure and services required to service the Gold Ridge Mine, which occupies a remote rural and formerly undeveloped area of Guadalcanal; the introduction of new industry, with a range of executive, professional, skilled, semi-skilled, and unskilled occupations; a greatly increased reputation in the international mining arena proving that mining companies can do business with Solomon Islands; and enhanced environmental skills, expertise and data collection development.

Conclusions

Countries such as Solomon Islands are populated by an indigenous culture that places an extremely high value on land ownership and land usage. A failure to take account of cultural sensitivities can even result in a failed mine, such as occurred in Bougainville, Papua New Guinea. The adopted procedures should be viewed as essential business practices. Sustainable development has encouraged new thinking in addressing the holistic nature of development and who precisely benefits in the longer term. Mining companies used to have a bad reputation in some parts of the world for operating profit-only motivated mines. However, particularly since the 1980s, most serious long-term players in the mining industry have realized that they must develop a range of practices that allows them to meet modern standards with respect to environmental protection, social justice and long-term post-mine closure economic benefits to the areas and countries in which they operate. These practices are particularly valued in a country such as Solomon Islands, which is a small island state, with a relatively low population, very limited economic and educational development and a fragile socio-cultural eco-system. Solomon Islands, is a young country, which has had a limited exposure to the mining industry and globally operating commercial companies. Patient careful planning with respect to consultation must be embedded within the mine development process from the earliest stages of exploration. Indigenous people, as well as relatively naive and young government systems, must be given time and access to information and transparent decision-making processes. Only through careful, honest and open dialogue and consultation can consensual decision making over life-changing projects such as mine developments be made and truly successful mine operations be brought into operation. The authors suggest that the consultation processes associated with the Gold Ridge Mine are a reasonable model to learn from.

Postscript

In spite of the best endeavours of the key stakeholders involved in the GRM and the successful operation of the mine, the mine closed in 2000. This was due to ethnic tension on Guadalcanal unconnected in any way to the mine. There are signs that the mine may reopen in the short–medium term.

References

BRUNTLAND, G. (ed) 1987. *Our Common Future: The World Commission on Environment and Development*. Oxford University Press, Oxford.

DELTA GOLD 2000. *Annual Report 2000*. Delta Gold Publications, Perth, Australia.

GOLD GAZETTE AUSTRALIA 1998. Solomon Islands prepare to welcome first gold mine. *Gold Gazzette Australia*, **4**(4), 15–33.

GOLD RIDGE MINING LIMITED. Monthly Report for May 2000. Gold Ridge Mining Limited, Int Rpts of GRML, Honiara, Solomon Islands.

GROVER, J. C. 1955. Gold Ridge, Guadalcanal, discovery of gold bearing bodies, implementation of sample assays and future prospect. *In: The Solomon Islands Geology, Exploration and Research, 1953–1956*, Ch. XII, C. F. Hodgson & Son Ltd, London, UK, 63–80.

HUGHES, A. V. 1997. *Financial Flows Interim Report: Gold Ridge Mine, Solomon Islands*. UN-ESCAP, United Nations, New York.

LUM, J. A., CLARK, A. L., COLEMAN, P. J. 1991. *Gold Potential of the SW Pacific, Papua New Guinea, Solomon Islands, Vanuatu, and Fiji*. SOPAC Publications, Suva, Fiji.

McGAVIN, P. A. 1993. *Economic Security in Melanesia: Key Issues for Managing Contract Stability and Mineral Resources Development in Papua New Guinea, Solomon Islands, and Vanuatu*. Research Report Series No. 16, Pacific Islands Development Program, East-West Centre, Hawaii.

NAITORO, J. H. 1995. *Report of the Social Environmental Study of Gold Ridge, Central Guadalcanal, Solomon Islands*. Solomon Islands Government Publications, Honiara, Solomon Islands, New York.

ORR, M. 1999. *The Approval of a Gold Mine in Solomon Islands, Environmental Planning Issues.*

Ross Mining NL, Ross Mining Internal Reports, Brisbane, Australia.

PYPER, R. 1996. *Geological Review and Valuation Report of Tenements at Guadalcanal, Sol. Is. and Vanua Levu, Fiji.* R. C. & J. M Pyper Geological Consulting & Contracting Services, QLD Australia.

ROSS MINING NL 1995–1999. *Annual Reports, 1995 to 1999.* Ross Mining NL, Ross Mining Internal Reports, Brisbane, Australia.

Solomon Islands Human Resources Development Report 2002 – Building a Nation, Vol. 1, Main Report, Mark Otter, Australia, Solomon Islands Government Publications, Honiara, Solomon Islands.

SHIELDS, D. J. & ŠOLAR, S. V. 2005. Sustainable development and minerals: measuring mining's contribution to society. *In*: MARKER, B. R., PETTERSON, M. G., MCEVOY, F. & STEPHENSON, M. H. (eds) *Sustainable Minerals Operations in the Developing World.* Geological Society, London, Special Publications, **250**, 195–211.

TOLIA, D. H. 2000. *The Environmental Management Planning for the Gold Ridge Mine,* The major course assignment for the Environment Management Course in Mining Industry Management, School of Mining Engineering, University of New South Wales, Australia.

WALSHAW, R. D. 1974. *A Geological Investigation of Gold Rudites at Gold Ridge, Guadalcanal.* Geological Survey Division, Ministry of Lands, Energy and Natural Resources, Sol. Is. Government. Bulletin 14.

Construction raw materials in Timor Leste and sustainable development

JORGE F. CARVALHO & JOSÉ V. LISBOA

Instituto Geológico e Mineiro, Estrada da Portela, Zambujal, 2721-866 Alfragide, Portugal
(e-mail: jorge.carvalho@ineti.pt)

Abstract: Timor Leste is the newest and one of the poorer nations in the world. One of its main challenges that could lead to poverty reduction is the reconstruction and maintenance of the infrastructures that were almost completely destroyed after its independence referendum. To achieve this, there is an imperative need for construction raw materials in a country where the extractive industry is scarce and artisanal. Available geological studies deal with the island's geology and tectonic evolution or its oil and gas potentialities. Very few broach other geological resources. A general study of the country's territory demonstrates that Timor Leste possesses large resources in clays, limestones and sand and gravel, which can support small- to large-scale raw material extractive industries. Some selected areas have been the target of more detailed study: Venilale and Aileu, with resources for structural ceramics and whiteware respectively, and Beheda, where a crinoid-rich limestone crops out, with potential for usage as ornamental stone. These resources are suitable for non-sophisticated small-scale mining operations that should be able to accomplish environmental and social liabilities. No public policy exists for the management of these mineral resources, which is essential for the sustainable development of Timor Leste.

Timor Leste is a newly independent country on the eastern part of the island of Timor that is located on the border of the Lesser Sunda archipelago (Indonesia). It has an area of about 15 000 km^2, which includes two islands, Atauro and Jaco, and the enclave of Oecussi located on the north coast of Western Timor. The country's capital, located on the north coast, is Dili. The Timor Sea separates the island from Australia (Fig. 1).

The territory is about 275 km long and 100 km at its greatest width, and is characterized by a varied morphology and ecology. A central range of hills and mountains, up to 3000 m, divides the country's northern region from the southern one. The climate is tropical, with distinct rainy and dry seasons. The natural combination of heavy rainfall during the rainy season (November to April) and a terrain morphology with slopes of around 40% occurring in approximately half of the territory's area, can result in a range of natural hazards (mainly floods, landslides and soil erosion).

The first known geological study is reported by Hirschi (1907). However, it was not until the post-World War II period that important regional studies were carried out, mainly on behalf of oil industry interests (Grunau 1953, 1956, 1957; Wanner 1956; Gageonnet & Lemoine 1957, 1958; Leme & Coelho 1962; Leme 1963, 1968). Several of these geological studies included mapping at different scales, but it was Audley-Charles (1968) who first produced a regional geological map of Timor Leste (scale 1:250 000) and formalized previous work into the presently recognized stratigraphy of the country. This map (Fig. 2) remains the main geological reference of Timor Leste. Geological research undertaken in Timor Leste by the University of London, Southeast Asia Research Group, was interrupted by Indonesian annexation. During this period, little geological fieldwork was carried out, except for that by Harris and his co-workers (Harris 1991; Prasetyadi & Harris 1996; Reed *et al.* 1996; Harris *et al.* 1998) and by the Indonesian Geological Research and Development Centre, which published the Dili and Baucau geological map sheets of Timor Leste at 1:250 000 scale (Bachri & Situmorang 1994; Partoyo *et al.* 1995). In recent years, significant geological studies have been carried out in eastern Timor (Harris & Long 2000; Charlton 2001, 2002; Charlton *et al.* 2002), mostly concerning its tectonic evolution and implications for oil exploration. Nevertheless, the geological knowledge of Timor Leste is incomplete, particularly in respect to its mineral resources.

From: MARKER, B. R., PETTERSON, M. G., MCEVOY, F. & STEPHENSON, M. H. (eds) 2005. *Sustainable Minerals Operations in the Developing World.* Geological Society, London, Special Publications, **250**, 161–184. 0305-8719/05/$15.00 © The Geological Society of London 2005.

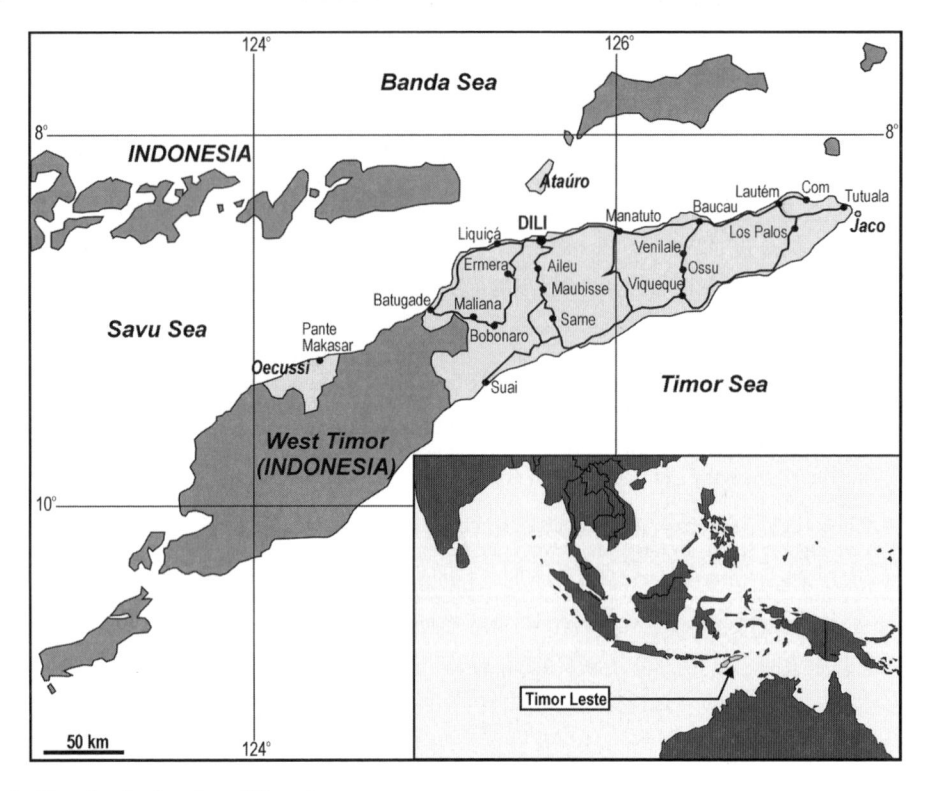

Fig. 1. Map showing location of Timor Leste.

Timor is a non-volcanic island with a complex geological history and is not yet well understood in the evolutionary context of the boundary between Asian and Australian tectonic plates. According to the recent studies, Timor represents a contractional wedge formed by mechanical accretion of underthrusted Australian continental-margin cover sequences that represent autochthonous, parauthochthonous, allochthonous and syn-orogenic olistrostrome units (Audley-Charles 1968; Charlton et al. 1991; Harris 1991, Harris & Long 2000; Charlton 2002).

The economic and social infrastructure of Timor Leste was severely damaged as a result of the devastating conflict that occurred after the independence referendum of 1999. International support was coordinated by the United Nations agencies in Timor Leste in order to restore peace and security, but also to help with reconstruction and economic revival. In this context, one of the contributions of the Portuguese Government was a preliminary assessment of the non-metallic mineral resources of Timor Leste, which is the basis of the results presented here.

Timor Leste's development will undoubtedly require the use of its natural resources. A key challenge for this new nation is how to develop it within the principles of sustainable development, thus contributing to the improvement of the well-being of the Timorese people.

The development strategy of Timor Leste

Timor Leste is one of the poorest countries in the world, with a per capita income lower than US$1 a day (UNDP 2000). This situation has a historical background that started with the settlement of a Portuguese trading post in 1562, the ensuing colonization period and later Indonesian military occupation after 1975. During the 500 years of colonization and foreign occupation, the economic structure was based on the exploitation of Timor Leste's natural resources, for short-term profits without a sustainable economic policy (Sandlund et al. 2001). The main commodities that were exploited included sandalwood, honey and bees wax. Since the beginning of the 20th century, a more sustainable policy led to a few phases of agricultural development based on coffee production (UNDP 2002). During Indonesian occupation, Timor Leste's fragile economy became increasingly based on urban

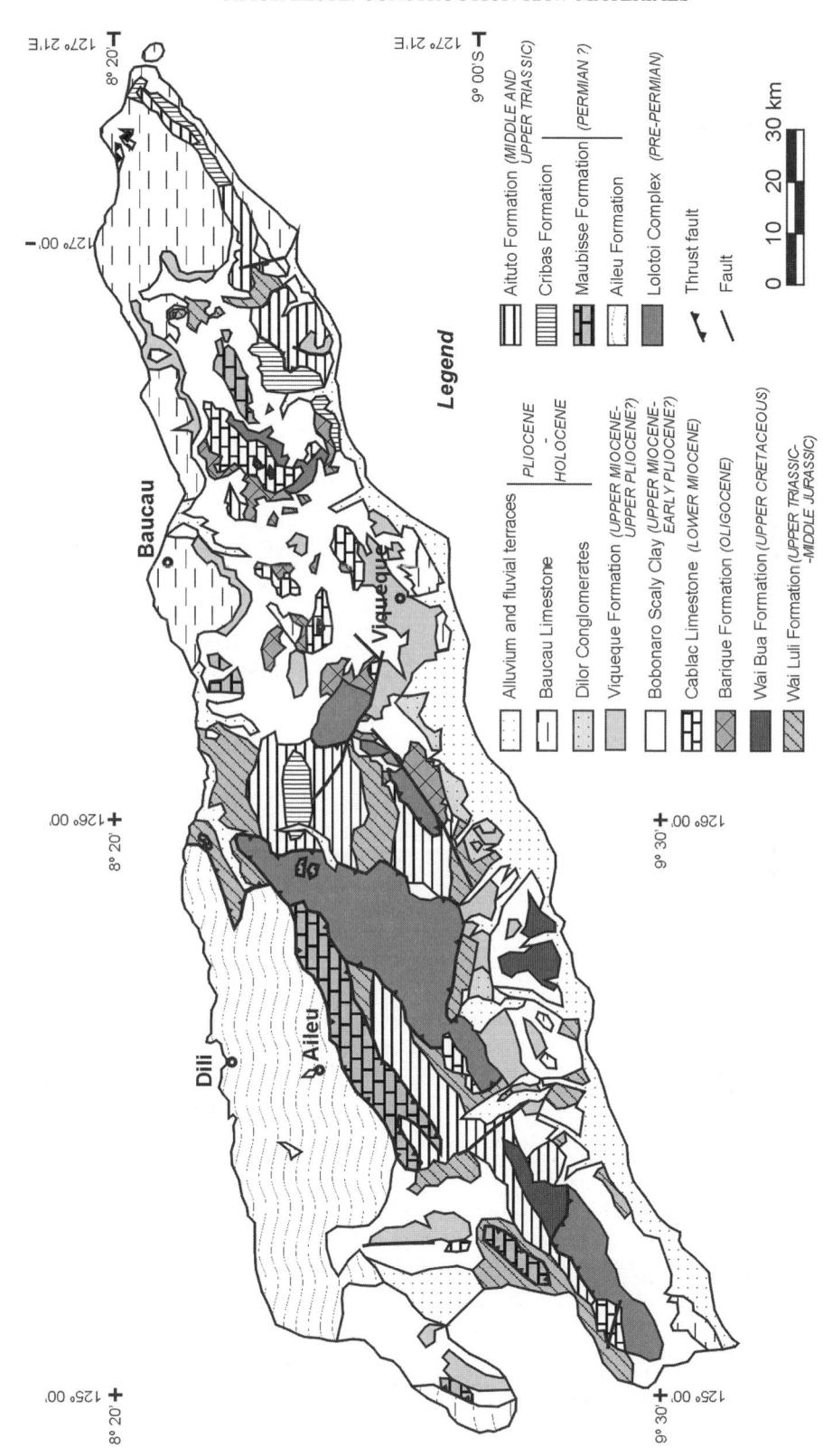

Fig. 2. Simplified geological map of Timor Leste (modified from Audley-Charles 1968).

services, resulting in an economic downturn in the contribution of agriculture to GDP from 60% in 1981 to about 25% in 1998 (UNDP 2002). Other economic activities such as furniture manufacturing and mining remained largely undeveloped.

The tragic and violent conflicts carried out by the Indonesian troops and anti-independence militias after the independence referendum in 1999 caused great changes in population settlement patterns, agricultural production and other microeconomic activities. Approximately 70% of all houses and infrastructure were destroyed and 75% of the entire population was displaced (UNDP 2000; Sandlund et al. 2001). Around one-quarter of the 28 000 civil servants in pre-conflict times were non-Timorese, occupying the top administrative positions. Their exit left the territory lacking skilled human resources (UNDP 2002).

According to statistical data provided in UNDP (2000, 2002) reports, Timor Leste's present-day population of about one million people is very young (48% below the age of 17 years), and is growing at a rate of 2.5%. More than three-quarters of the population live in rural areas from subsistence agriculture without access to basic social services such as drinking water, electricity, and health and sanitation services. Official unemployment figures indicate that 16% of the population are out of work, but according to CIA's World Fact Book (WFB 2004), this figure is closer to 50%. Half of the Timorese people are unable to read or write.

At present, Timor Leste's exports are almost totally linked to coffee, which is currently being reestablished. This situation is expected to change shortly, with revenues from the exploitation of oil and gas resources in the Timor Sea. Most commodities, from cooking oil to cement, have to be imported, which is the reason why the future of Timor Leste depends on present international support.

Timor Leste became an independent and internationally recognized new nation in May 2002. Its government presented a strategic plan regarding the nation's development – the National Development Plan (NDP 2002) – with two main goals: poverty reduction and sustainable economic growth. The Plan is heavily focused on the sustainable development principles advocated on United Nation's Agenda 21 and the Johannesburg's World Summit on Sustainable Development.

In connection to the Millennium development goals, a policy for poverty reduction is outlined based on a fast but sustainable economic growth supported not only by future oil and gas revenues, but also by the exploitation of economic opportunities based on the use of other resources, such as agriculture, fisheries and forestry. Agriculture is highlighted as a key growth area, especially if productivity improvements are achieved within small-scale agro-units.

Timor Leste' strategy on mineral resources

The NDP's outlined strategy for mineral resources in Timor Leste is mainly described in a report entitled 'Natural and Mineral Resources Inventory, Policy and Development Strategy for East Timor', which was prepared by the United Nations agency ESCAP (Economic and Social Commission for Asia and the Pacific). It is published as 'Exploring Timor-Leste – Mineral and Hydrocarbon Potential' (ESCAP 2002). This important document presents a draft legal framework regarding the management of Timor Leste's mineral resources. In addition, it provides an overview of the general characterization of the resources and their potential, based on known mineral occurrences. Special emphasis is given to resources that can yield important economic revenues, such as onshore oil & gas, chromite, copper and gold. To ensure a sustainable framework for mineral development, the priorities addressed in the NDP of Timor Leste's Government are:

- the establishment of a legal framework and an institutional capacity building programme for medium- to long-term management of the mining sector; and
- the creation of investment opportunities.

The Government has identified one of the short-term paths to economic growth and consequent poverty reduction as the infrastructure reconstruction maintenance development. The NDP states that an effective physical infrastructure system is crucial for agricultural productivity, business investment, and is instrumental to human development. Industrial minerals play a key role in this development, providing raw materials essential for construction, such as aggregates and common clays for bricks and roof tiles manufacturing. However, the importance of these mineral resources, namely with respect to resource availability, accessibility and environmental sustainability, is stated neither in the government's NDP, nor in the several strategic documents that have been published by United Nations agencies dealing with Timor Leste development, such as 'Building Blocks for a Nation' (UNDP 2000) and 'Timor-Leste: Programme Package

Document for Sustainable Human Development' (UNDP 2003). One of the most important documents is 'The Democratic Republic of Timor Leste – Public Expenditure Review' (World Bank 2004), where several pages address the needs and economic assessment of road reconstruction and maintenance without mentioning raw materials requirements. This view, that construction mineral resources 'come from heaven', is not just a Timor Leste problem, but is common worldwide (Perez 2001; Wellmer & Becker-Platen 2002; Cárdenas & Chaparro 2004) and fails to embrace the fundamental concepts of sustainable development.

Construction raw materials in Timor Leste

Timor Leste's geology is characterized by considerable lithological diversity, favourable for the occurrence of a wide variety of mineral resources. Nevertheless, taking into account the immediate infrastructure reconstruction and development needs, attention is focused on construction raw materials, and particularly on the identification of target sites that in the short term can support the supply of these raw materials. Table 1 summarizes the results of a regional geological field survey supported by the mapping of Audley-Charles (1968) and presents the mineral resource potential of Timor Leste. All the geological units referred to from this point forward were defined by Audley-Charles (1968), unless otherwise mentioned.

Old and present-day mining activities represent the best evidence for mineral potential. Mining activity in Timor Leste has always been very incipient. Evidence of old exploitations found in Timor Leste comprise the following.

- Alluvium clays exploitation in the Dili's Fatumeta quarter, where a small deactivated two-kiln brick- and tile-making plant still stands. Nowadays, bricks and tiles are sporadically produced there by artisanal means, being fired in holes that are used as kilns.
- Brick- and tile-making plant near Aileu town, which has never worked due to the 1999 conflict; it was intended to use whitish clayey raw material, which crops out in the surrounding area.
- Inactive exploitation of marble boulders in a colluvium deposit in the Manatuto district, 40 km east of Dili, for ornamental purposes.
- An inactive kiln in Balibo (Maliana district) village, about 70 km southwest of Dili, where coral reef debris was used for lime production.

The present-day mining activity is limited to:

- sand and gravel extraction for road and building construction exploited by artisanal means in flat downstream areas of the main rivers;
- small artisanal limestone crushing plant and a modern quarry and stone crushing plant, managed by a private company near Dili, where intrusive igneous basic rocks are exploited;
- clay-rich material from alluvium deposits used countrywide for the manufacture of sun-dried bricks.

Ceramic raw materials

Brick- and tile-making plants can be of major importance for the reconstruction needs and further development of Timor Leste. The use of indigenous construction materials can replace uneconomic practices, such as the present-day use of heavy cement bricks made with imported cement and imported zinc plates used for climatically inappropriate roofing purposes. The industry would also create much needed job opportunities and provide training of skilled human resources.

Critical to the development of a ceramic industry in Timor Leste is an assessment of the country's potential for ceramic raw materials, in particular, common clays. This assessment must ensure that the lithologies are of suitable quality, uniformity and thickness and are free of major tectonic disturbances, in addition to having a favourable location in relation to consumer centres and accessibility.

Potential lithostratigraphical units for the supply of ceramic raw materials, in order of importance, are the Bobonaro Scaly Clay, Aileu Formation, Suai Formation and Ainaro Gravels (Table 1).

The Bobonaro Scaly Clay (Audley-Charles 1965), the Bobonaro Complex (Rosidi *et al.* 1979) or the Bobonaro Mélange (Harris *et al.* 1998) are all different names for a clay-rich unit, which is widespread all over the Timor Leste's territory, covering about 60% of its area. This unit is a tectonic/sedimentary mélange of Upper Miocene to Early Pliocene age (Audley-Charles 1968; Harris *et al.* 1998), with considerable lithological uniformity. It consists of a clay matrix in which unsorted blocks from structurally and stratigraphically overlying units are found. The clays vary widely in colour, but commonly have a distinctive scaly clay fabric. Clay mineralogy varies throughout the unit, but two main mineral assemblages occur: illite and smectite, and illite, kaolinite and chlorite (Harris *et al.* 1998).

Table 1. *Exploitable non-metallic raw materials and their potential application*

Formation*	Age	Thickness (m)	Lithology	Raw Material	Possible application
Suai Formation	Holocene and Pleistocene	1000	Unconsolidated rudites and arenites ranging from fine silts to pebbly gravels	Aggregates, common clays	Building industry, ceramics industry
Poros Limestone	Holocene and Pleistocene	20	Pale-brown to cream limestone that weathers grey. It is hard, thin bedded and rich in lacustrine gastropods and algae	–	–
Baucau Limestone	Holocene and Pleistocene	100	Hard, cavernous, massive white coral-reef limestone, weathers to a pale grey colour	Biocalciclastic limestone	Chemical industry, lime
Ainaro Gravels	Pliocene to Holocene	100 (?)	Stranded alluvium terraces	Kaolin, aggregates, common clays	Building industry, ceramics industry
Dilor Conglomerate	Upper Miocene to Upper Pliocene	300	Poorly sorted sandy conglomerate with a dark red lateritic crust	–	–
Lari Guti Limestone	Upper Miocene to Upper Pliocene	75	Sequence of yellow calcarenites and thin coral reef rocks	Biocalciclastic limestone	Chemical industry, lime
Viqueque Formation	Upper Miocene to Upper Pliocene	800	Massive white marl and grey claystone interbedded with a few chalky limestones; siltstones and sandstones upwards	Marls, micritic limestones	Cement industry, chemical industry (?)
Bobonaro Scaly Clay	Upper Miocene to Early Pliocene (?)	>2000	Soft and variegated scaly clay with exotic blocks and lenses of rocks of all ages and sizes in it; clay matrix colour varies much, but predominantly grey	Common and special clays (specially bentonitic (?) clays)	Ceramics industry, chemical industry (?)
Cablac Limestone	Lower Miocene	600	Hard, massive limestones of several types: calcilutites, oolitic limestone, calcarenite and intraformational conglomerate	Limestone (sparry, micritic, brecciated), aggregates	Ornamental stone industry, building industry
Barique Formation	Oligocene	300	Basic tuffs (fragments of basalts and serpentinites), feldspathic dacitic tuffs, alteration usually severe; minor interbedded foraminiferal quartz-sandstones. Andesites with zoned feldspars are common	–	–

TIMOR LESTE: CONSTRUCTION RAW MATERIALS

Formation	Age	Thickness (m)	Lithology	Raw materials	Applications
Dartollu Limestone	Middle and Upper Eocene	100	Thick bedded, brown biocalcarenites containing carbonate algae and foraminifera or echinoderm fragments and foraminifera	Limestone (sparry, micritic, brecciated), aggregates	Ornamental stone industry (?), building industry
Seical Formation	Middle Eocene and Lower Cretaceous	100	Radiolarites, radiolarian shales, cherts and marls	Radiolarites	Chemical industry, abrasives, filters, absorbents
Borolalo Limestone	Maestrichtian to Campanian	200	Thickly bedded calcilutites; cherts occur as red or black nodules or veins	Micritic limestone, aggregates	Ornamental stone industry (?), building industry
Wai Bua Formation	Maestrichtian to Aptian	500	Radiolarian marls and shales, bedded coloured cherts, radiolarites, biocalcarenites	Radiolarites	Chemical industry, abrasives, filters, absorbents
Wai Luli Formation	Upper Triassic to Middle Jurassic	1000	Marls, calcilutites, micaceous shales and quartz arenites; basal units spotted blue-grey marls and calcilutites bearing ammonites	Marls, micritic limestones bearing ammonites	Cement industry (?)
Aitutu Formation	Norian to Ladinian	1000	Radiolarian calcilutites, shales and carbonate-rich shales and sandstones	–	–
Cribas Formation	Upper Permian	500	Shales, micaceous shales, silty shales with carbonate and clay-ironstone nodules	–	–
Atahoc Formation	Lower Permian	600	Black pyritic shales, silty-shales, quartz sandstones calcilutites and carbonate nodules	–	–
Maubisse Formation	Permian	900	Well-bedded limestones (biocalcarenites) and massive reefs; limestones are coloured red, pink, white and grey and very rich in reef fauna and debris. Interbedded conglomerates contain clasts of eruptive rocks and tuff	Limestone (sparry, micritic, brecciated), aggregates	Ornamental stone industry, building industry, lime
Aileu Formation	Permian (?)	>1000	Light-coloured shales, phyllites, slates, igneous rocks; in the north coast schists, volcanics, amphibolites, serpentinites and diorite	Marble, kaolin, common clays, igneous aggregates (andesite, basalt, gabbro)	Ornamental stone industry, ceramics industry, building industry
Lolotoi Complex	Pre-Permian (?)	1300 (?)	Quartz-mica-phyllites, quartz-mica-schists, black schists with quartz; metagabbro, dolerite and gneiss, strongly fractured are also present	Igneous aggregates (gabbro, gneiss, quartzite, dolerite)	Building industry

*Main lithostratigraphical units in East Timor as defined by Audley-Charles (1968).

The Aileu Formation or Aileu Metamorphic Complex (Charlton 2002) occupies a single large massif in northwestern Timor Leste, consisting of a sequence of deformed and metamorphosed pelitic, psammitic, basic and carbonate-rich rocks. Earlier studies suggested a Permian age for the metasedimentary sequence (Gageonnet & Lemoine 1958; Leme 1968), but Mesozoic fossils have been reported (Brunnschweiler 1977; Harris & Long 2000; Charlton 2002). In the southwest, clay alteration of argillaceous schist has produced ceramic-grade material.

The Suai Formation and the Ainaro Gravels are Quaternary (Audley-Charles 1968) thick coastal unconsolidated sediments, including alluvium deposits, and stranded alluvial terraces, respectively.

Geological mapping and preliminary characterization of the raw materials were carried out in two areas, Venilale and Aileu, representing target sites where ceramic raw material can be provided in the short term.

Venilale clayey deposits

In the Venilale area, 25 km south of Baucau (the second largest city in Timor Leste) the Bobonaro Scaly Clay occurs in extensive and massive outcrops. In the area under investigation (Fig. 3), fresh exposures of clayey deposits were identified (Fig. 4a). These clays are consistently soft with a variegated colour and frequently display a scaly texture (Fig. 4b). The predominant colour is light grey to dark reddish-brown and dark olive-green. Fragments of exotic blocks occur embedded in the clay matrix, chaotically distributed and randomly orientated, with a size range from a few millimetres to 60 cm in diameter. Its nature is highly variable, although reddish and brown cherts, probably related to the Wai Bua Formation, prevail. Other exotic material found in the area includes shale fragments, probably from the Wai Luli Formation and, less frequently, crinoid-rich Permian limestones of Maubisse Formation.

Three channel samples were collected, representative of the prevailing clayey facies found in the study area. All three samples consist of fine-grained sediments falling in three distinct fields of the Shepard's diagram (Fig. 5; Shepard 1954): clay, silty clay and clayey silt. On the basis of their granulometry, these materials may be adequate for the manufacture of structural clay products,

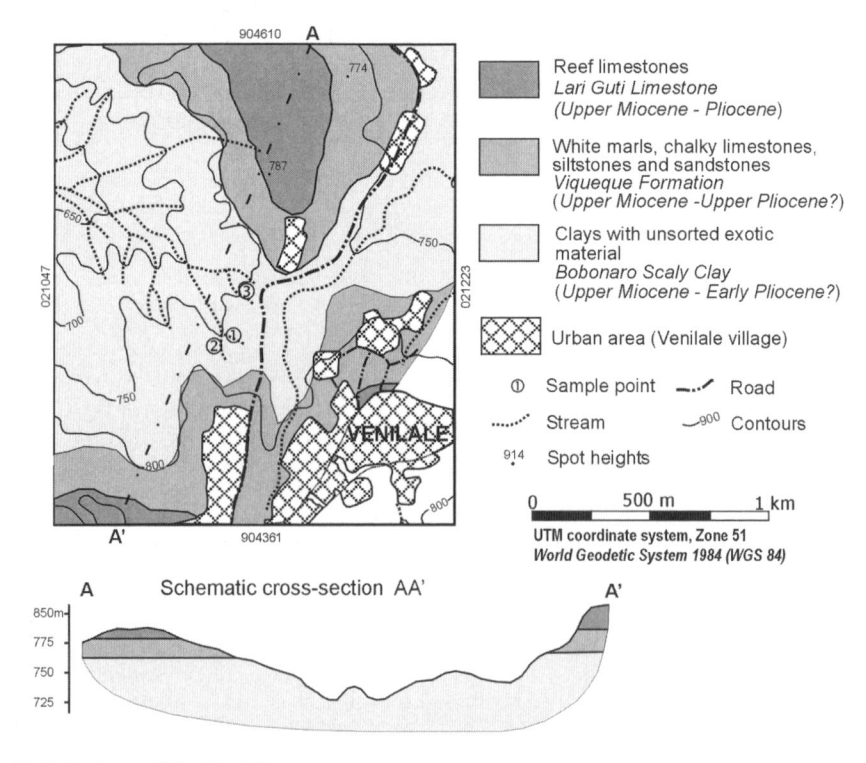

Fig. 3. Geological map of the Venilale area.

Fig. 4. (a) Outcrops of Bobonaro Scaly Clay near Venilale, in the study area; (b) Detail of scaly clay layer, green and reddish coloured.

according to Winkler's diagram (Fig. 6; Winkler, 1954), except for sample Ven1; in order to be used as a ceramic body for brick manufacture, this clay requires blending with a non-plastic material.

Mineralogically (Table 2), clay minerals clearly prevail over non-clay minerals, except for sample Ven1, which is characterized by a high amount of quartz (59%). Samples Ven2 and Ven3 are compositionally similar, except for smectite content (20 and 12%, respectively), with mineral assemblages comprising illite, kaolinite, quartz and smectite. Associated non-clay minerals include K-feldspar with smaller quantities of plagioclase, gypsum and hematite. The mineralogical composition is reflected in its chemical analysis. As would be expected, the lithotype Ven1 has the highest silica content (78.74%) and lowest alumina content (8.53%). The major oxide content of the two other lithotypes is very similar (Table 3). The red firing colour (900°C under an oxidizing atmosphere) is consistent with the Fe_2O_3 (5.07%) and TiO_2 (0.54%) content. Although these are preliminary data, due to the smectite content these clayey raw materials cannot be utilized with firing technology for the manufacturing of structural clay products (common and hollow bricks) and tiles for several reasons. In particular this is due to the occurrence of drying cracks, deformation during firing and small efflorescence. In order to overcome these drawbacks caused by the samples' high plasticity, the clays must be blended in the correct proportions with a non-plastic material.

The mineralogy of the sampled material does not confirm previous studies (ESCAP 2002), which refer to bentonitic clays in this particular area. Nevertheless, it is worth considering the occurrence of bentonitic clays in the area,

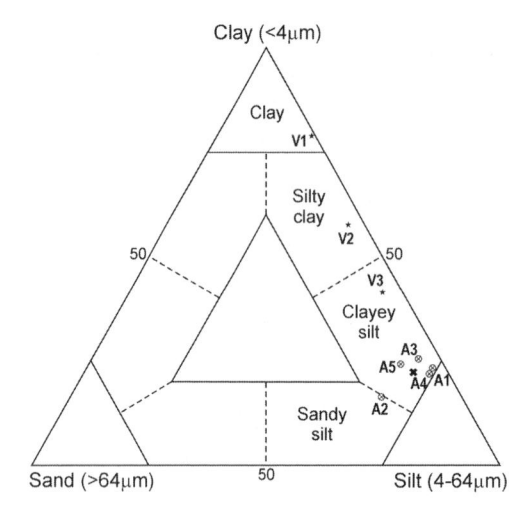

Fig. 5. Grain size classification diagram (Shepard, 1954): A1 to A5, Aileu samples; V1 to V3, Venilale samples; (×) average value.

which, although probably limited in distribution, should be assessed by sampling.

Aileu silt-rich deposits

The Aileu area, about 20 km south of Dili, is located in a valley, where thick coarse and reddish coloured river terraces crop out. Potential economic deposits of ceramic raw materials, characterized by light-coloured silt-rich sediments,

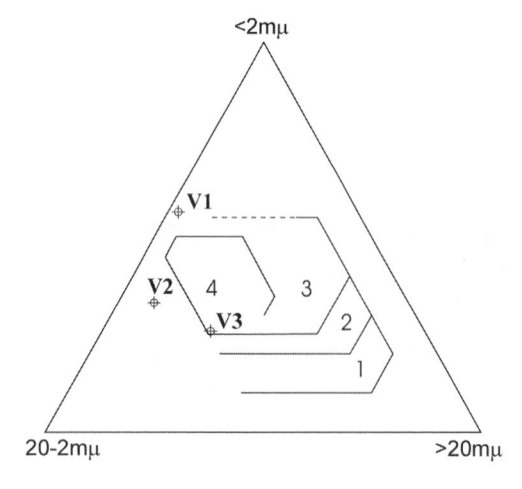

Fig. 6. Winkler diagram (Winkler 1954) for the technological classification of bodies for structural clay products: (1) Solid bricks; (2) vertically perforated bricks; (3) roofing tiles; (4) thin-walled hollow bricks. Venilale samples are V1 to V3.

which occur interbedded within the terraces, were studied. They do not represent a structural red ceramics target but its assessment was included due to its potentialities for whiteware ceramics under optimum accessibility conditions, which may trigger the development of a local industrial nucleus. The terraces and interbedded silt-rich deposits are confined by two faults striking WSW–ENE and other inferred fractures, which partially limit them, tectonically defining a graben, as shown in the mapped area presented in Figure 7.

Terrace composition is strongly related to the underlying bedrock, the Aileu Complex. In the study area it encompasses a series of light-coloured shales frequently interfingered with quartz veins. The targeted silt-rich deposits (maximum thickness of 25 m) occur in tabular layers, 0.5–3 m thick, encompassing silt, fine grain sand and clay (Fig. 8). Five channel samples representative of the deposit's predominant facies were collected for preliminary characterization studies. On the basis of their granulometry, studied materials may be classified as clayey silts, with the exception of sample Aileu2, which falls on the boundary between sandy silt and clayey silt fields of Shepard's diagram (Fig. 5; Shepard 1954).

Mineral assemblages of samples Ail1, Ail3 and Ail4 are quite similar, and samples Ail2 and Ail5 have the higher non-clay minerals content; illite is the prevailing clay mineral in all samples (Table 4). The high amounts of SiO_2 (64.89–72.85%), K_2O (3.39–4.18%) and fairly high Al_2O_3 (16.12–21.27%) reflect the quartz and illite contents of these samples; loss on ignition (LOI) values are concordant with the clay minerals' composition, mainly illite (Table 5).

Chemical analyses of the samples were plotted on a triangular diagram (Fabbri & Fiori 1985) to test the possible use of these raw materials as ceramic products. All samples fit into the white stoneware field, although relatively close to the edge, especially samples Ail3 and Ail2, the former with the highest total oxides content and the latter with the highest silica content (Fig. 9). The Atterberg limits obtained for whole samples show that apart from for sample Ail2, the remaining samples have a satisfactory to optimum extrudability and have a medium plasticity (Fig. 10). After firing (900°C under an oxidizing atmosphere), sampled clays behave in a way consistent with their mineralogical composition, forming a final product with a pearl-fired colour, homogeneous and without structural flaws.

The granulometrical, mineralogical and chemical studies as well as the clay's

Table 2. XRD mineralogy of Venilale samples estimated by XRD analysis (wt%)

Sample	Smectite	Illite	Kaolinite	Quartz	KF	NaF	Gypsum	Hematite	Calcite
Ven1	14	11	9	59	4	2	1	–	vest.
Ven2	20	28	24	21	4	1	1	2	vest.
Ven3	12	29	25	25	5	2	1	1	vest.

KF = K-feldspar, NaF = Na-feldspar.

Table 3. Chemical analysis (wt%) of Venilale samples determined by XRF

Sample	SiO_2	Al_2O_3	Fe_2O_3	MnO	CaO	MgO	Na_2O	K_2O	TiO_2	LOI
Ven1	78.74	8.53	3.62	0.10	0.37	1.19	0.35	1.34	0.45	4.94
Ven2	60.65	16.51	5.73	0.25	1.79	1.87	1.16	2.87	0.58	8.29
Ven3	61.69	16.80	5.85	0.26	1.82	1.86	1.18	2.92	0.60	6.61

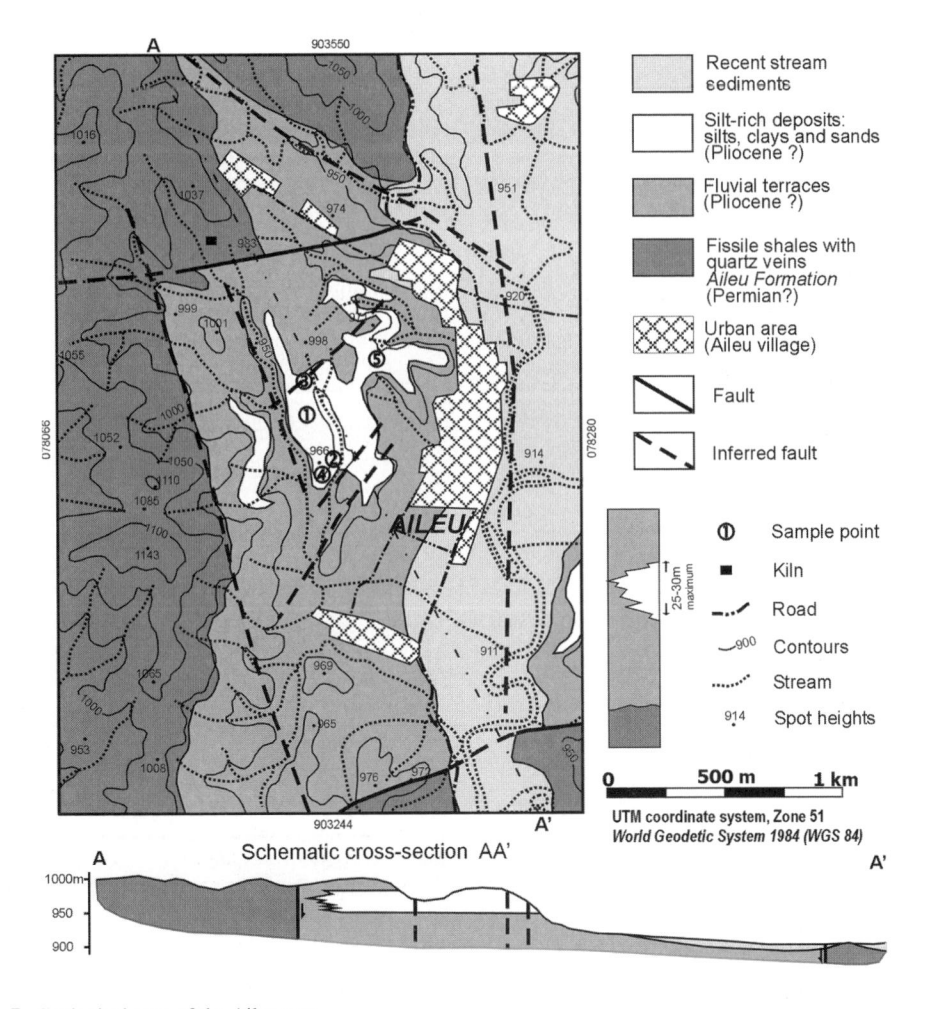

Fig. 7. Geological map of the Aileu area.

Fig. 8. Detail of the light coloured clay deposits.

workability, firing colour and final product characteristics, suggest that provisionally the raw material from the Aileu silt-rich deposits may be used for ceramic whiteware manufacture or for structural ceramic products if blended.

Other clay deposits in the Aileu area

Villages built on kaolin and a kaolin belt in Aileu region are mentioned in ESCAP (2002). In fact, the surrounding area of Aileu, especially towards the north of the city, contains large resources of *in situ* clay-rich deposits, resulting from the alteration of shale or alluvium deposits. Although vast quantities of clay-rich material occur, the *in situ* deposits are characterized by embedded quartz fragments, resulting from a complex interfingering of quartz veins frequently found in the Aileu Formation's shales. Moreover, the occurrences are scattered and too small for the development of an industry with modern firing technology. Clays in alluvium deposits have unfavourable exploitation conditions, due to their topographic locations near stream channels, exposing them to the risk of seasonal flooding. Nevertheless, they are exploited by

Table 4. *Mineralogy of Aileu samples estimated by XRD analysis (wt%)*

Sample	Ill/Verm	Illite	Kaolinite	Quartz	K-Feldspar
Ail1	2	47	24	26	1
Ail2	1	32	14	51	2
Ail3	4	50	21	24	1
Ail4	2	38	20	38	2
Ail5	1	31	16	50	2

Ill/Verm = mixed-layer illite/vermiculite.

Table 5. *Chemical analysis (wt%) of Aileu samples determined by XRF*

Sample	SiO_2	Al_2O_3	Fe_2O_3	MgO	K_2O	Na_2O	TiO_2	LOI
Ail1	67.51	19.41	1.90	0.79	3.72	<0.2	0.84	5.63
Ail2	72.85	16.12	1.60	0.69	3.43	<0.2	0.79	4.42
Ail3	64.89	21.27	1.72	0.95	4.18	<0.2	0.85	5.93
Ail4	68.78	18.73	1.40	0.80	3.62	<0.2	0.87	5.43
Ail5	71.67	16.90	1.47	0.76	3.39	<0.2	0.83	4.98

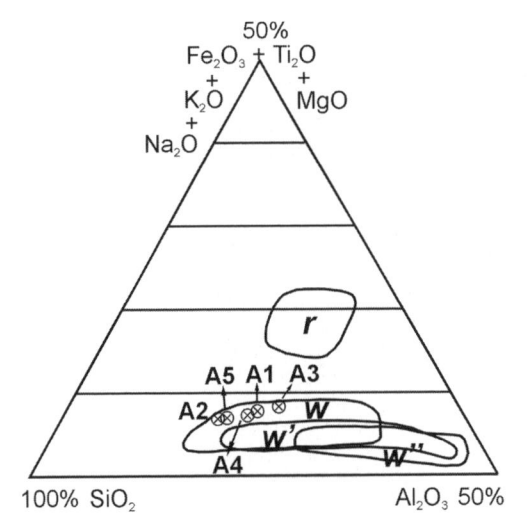

Fig. 9. Chemical composition of studied raw materials in comparison with red stoneware (r) and white stoneware (w = German; w′ = English; w″ = French) application fields; Aileu samples are A1 to A5.

local people for the artisanal making of sun-dried bricks. It can be concluded that the Aileu region is an important source of clay raw materials and warrants further investigation. Although the expected deposits are small and widely scattered, they may in time support an industry based on the blending of different types of clay to ensure adequate quality and sufficient reserves.

Ornamental stones

Ornamental stone exploitation has long been important in countries such as Italy, Spain and Portugal and, more recently, in developing countries such as Brazil, India and China. In general, ornamental stones are relatively high-value products. The establishment of an ornamental stone mining industry in Timor Leste could play an important role in its economic and social development. However, it is important that large-scale environmental impacts are avoided and mitigated against, in particular degradation of the landscape, as this may impact on the development of a tourist industry.

Favourable criteria for ornamental stone resources include high lithological thickness, lithological uniformity, and moderate fracturing grade. In Timor Leste, there are several geological units, such as the Aileu, Maubisse, Borolalo Limestone and Cablac Limestone formations (Table 1) that meet (at least partially) those broad criteria and thus are potentially suitable for the ornamental stone extractive industry. The Beuah marbles (Aileu Formation) and the Beheda limestones (Maubisse Formation) were selected for follow-up investigations.

Beuah marbles

Owing to the existence of an old marble quarry near Beuah village in Manatuto district, east of Dili, during the last years of Indonesian

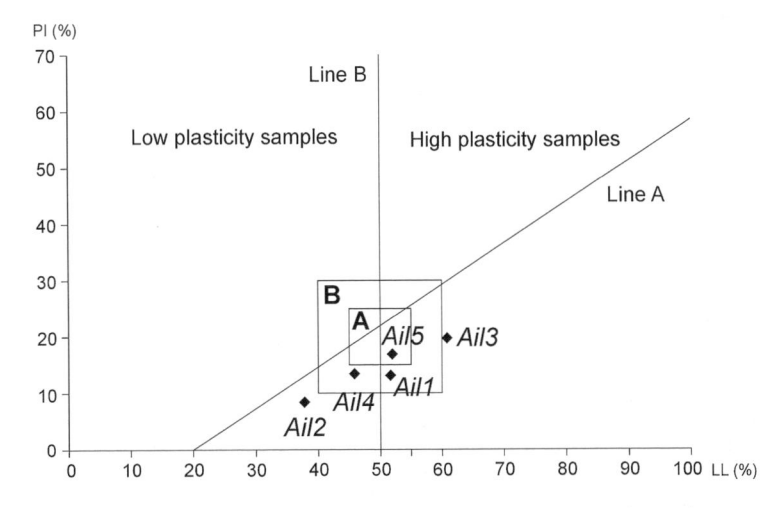

Fig. 10. Casagrande diagram with domains (Gippini 1969) relative to extrudability of ceramic raw materials: A, optimum; B, satisfactory; empirical boundary 'Line A' separates inorganic clays (above the line) from inorganic silts and organic soils; 'Line B' separates raw materials with low plasticity from those with high plasticity; PI, Atterberg plasticity index; LL, Liquid limit.

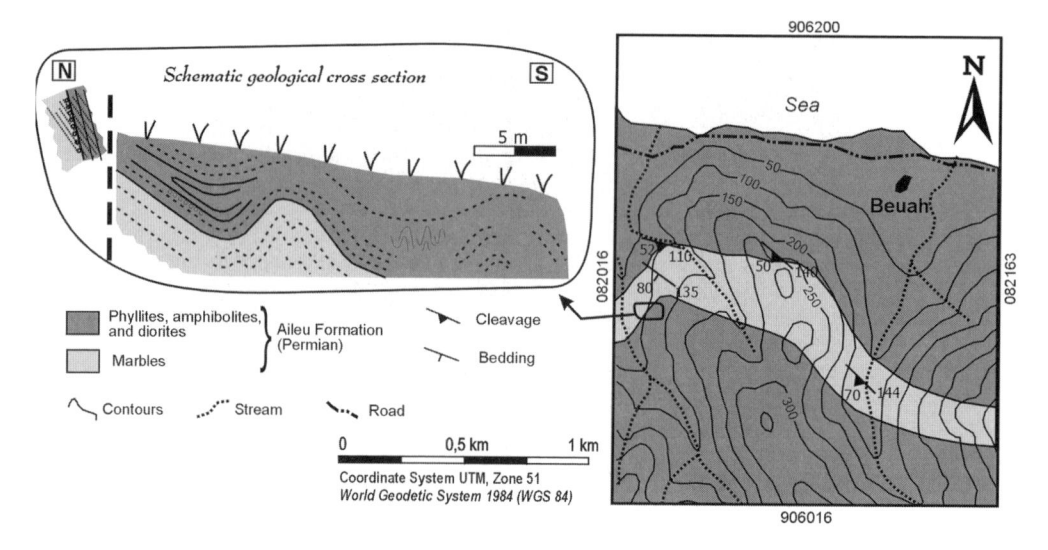

Fig. 11. Geological map of Beuah area.

occupation, marbles are referred to as one of the potential economically exploitable mineral resources of Timor Leste (ESCAP 2002; WFB 2004). They represent a minor lithological component of the northeastern multiply-deformed Aileu Metamorphic Complex sequence, occurring in layers usually less than 100 m thick (Berry & Grady 1981).

A preliminary geological study of the area where exploitation took place (Fig. 11) shows that marbles crop out along a belt about 200 m wide, which corresponds to a deposit with an apparent thickness of approximately 50 m. However, the real deposit thickness is difficult to estimate due to the polyphase deformation: at least two phases of folding can be recognized (Fig. 12). Marble outcrops in the study area seem to represent the inverted flank of a north-verging major fold, although insufficient data did not allow a conclusive characterization of the structure.

The marbles cropping out in the study area show significant texture and colour variations. The exploited lithotypes are typically light coloured (white, pale pink or yellow) and fine grained with few grey streaks of dark schistose material (Fig. 13). However, such lithotypes occur mainly within the central area. Laterally, there is a variation to a greyish dark coloured facies with numerous streaks of dark schistose material crisscrossing the marble, which is fine to coarse grained. The presence of interbedded fine layers (about 1 to 3 cm thick) of calcsilicate rocks (Fig. 14), as well as coarse, weathered, red garnets reduce the quality of the marble.

Marble quarrying in the area was based on the exploitation of large boulders from a colluvium deposit. Their provenance is a steep area to the south, where the marbles crop out at 300 m in

Fig. 12. Structural aspects of Beuah marbles. Two phases of deformation can be recognized in the folded structure.

Fig. 13. Remaining marble blocks from Indonesian exploitation and detail of the most commonly used lithotype.

elevation. Owing to their inaccessibility, the extractive activity never focused on these *in situ* outcrops. Attempts were made to exploit the deposits situated in the valleys east and west of the above mentioned central area, but the exploitation was abandoned as a consequence of the lateral variations to dark coloured facies and structural complexity. All things considered, it is likely that these deposits may prove unfavourable for exploitation. Nevertheless, as evidenced by Berry and Grady (1981), there are much larger marble occurrences to the south of this area that warrant assessment.

Beheda limestones

In the surrounding area of Beheda village (10 km west of Manatuto), widespread outcrops of the Maubisse Formation reveal the economical potential for ornamental stone extraction. Moreover, proximity to the main road between Dili and Baucau and good accessibility conditions, makes this area very attractive for the extractive industry.

The unmetamorphosed Permian Maubisse Formation is a sequence of carbonate-rich and volcanic rocks, where fauna-rich massive limestones prevail at the base of the unit (Audley-Charles 1968). Its geological origin and settlement has been the object of controversy, including its association with the Aileu Formation (Barber & Audley-Charles 1976; Carter *et al.* 1976; Barber *et al.* 1977; Charlton *et al.* 2002). In the Beheda area, the Maubisse Formation is represented by crinoid-rich limestones that crop out as a *klippe* overlying parautochthonous Mesozoic terrains of the Wai Luli Formation, which mainly consists of shales, marls and sandstones (Fig. 15).

The mapped Maubisse Formation's limestones occur in an elongated area oriented NE–SW, 500 m in width and 1500 m in length. The outcrops are large, occurring in blocks as shown in Figure 16, the largest of which are approximately 50 m^3 with very few fractures. Owing to the

Fig. 14. Greyish marble outcrop with thin calcsilicate layers.

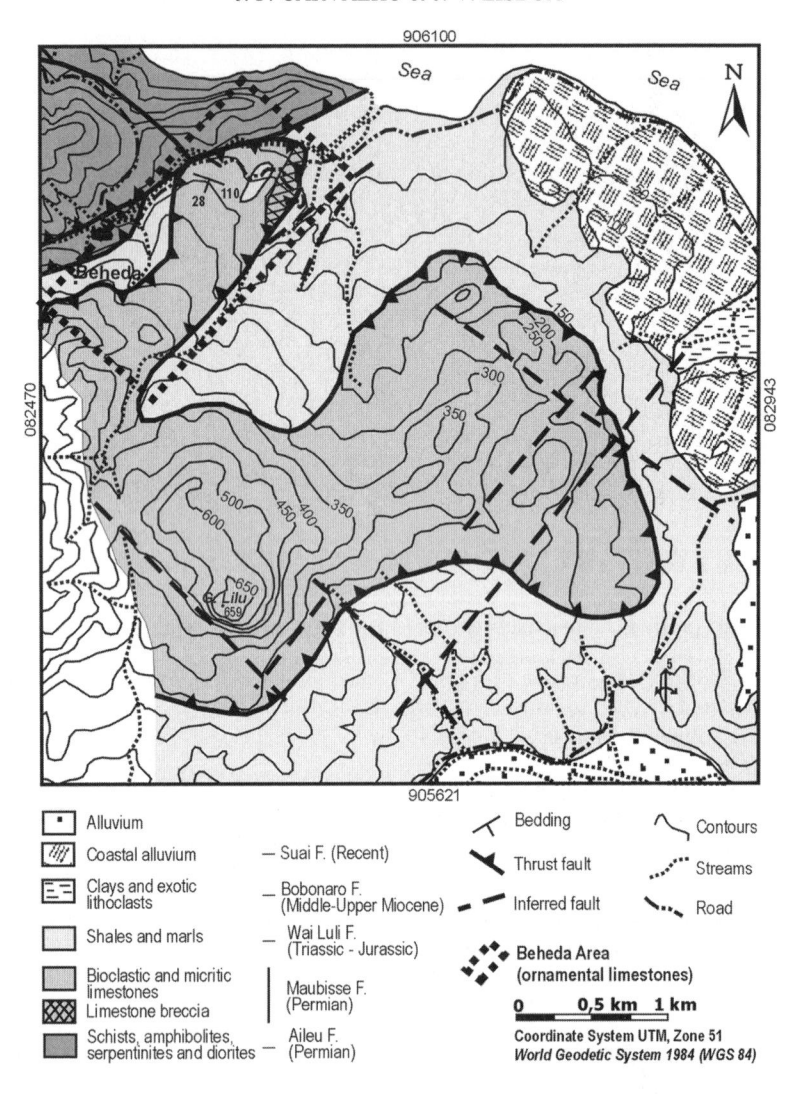

Fig. 15. Geological map of Beheda area.

massive structure of the limestone, it becomes difficult to estimate its thickness, which is probably between 3 and 6 m. Beds dip gently from 10 to 25° SSW. Two main facies coexist:

- A prevailing facies of white to very light coloured biocalciclastic limestones (biocalcarenites), medium to coarse grained with sparry cement. White coloured crinoids are the most abundant bioclastic elements. Their average size is approximately 1 cm, but they reach up to 5 cm (Fig. 17).
- Micritic (calcilutites) to microsparritic limestones, light to dark grey in colour, with disseminated fine bioclasts and calciclasts. Large bioclasts can also occur, as well as abundant translucide calcite veins (<2 mm thick). The biocalciclastic limestones abruptly and conformably overlay this facies.

Both facies are potentially suitable for use as ornamental stone, in particular the crinoid-rich limestones. Evaluation of the resources in the study area considered geological structure and the raw material's final use. Considering the inherent restrictions to the yield of a typical ornamental stone exploitation, a probable resource of 300 000 m^3 is inferred by simple geometric

Fig. 16. Huge limestone outcrop in Beheda area.

calculations. Although this value does not support long-term modern industrial exploitation, the country's economic and social situation is rather favourable for small-scale operations, since they do not require sophisticated machinery. These limestones can also be used as raw materials for lime or cement production. In these circumstances the available resources are much larger. In a sustainable approach, aggregates should always be considered as a subproduct of any ornamental stone exploitation.

The area warrants further detailed investigation and future work should also consider a thematic geological mapping approach for the Maubisse Formation's extensive outcrops, south of the study area. These could provide large volumes of raw material suitable for the production of lime, cement, crushed-stone aggregates or, possibly, ornamental stone.

Aggregates

Aggregates are used as construction raw materials, with or without a binder. Road base or road surfacing material and macadam are the major envisaged uses without a binder in Timor Leste. Aggregates for cement and bituminous concrete in road construction and repair, and in

Fig. 17. Detail of crinoid-rich limestone from Maubisse Formation in Beheda area.

residential and public building construction are the major likely uses with a binder. Other important uses include cement and lime manufacture and soil correctives in agriculture. Despite their low bulk value, aggregates are among the most important commodities, being major contributors to the economic and social well-being of nations and are key to sustainable development (Wellmer & Becker-Platen 2002; Cárdenas & Chaparro 2004; Langer & Tucker 2004).

Aggregates are a very important commodity in Timor Leste, playing an increasingly important role in the immediate future development of the country. Extensive rock deposits, with the potential for aggregate resources, are found all over the island. Sources for crushed stone encompass rocks of diverse lithostratigraphical units. The most important ones are the Aileu Complex (igneous and metamorphic rocks), the Cablac Limestone and the Maubisse Formation (limestones and dolostones). The main source of superficial sand and gravel aggregates in Timor Leste is stream channel deposits within the Suai Formation.

Crushed stone

Intermediate to ultrabasic intrusive hard rocks are relatively common in the northern part of the Aileu Complex. They are an excellent source of crushed stone aggregates, which could be an important resource of high-quality raw material suitable for road metal (road-base aggregates), as well as for armour stone for harbour and pipeline protection.

In Dili's coastal region, and to the west of the town, extensive outcrops of these rocks are common with good accessibility. Examples of such areas include Cristo Rei and Comoro River (in the surroundings of Dili) and Maubara and Tibar-Liquiça (some kilometers to the west of Dili). In the latter area an important but now inactive quarry once exploited a competent dark-green igneous basic rock that crops out for more than 400 m along the coastal road. Although strongly fractured, these fine to medium grained rocks are not significantly weathered, making them suitable for crushed stone production. Resources are extensive, but an assessment of the material's physical and chemical properties has yet to be undertaken. The area is important due to its proximity to the country's capital, the major consumer and trade centre.

In terms of accessibility and proximity to other main villages, the most relevant limestone and dolostone occurrences are the Maubisse Formation outcrops in Maubisse district and the Cablac Limestone Formation outcrops in Viqueque district. As mentioned elsewhere, the Maubisse Formation lower sequence is characterized by prevailing massive limestones, which have a high potential for ornamental purposes, as is the case for the outcrops near Beheda village. Thus, aggregate production from Maubisse Formation limestones must be regarded always as a byproduct of the eventual exploitation for ornamental stone, contributing to a prudent use of this natural resource.

The Cablac Limestone Formation has been dated as Lower Miocene (Audley-Charles 1968) and is mostly composed of thick bedded, hard, massive limestones that commonly occur in mountain tops as steep outcrops (e.g. Baucau, Viqueque and Covalima districts). Although the general study carried out on these limestones suggested a high economic potential for ornamental purposes, they are also suitable for crushed stone aggregates. These rocks have high potential for using as concrete aggregate, as they are hard and clean. In addition, their use in cement and lime industries should not be disregarded. In the Ossu village area (Viqueque district), a conglomeratic facies of this formation is exploited in a small quarry about 2.5 km north of Ossu, for road building purposes (Fig. 18).

Other predominantly carbonate-rich units, but with more restricted uses are the Baucau Limestone and the Viqueque Formation. The Baucau Limestone Formation (Baucau and Lautem districts) is referred to as a fossilized coral reef and associated carbonate-rich deposits dated as Pleistocene to Holocene in age (Audley-Charles 1968). These are exploited in small artisanal works and crushed to obtain a sand grain size aggregate, which is used locally as filler in the production of cement made bricks. The purity of these limestones makes them an optimal raw material for lime production. The facies composed of carbonate-rich rocks in the Viqueque Formation (mainly in the Viqueque district), dated as Upper Miocene to Pliocene (Audley-Charles 1968) is used for the same purpose. These rocks, particularly a white marl distinctive facies that crops out extensively in the Viqueque's surrounding area, might possibly have another potential use as a cement raw material. Nevertheless a detailed assessment of their chemical and physical properties needs to be undertaken.

Sand and gravel

Owing to the climatic, topographic and tectonic conditions of the Timor island, alluvium deposits are thick and vast. The torrential weather regime produces wide stream channels with thick

Fig. 18. Small quarry in Ossu area, where a conglomeratic facies of Cablac Limestone Formation is exploited.

unconsolidated deposits of mostly unsorted material, even in upstream areas. Their nature is very diverse depending on the source lithologies.

The most important river courses in the north coast occur within the Aileu Complex. The deposits are characterized by the presence of clasts, ranging in size from pebbles to metric boulders of schistose rocks, gabbros and quartz within a clay matrix. In the east of the country, where argillaceous and carbonate-rich rocks of Triassic to recent aged formations predominate (Fig. 2 and Table 1), sand and gravel deposits are not as thick, and are mostly associated with talus deposits and storm waves accumulations of coral reef rocks in beaches. The main sand and gravel exploitations are located near the coast road between Liquiça and Manatuto. The large valley of the Comoro River, located in the Dili area, is the main aggregate source (Fig. 19). The alluvium deposit is a poorly sorted material composed of greyish schistose and quartz clasts of sand and gravel, with larger clasts, mainly of igneous mafic rocks. The material is embedded in a silt and clay matrix. The presence of elongate and/or flaky particle shape of foliated

Fig. 19. Sand and gravel exploitation in Comoro River, Dili. Note the numerous small pits dug by local miners.

rock and of mafic minerals, iron oxides and clay reduces the quality of this raw material. It is, however, exploited by a large number of artisanal miners, who sort out sediments of different size grades by hand screen. Despite the poor aggregate quality of this material for use in the building industry, its consumption is primarily as sand for concrete and cement bricks. Gravel is used in road building and repair.

The sand and gravel resources of Timor Leste are extensive. In addition to the Comoro River, other river valleys have large and thick alluvium deposits, such as the Laclo Norte River near Manatuto, the Laleia River between Laleia and Vemasse, the Lois River between Balibó and Maliana, and the Cua River near Viqueque. The artisanal procedure for exploiting these deposits is not expected to cause relevant modifications to the streamline dynamics. However, considering the country's expected development and the consequent technological enhancements in industry, the exploitation process should be reformulated in order to avoid pervasive environmental disturbances.

Discussion and conclusions

Timor Leste is the poorest country of the Asia-Pacific region due to the devastating and violent incidents that occurred in 1999. To promote the economic development of the country, a strategy of infrastructure reconstruction has been outlined, for effective impact on poverty reduction. However, it is impossible to plan for major infrastructures such as roads and bridges or houses and public buildings without considering the mineral resources and raw materials required to construct them.

This work has focused on the identification of mineral resources suitable for use as construction raw materials, namely aggregates, common clays and building stones. In Timor Leste, these resources are crucial, both directly and indirectly for infrastructure reconstruction, maintenance and future developments, thus contributing to and determining its sustainable development.

In order to support an emergent extractive industry, a few target areas were identified, although many favourable areas for construction raw material certainly exist. In Venilale, a village with good access to Baucau, the second largest city in Timor Leste, large resources of common clays occur over a large area. Although the results of preliminary sampling and characterization were not favourable due to the materials' high plasticity, these clayey materials could support brick- and tile-making industries, if properly blended with a non-plastic raw material. A regional sampling programme could be successful in identifying areas with more adequate ceramic properties, although the formation has a characteristically high smectite content (Audley-Charles 1965, 1968; Harris et al. 1998).

The ceramic industry is perhaps the industry that could most easily be developed due to the traditional skills that still remain as a result of the old ceramic plants in Dili. Besides providing raw materials for the building industry, it could provide employment opportunities, training of skilled human resources, and encouragement of an internal trade economy, thus contributing to the overall economic development of the country. In the longer term, bricks and roof tiles could be exported to neighbouring countries, particularly to Indonesia.

In the surrounding area of Manatuto, located half way between Dili and Baucau, large and massive limestone outcrops may be used in the ornamental stone industry. They may represent an important commodity for Timor Leste not only for internal consumption, but also for exporting purposes if extensive deposits are to be found. Although the immediate requirement of Timor Leste is not the embellishment of its buildings, ornamental stones are an important commodity as they can perform a structural role in the building industry, replacing other materials that otherwise must be imported. With regard to eventual exportation of this commodity, the benefits for the economy are obvious. Nevertheless, this option must be supported by an adequate market study, particularly focusing on potential markets in neighbouring countries.

A major issue that must be addressed regarding the establishment of an ornamental stone's mining industry in Timor Leste is the lack of technical means and human skill resources required to exploit, saw and polish large blocks of stone. Intervention by foreign companies will be crucial for the development of this industry and for training local counterparts. This area is also the stepping-stone to areas where the Maubisse Formation crops out (east of Luro, Tapau and Balibo areas and south of Vemasse), which can be considered potential targets for ornamental stone exploration.

Environmentally, the extraction of ornamental stones is one of the most problematic in relation to landscape degradation. This is due to the often common, large accumulations of residues or waste. To ensure the prudent use of this resource, the byproducts of ornamental stone production should, as far as economically possible, be used for other purposes such as crushed stone aggregates.

The importance of aggregate resources across the world has long been recognized, to support both developing and developed economies. Timor Leste has extremely large aggregate resources, in particular river sand and gravel deposits. Other important aggregate resources include crushed stone aggregates from hard, basic intrusive rocks near Dili, the main consumer centre. Nevertheless, owing to the country's geological diversity, almost all major consumer centres are within close proximity to hard rock resources suitable for crushed stone aggregates and/or in the vicinity of sand and gravel alluvium-rich stream channels. Irrespective of the resources type, the key concern is ensuring that these resources are exploited in a sustainable manner, economically, environmentally and to the maximum social benefit.

Even in developed countries, geological information and interpretation has been used to a limited extent in exploration and exploitation of construction raw materials as a result of the resources' near surface relationship and worldwide abundance. This has indirectly and negatively affected the integration of the extractive industry into land use planning in many developed countries, particularly in Europe. The concerns of accessing mineral resources by the extractive industry can be avoided if adequate land use planning methodologies are adopted in Timor Leste regarding not only agriculture and tourism as sustainable development factors, but also mining.

The emergence of a large number of unsustainable small artisanal mining activities in Timor Leste, proximal to the larger consumer centres, is a response to the country's infrastructure reconstruction raw material needs. Most of them are exploiting sand and gravel in large stream channels without any environmental considerations or work security issues. Although aggregates, common clays and ornamental stone are essential raw materials for the reconstruction of the country's infrastructure, their commonly adopted exploitation methodologies can, if not planned and managed in a sustainable manner, result in large visual impacts on the landscape and, consequently, a negative opinion of the mining industry.

A short-term policy adapted to the country's raw material requirements and to the current lack of human skill resources and technological processing capabilities is a major challenge that must be addressed in order to create mining investment opportunities within a framework of sustainable development. Perhaps one of the ways forward is to support present-day artisanal mining activity through the promotion, development and regulation of small-scale mining industries. In the short to medium term, this would satisfy market demands, contribute to unemployment reduction (as it could be carefully planned to employ large numbers of Timorese personnel) and reduce the country's dependency on imported goods.

These small-scale industries should not be confused with the small-scale artisanal mines that are typified worldwide by an absence or low degree of mechanization, precarious safety standards, poorly skilled and paid personnel, and illegality due to the inexistence of mining rights. Instead, the well-known issues associated with these environmentally and socially problematic mines should be considered as business practices to avoid. The Berlin II Guidelines for Mining and Sustainable Development (UNEP 2002) is a starting point for the implementation and control of a sustainable development indicator system for the extractive industry. However, care must be taken that such a system should take into account the amendment and customization of indicators to the specificities of Timor Leste.

In this context, Timor Leste's government can play an important role in ensuring mineral resources are acknowledged as a key determinant of the country's sustainable development, but also key to the promotion of sustainable development principles in the mining industry. Institutional capacity in the fields of Earth Sciences and mineral resources management is the key to the discovery and promotion of new deposits, to the integration of the mining industry into land use planning policies and to the promotion of extractive activities with minimal environmental impact.

Our thanks are given to A. Oliveira, C. Carvalho and J. Grade, from INETI's Laboratory in Oporto, for their analytical and technical support. We also thank D.P.S. Oliveira and F. McEvoy for an early review of the manuscript.

References

AUDLEY-CHARLES, M. G. 1965. A Miocene gravity slide deposit from Eastern Timor. *Geological Magazine*, **102**(3), 267–279.

AUDLEY-CHARLES, M. G. 1968. The geology of Portuguese Timor. *Memoirs of the Geological Society of London*, **4**, 76.

BACHRI, S. & SITUMORANG, R. L. 1994. *Geological Map of the Dili Quadrangle 2406–2407*, scale 1:250 000. Geological Research and Development Centre, Bandung.

BARBER, A. J. & AUDLEY-CHARLES, M. G. 1976. The significance of the metamorphic rocks of Timor in the development of the Banda Arc, eastern Indonesia. *Tectonophisics*, **30**, 119–128.

BARBER, A. J., AUDLEY-CHARLES, M. G. & Carter, D. J. 1977. Thrust tectonics in Timor. *Journal of the Geological Society of Australia*, **24**, 51–62.

BERRY, R. F. & GRADY, A. E. 1981. Deformation and metamorphism of the Aileu Formation, north coast, East Timor and its tectonic significance. *Journal of Structural Geology*, **3**(2), 143–167.

BRUNNSCHWEILER, R. O. 1977. Notes on the geology of eastern Timor. *BMR Bulletin of Australian Geology and Geophysics*, **192**, 9–18.

CÁRDENAS, M. & CHAPARRO, E. 2004. Industria minera de los materiales de construcción. Su sustentabilidad en América del Sur. División de Recursos Naturales e Infraestructura de la Comisión Económica para América Latina y el Caribe (CEPAL). Naciones Unidas, Santiago de Chile. Serie Recursos Naturales e Infraestructura, **76**, 84.

CARTER, D. J., AUDLEY-CHARLES, M. G. & BARBER, A. J. 1976. Stratigraphical analysis of islandarc-continental margin collision in eastern Indonesia. *Journal of the Geological Society, London*, **132**, 179–198.

CHARLTON, T. R. 2001. Permo-Triassic evolution of Gondwanan eastern Indonesia, and the final Mesozoic separation of SE Asia from Australia. *Journal of Asian Earth Sciences*, **19**, 595–617.

CHARLTON, T. R. 2002. The structural setting and tectonic significance of the Lolotoi, Laclubar and Aileu metamorphic massifs, East Timor. *Journal of Asian Earth Sciences*, **20**, 851–865.

CHARLTON, T. R., BARBER, A. J. & BARKHAM, S. T. 1991. The structural evolution of the Timor collision complex, eastern Indonesia. *Journal of Structural Geology*, **13**(5), 489–500.

CHARLTON, T. R., BARBER, A. J. ET AL. 2002. The Permian of Timor: stratigraphy, palaeontology and palaeogeography. *Journal of Asian Earth Sciences*, **20**, 719–774.

ESCAP 2002. *Exploring Timor-Leste – Mineral and Hydrocarbon Potential. Economic and Social Commission for Asia and the Pacific*, ST/ESCAP/2243, 87 p http://www.undp.east-timor.org/documentsreports/environment/index.html

FABBRI, B. & FIORI, C. 1985. Clays and complementary raw materials for stoneware tiles. *Mineralogica et Petrographica Acta*, **29 A**, 535–545.

GAGEONNET, R. & LEMOINE, M. 1957. Note préliminaire sur la géologie du Timor Portugais. *Garcia de Orta* **5**(1), 153–163.

GAGEONNET, R. & LEMOINE, M. 1958. Contribution à la connaissance de la géologie de la province portugaise de Timor. Junta de Investigações do Ultramar. *Estudos, Ensaios e Documentos*, **48**, 138.

GIPPINI, E. 1969. Contribution à l'étude des propriétés de moulage des argiles et des mélanges optimaux de matières premières. *L'Industrie Céramique*, **619**, 423–435.

GRUNAU, H. R. 1953. Geologie von Portugiesisch Ost-Timor. Eine kurze Übersicht. *Eclogal Geologieae Helvetiae*, **46**, 29–37.

GRUNAU, H. R. 1956. Zur geologie von Portugiesisch Ost-Timor, Mitteilungen Naturf. *Gesellschaft Bern*, **13**, XI–XVIII.

GRUNAU, H. R. 1957. Geologia da parte oriental do Timor Português, Nota abreviada. *Garcia de Orta*, **5**(4), 727–737.

HARRIS, R. A. 1991. Temporal distribution of strain in the active Banda orogen: a reconciliation of rival hypotheses. *Journal of Southeast Asian Earth Sciences, Special Issue: Orogenesis in Action – Tectonics and Processes at the West Equatorial Pacific Margin*, **6**(3–4), 373–386.

HARRIS, R. A., SAWYER, R. K. & AUDLEY-CHARLES, M. G. 1998. Collision melange development: geologic associations of active melange-formation processes with exhumed melange facies in the western Banda orogen, Indonesia. *Tectonics*, **17**, 458–479.

HARRIS, R. & LONG, T. 2000. The Timor ophiolite, Indonesia: Model or myth? *Geological Society of America Special Paper*, **349**, 321–330.

HIRSCHI, H. 1907. Zur Geologie und Geographie von Portugiesich Timor. *Neus Jahrbuch für Mineralogie, Geologie und Paläontologie*, **24**, 460–473.

LANGER, W. H. & TUCKER, M. L. 2004. *Specification Aggregate Quarry Expansion – A Case Study Demonstrating Sustainable Management of Natural Aggregate Resources*. U.S. Geological Survey Open-File Report 03–121.

LEME, J. C. A. 1963. The eastern end geology of Portuguese Timor (a preliminary report). *Garcia de Orta (Lisbon)*, **11**(2), 379–388.

LEME, J. C. A. 1968. Breve ensaio sobre a Geologia da Província de Timor. Curso de Geologia do Ultramar, 1, Junta de Investigações do Ultramar, Lisboa, 106–161.

LEME, J. C. A. & Coelho, A. V. P. 1962. Geologia do encrave de Ocússi (Província de Timor). *Estudos Agronómicos (Lisboa)*, **3**(3), 119–132.

NDP 2002. *East Timor National Development Plan*. Prime Minister and Cabinet, Timor-Leste Government, http://www.pm.gov.tp/ndp.htm

PARTOYO, E., HERMANTO, B. & BACHRI, S. 1995. *Geological Map of the Baucau Quadrangle 2057, scale 1:250000*. Geological Research and Development Centre, Bandung.

PEREZ, B. C. 2001. *As rochas e os minerais industriais como elemento de desenvolvimento sustentável*. CETEM/MCT, Rio de Janeiro; Série Rochas e Minerais Industriais, **3**, 37.

PRASETYADI, C. & HARRIS, R. A. 1996. Hinterland structure of the active Banda arc-continent collision, Indonesia: constraints from the Aileu complex of East Timor. *Proceedings of the 25th Annual Convention of the Indonesian Association of Geologists*, 144–173.

REED, T. A., SMET, M. E. M., HARAHAP, B. H. & SJAPAWI, A. 1996. Structural and depositional history of East Timor. *Proceedings Indonesian Petroleum Association*, 25th Annual Convention, Jakarta, **25**(1), 297–312.

ROSIDI, H. M. D., SUWITODIRDJO, K. & TJOKROSAPOETRO, S. 1979. *Geologic Map of the Kupang-Atambua Quadrangles, Timor, scale 1:250,000*.

Geological Research and Development Centre, Bandung, Indonesia.

SANDLUND, O. T., BRYEESON, I., CARVALHO, D., RIO, N., SILVA, J. & SILVA, M. I. 2001. *Assessing Environmental Needs and Priorities in East Timor: Issues and Priorities.* Report UNOPS/NINA-NIKU, Trondheim, 45.

SHEPARD, F. P. 1954. Nomenclature based on sand-silt–clay ratios. *Journal of Sedimentary Petrology* **24**, 151–158.

UNITED NATIONS DEVELOPMENT PROGRAMME (UNDP) 2000. *Building Blocks for a Nation. A Common Country Assessment for East Timor*, http://www.undp.east-timor.org/documentsreports/index.html or www.undp.org/rbap/Country_Office/CCA/Cca-EastTimor2000.pdf

UNITED NATIONS DEVELOPMENT PROGRAMME (UNDP) 2002. *East Timor Human Development Report 2002. United Nations Development Programme*, http://www.undp.east-timor.org/documentsreports/index.html

UNITED NATIONS DEVELOPMENT PROGRAMME (UNDP) 2003. *Timor-Leste: Programme Package Document for Sustainable Human Development*, http://www.undp.east-timor.org/documentsreports/ppd.html

UNITED NATIONS ENVIRONMENT PROGRAMME (UNEP) 2002. *Berlin II: Guidelines for Mining and Sustainable Development*, http://www.mineralresourcesforum.org/workshops/Berlin/

WANNER, J. 1956. Zur stratigraphie von Portuguesisch Timor. Zeitschrift der Deutschen. *Geologischen Gesellschaft*, **108**, 109–140.

WELLMER, F. W. & BECKER-PLATEN, J. D. 2002. Sustainable development and the exploitation of mineral and energy resources: a review. *International Journal of Earth Sciences (Geologische Rundschau)*, **91**, 723–745.

WFB 2004. *The World Factbook 2004. Central Intelligence Agency*, http://cia.gov/cia/publications/factbook/geos/tt.html

WINKLER, H. G. F. 1954. Bedeutung der korngrossenverteilung und des mineral-bestandes von tonen fiir die herstellung grobkeramischer erzeugnisse. *Berichte der Deutschen Keramischen Gesellschaft*, **31**(10), 337–343.

WORLD BANK 2004. The *Democratic Republic of Timor Leste – Public Expenditure Review*. Report 27886-TP, 103.

Capacity building of developing country public sector institutions in the natural resource sector

M. H. STEPHENSON[1] & I. E. PENN[2]

[1]*British Geological Survey, Keyworth, Nottingham NG12 5GG, UK (e-mail: mhste@bgs.ac.uk)*
[2]*12 Peacock Close, Ruddington NG11 6JF, UK*

Abstract: The natural resources of developing countries, particularly in a post-conflict situation, are the key to creating wealth, getting people back to work, and to improving security. However, public sector institutions like geological surveys, and government departments such as mines, energy and water ministries often need help in their vision to promote and sustainably develop their natural capital, as well as to protect the lives and livelihoods of people affected by development. Some have few physical resources, and a poorly trained and motivated workforce; others may be housed in buildings that have borne the brunt of prolonged fighting and a long period of neglect. In many developing countries, such institutions have a rather inward-facing colonial-style civil service culture that lacks the ability to liaise and engage with modern multinational investors. Unfortunately, donor organizations that seek to build the capacities of institutions do not build sufficient 'project ownership' and fail to incorporate into their plans the culture of the organization, or fail to integrate parts of multidisciplinary projects. Development projects supported are often perceived to reflect donor agendas rather than the needs of the recipient institution. Using experience in a number of developing country and post-conflict contexts, a methodology to plan and integrate capacity building has been developed, to help employees and management, and donor organizations, deal with these difficulties. Through training tuned to business need, institutions will develop appropriate IT and communication skills, while at the same time developing corporate understanding of the private sector, which is needed to interact successfully with it. Stakeholder analysis gauges the organization's strengths and weaknesses and ensures coordination of aid, which takes account of the local social, political and business context. The methodology will also establish a system allowing regular cyclical business/training review, so that the institutions can adapt to further change.

Building the institutions of an effective, democratic, modern state is the most important condition for sustainable development (DFID 2001).

Poorly functioning public sector institutions and weak governance are major constraints to growth and equitable development in many developing countries (World Bank Public Sector Group 2000).

Organizations such as the UK Department for International Development (DFID) and the World Bank are committed to building institutions that are vital to the economic success and security of post-conflict countries. Such institutions include geological surveys and energy/mines ministries and departments, which deal with acquisition of environmental data, regulation of natural resource development, and policy advice to government. Successful institutions of this type allow the indirect trickle down of wealth to poor people through increased investment, and direct employment and service provision. They also connect the citizens in a democratic society with the decisions that allow the development of natural resources and the protection of environmental and human interests. However, many institutions need to do more to promote sustainable resource development, while at the same time adequately protecting the human and natural environment (World Bank Public Sector Group 2000; DFID 2001). The donor community has also recognized that capacity-building projects, which try to help, have sometimes supported disparate activities that are not consistent with integrated development (Carlsson *et al.* 1998; World Bank Public Sector Group 2000). Instead of working together, donor organizations often compete for national or organizational prestige or with an eye for downstream contracts (DFID 2001).

To improve capacity building, a holistic approach is needed that understands the business

From: MARKER, B. R., PETTERSON, M. G., MCEVOY, F. & STEPHENSON, M. H. (eds) 2005. *Sustainable Minerals Operations in the Developing World.* Geological Society, London, Special Publications, **250**, 185–194.
0305-8719/05/$15.00 © The Geological Society of London 2005.

of the organization and its relationship with stakeholders, as well as the motives and methods of aid donors. The aim of such an approach would be to maximize the benefit of aid, make the organization better at its job, and make it able to change with its changing business.

The British Geological Survey (BGS) has a long history of capacity building in geological surveys and energy/mines ministries and departments in the developing world and post-conflict situations (Reedman et al. 1996; O'Connor 1996). In recent work, the relationships between such public sector bodies, their stakeholders, and large aid donors, have been studied in detail, particularly in countries of central Asia, southeast Asia and the Pacific (Stephenson et al. 2002; Stephenson 2003; Stephenson & Penn 2003). The results of this and other unpublished research, and the foundation of holistic capacity building are presented here. First, the methods of investigating relationships with internal and external stakeholders, and some general findings, are described; this is followed by analysis of the methods of donors, and how they affect capacity building.

Relationships with external stakeholders

Geological surveys and energy/mines ministries serve a wide range of external stakeholders including government departments at national, regional, local and international scales, NGOs, environmental, industrial and regulatory organizations, natural resource-based private industry and hazard mitigation bodies (Reedman et al. 1996; Findlay 1997). Within government, the organization's key role is in advising other departments and developing policy on economic, social and environmental issues. The key role in relation to private sector clients including mining, hydrocarbon, civil engineering, construction and privatized utilities companies (Reedman et al. 1996; Findlay 1997) is to develop information packages that inform on resource priorities, promote resource wealth, and attract sustainable inward investment. The organization may also play a crucial role in negotiations between government, industry and landowners with the objective of maximizing economic and social benefits to as wide a stakeholder community as possible (e.g. Power & Hagen 1996). Mining and hydrocarbon activity is undertaken within licence blocks that are granted, processed and monitored by the organization. It must therefore deal with the numerous applications, renewals, adjustments and queries that occur in the day-to-day operation of the commercial sector. Specialist personnel in the organization may advise on policy for natural resource development and large-scale construction, or may be responsible for maintaining occupational health/safety and environmental standards on mine, oil and gas installations, and construction sites.

Most organizations of this type also play a crucial role in investigating and assessing natural hazards (Reedman et al. 1996) and developing hazard mitigation plans. Such plans have to be distributed among poor people, who often have little education and understanding of geological hazards, and who are at risk. The information may be in the form of booklets, seminars, worksheets and talks and contain advice on personal and family protection, building and agriculture, and escape in hazardous conditions. Other direct contact with poor rural people involves provision of technical training and safety advice in small-scale artisanal mining. In many areas of the South Pacific and Africa, this is an important subsistence-level activity for large numbers of rural people.

Methodology of investigation

The purpose of the investigation is to understand the relationship between the organization and its external stakeholders, and its strengths and weaknesses as perceived by external stakeholders, and internally by senior managers. Data are collected internally by researching relevant literature including business plans, corporate organograms, training plans, mission statements and job descriptions, and by interview with senior managers. External data are gathered by discussing the organization's performance and structure with a representative sample of its clients. Interviews within the organization are undertaken to determine whether senior managers understand the business needs of their clients, while interviews with clients are designed to determine how well the organization serves them. The interviews focus on the following areas:

(1) Current state of the mining/energy sector and likely future developments;
(2) Functions of the organization;
(3) Organization's relationship with its clients;
(4) Types and levels of products/services delivered to clients;
(5) Level of internal and external awareness of the organization's corporate plan and objectives; and
(6) Types of training required.

Information from this study builds a comprehensive picture of the organization's strengths and weaknesses, and informs the training needs analysis and internal stakeholder analysis that follows.

Results of investigations: external stakeholders' views

It is not within the scope of this paper to discuss details of individual organizations; however, research has shown a number of themes common to several organizations in developing countries and post-conflict situations. Private sector companies want improvements in the day-to-day management of the resource sector, and are critical of the lack of timely delivery of data or decisions, as well as a perceived old-fashioned civil service culture not used to the demands of the commercial world. An example is tenement and licence block management in minerals and hydrocarbons development. Companies need to make numerous and frequent applications, renewals, adjustments and queries in their day-to-day operations, but these take too long to process, often because there is no computerized system. Exploration decisions may be delayed by lack of geological mapping data available quickly enough or in appropriate form. Companies also greatly value the coordination and liaison service between mining/energy companies, local landowners, and local government, which is vital to the smooth operation of the sector (e.g. Power & Hagen 1996). Insufficient budget or relevant training for staff leads to decline in this service, and introduces significant project risk. Similarly, insufficient budgets reduce a public sector organization's ability to carry out occupational health and safety surveys on mine and well sites. In some cases the company being inspected is obliged to finance inspections, which may compromise the independence and impartiality of inspectors.

Natural resource development has an impact on the economic and social development of most post-conflict countries and, as a result, the geological survey or mines/energy department regularly interacts with other government agencies. These agencies echo some of the concerns of the private sector since accurate and timely delivery of data is required by policy, environment, treasury and finance departments of the government in their regulatory, planning and promotion activities with major mining/hydrocarbons companies as well as the small-scale artisanal mining sector. Despite the importance of the mining/energy sector in developing and post-conflict economies, many geological surveys and mines/energy departments are regarded as having a low profile within government. Often, government clients are unaware of the organization's business plan or mission statement. This insularity sometimes leads to a perception in other parts of government that the organization is more accountable to the interests of the private sector than to citizens.

The general public receives hazard mitigation information from geological surveys and mines/energy departments, and may be represented by them during closure negotiations over mines and oil/gas installations. The theme of poor communication is taken up here also. Local geohazard and hydrogeological information is often not easily accessible or is in hard copy format. Perhaps more important is that there is a generally poor understanding of the geological survey and mines/energy department role in the natural resources sector. Many citizens are not aware of the importance of the mining/energy sector to the economy. Often they regard the sector as being controlled by expatriate companies exploiting the country's natural wealth, which may lead to poor relations between investors and citizens directly affected by development. Many organizations therefore need to demonstrate their role as administrator of the sector on behalf of the people, in publications, meetings and national fora.

In summary, a survey of external stakeholders' views indicates that many public sector organizations in the natural resource sector do not have the manpower and skills to cope with the volume and complexity of their task. They are unlikely to have a digital system of records to allow swift retrieval and transfer of data, which will allow, for example, facilitation of licensing of blocks for oil exploration. In addition, many staff, coming from a traditional civil service background, have little understanding of the motives and working practices of modern international mining, hydrocarbons or construction companies. Moreover, due to English being their second language, they may be poorly equipped to communicate scientific and technical information to customers or simple information on geohazard mitigation.

Relationships with internal stakeholders

Internal stakeholders of geological surveys and mines/energy departments comprise all employees from management, scientific staff, information technologists, human resource staff, to ancillary and support staff. The chief method of gathering data on internal relationships is training needs analysis and study of managerial methods and structures. Prior to training needs analysis, personnel data in the form of staff lists, job descriptions, employment histories, training records and information from staff appraisal are collected. Information gathered is used to design a questionnaire, which forms the basis of interviews with all staff. The main purpose of the questionnaire is (1) to guide the interview process; (2) to elicit training needs from each staff member, and (3) to determine the position of the staff member within the managerial structure. Different forms of questionnaire are used for non-managerial and managerial staff; this allows views of training needs and business methods from different perspectives. All information gathered is treated as confidential, and interviewees are encouraged to 'speak their mind'.

The purpose of interviews is to:

(1) Develop a relationship with the interviewee and gather information beyond the guide questionnaire;
(2) Understand learners' needs, backgrounds, preferences and abilities;
(3) Assess factors such as education, age, gender, and mother tongue and how they affect potential for learning;
(4) Determine levels of previous training (gained within or without the organization);
(5) Obtain a detailed job description for each employee and compare the documented skills and abilities against requirements of the job;
(6) Explain the rationale for, and instil a feeling of ownership in, the training plan;
(7) Determine a range of training needs outside the technical, including workplace skills, confidence, assertiveness, skills in written and oral expression.

Interviews are conducted by a training/business expert with relevant specialism; a human resources specialist is required for interview of non-technical, clerical and managerial staff, while a geological, mining or oil/gas specialist is required for interviews with scientific and technical staff. Training needs can be compiled for each individual, by 'functional group' (branch or division of the organization), and then corporately, depending on circumstances. In the event of comprehensive corporate re-organization, which is sometimes the object of large-scale aid programmes, a training plan cannot be linked with individuals, since their jobs may change or be discontinued.

Reults of investigations: internal stakeholders

As well as difficulties relating to their outward facing nature, public sector organizations need to address the concerns of their internal stakeholders. Research showed a number of common themes, including inappropriate managerial and communication methods, business structures or cultures, which prevent efficient function. For example, few organizations have business plans and clear business goals. They may also have only a relatively poor understanding of the links between institutional function, actual and required skills, stakeholder relationships, and developing strategies for evolving institutional function to meet changing demands.

Apart from the organization's formal structure, its corporate culture may be counterproductive, being a product of a colonial past, modified by piecemeal introduction of later developments, reflecting contact with more modern organizations. The result, as a whole, is a patchwork of old and new practices typical of public sector organizations in transition. Such organizations contain a group of graduate professional and technical experts who are supported by an administrative class, who administer the specialist training required by the experts. Training and business development tends to be sporadic, inconsistent with business need, and heavily weighted towards science and technical matters. Expatriate scientists and technicians are liable to leave, taking their expertise with them. They may be replaced, through localization policy, by national officers not yet able to operate at a higher level; or the skills deficit is remedied by costly overseas scientific training. The latter is expensive in terms of fees to be paid, but also in terms of staff time lost. In addition, many overseas-trained officers leave by finding jobs in the private sector.

A worldwide requirement for all geological surveys and mines/energy departments is the ability to work in a more commercial environment where staff take risks, make commercial decisions and respond more effectively to external stakeholder requirements. A serious bureaucratic impediment to such staff development is the rigidly fixed nature of job descriptions in many organizations. This makes poor business sense, because it assumes jobs are fixed or even permanent and encourages employees to think that a job 'is for life'.

The way aid is given

> Development is complex... for poverty elimination to be achieved, and for development to be sustainable... there must be a dynamic balance between policies and actions which promote sustainable livelihoods, human development and the better management of the natural and physical environment... (UK Government 1997)

Preparatory to capacity building, economic and sector analysis is carried out by donor and lending organizations like the World Bank, DFID and the Asian Development Bank (ADB), to define the nature and scope of projects and to ensure that aid recipients (or borrowers) have a strong sense of ownership of projects (e.g. World Bank Partnerships Group 1998). However, preparatory analytical work is often not done 'in country' or in partnership with other donors and, most importantly, with institutions of the recipient country (World Bank Partnerships Group 1998); similarly it is often based on models of best practice rather than on the particular cultural or business setting of the organization. World Bank experience has shown that among recipients with weak project ownership, there is a perception that projects are driven by donors and that final project decisions rest with donors and not with the recipient government (Kogbe 1991).

Weak project ownership and integration has led to 'project proliferation', where a large number of donor agencies at work in a particular sector increase demands on recipients' budgets for local and recurrent costs, and form a burden on the weak administrative capacities of recipients (e.g. Carlsson et al. 1998). Tanzania is said to have had some 2000 projects from some 40 donors operating in the 1990s (World Bank Partnerships Group 1998). There is a growing perception that coordination of development aid is therefore essential (UK Government 1997), so that the development agenda is designed and circumscribed properly, and so that its implementation can be monitored.

BGS experience in geological surveys and mines/energy departments reflects some of the concerns documented by the World Bank and DFID, particularly in the poor integration of projects. An example is the aid project portfolio for the Ministry of Mines and Industries (MMI) in Afghanistan. A small team from BGS visited the MMI in November 2002, to study the present plan for capacity building in the MMI, and to suggest other areas for development and funding. The areas of major donor support as of November 2002 are shown in Table 1; consideration of these and of the likely business needs of the organization indicated that a number of critical development areas were not provided for. These included mineral evaluation and geological mapping, a comprehensive digital information system, and groundwater scoping studies. Perhaps the most important area for development was in human resources. The lack of projects developing the managerial, training and organizational aspects of the institution reflects a bias seen in other capacity building elsewhere in the developing world, where most aid is concerned with technical and scientific areas. BGS has noted similar aid projects that allowed for hugely increased training budgets destined for a tiny proportion of technical and scientific employees in an environment where skills across the board were required for all employees at various levels.

Finally, inflexible or conditional funding, which is often a characteristic of aid, may be counterproductive. Funds destined for the procurement, running and maintenance of highly technical equipment (for example, seismological monitoring equipment) are sometimes made ineffective because more basic equipment, for example, generators or air-conditioning, on which the technical equipment relies, cannot be claimed legitimately through aid budgets. In these circumstances it is possible to have rooms full of state-of-the-art scientific equipment that cannot be used because there is no electricity.

In conclusion, the haphazard nature of the aid effort in public sector organizations in some developing countries does not come about by wilful neglect in the development process, but rather reflects the way that aid is a compromise between what the donor wants to provide for its own policy and strategy reasons, and what the recipient country/institution requires. It also probably reflects the lack of local coordination between aid projects on the ground, the lack of 'project ownership' in the recipient organization, and the onerous responsibility of administering complex multidonor aid projects. However, this disorganization is clearly counterproductive and wasteful of funds, and needs to be remedied.

Holistic capacity building: addressing the problems

The foregoing has described a number of the problems that face public sector organizations, both from the way that aid is applied in capacity building and from the difficulties within organizations, and between organizations and their external stakeholders. These problems are summarized in Figure 1; the diagram shows that to be effective, capacity building must act on aid projects and on the organization.

Stakeholders

The methodology described here allows the business and capabilities of organizations to be assessed and measured. In response to this, a snapshot holistic training plan is developed to

Table 1. *Major areas of donor support, as of November 2002, for the Ministry of Mines and Industries (MMI) in Afghanistan*

Funding source	Activities to be supported	Outcomes
World Bank	• Policy and legislation • Institution building	• New petroleum/mining sector policy • New petroleum/mine law • Model investment contract • New taxation regime for mining/petroleum • Institutional restructuring plans
GTZ	Various IT projects and training projects, and rebuilding/refurbishment of the MMI building	No detailed information on outcomes
Asian Development Bank	Establish gas regulatory framework	Effective regulatory system for the gas sector in Afghanistan to ensure efficient development and utilization of natural gas as the domestic market expands
USGS	Assessment of Afghanistan oil and gas resources	• Review of geochemical, seismic, tectonic exploration and production data from domestic and international sources • Characterization of petroleum geology, illustrated as a GIS • Assessment of undiscovered oil and gas • Dissemination of results via USGS web and meetings in Afghanistan
AACA	Basic IT and English language training	No detailed information on outcomes
JICA	• Consultancy study of analytical geoscience laboratory equipment needs • Contingency funding of acquisition, installation and training for some laboratory equipment	No detailed information on outcomes

solve 'present' problems. This will constitute a detailed training plan for each 'job' or employee within the organization, which is linked to a branch or division training plan, and an overall training strategy for the entire organization.

A training plan takes the training need and schedules it against the business need at individual, branch and corporate levels. Since both the business needs of an organization and its personnel continually change, the organization's training needs and consequent training plan are subject also to continual change. The plan has therefore to be continuously monitored and changed by a local training coordinator who understands the business needs of the organization as well the training solutions required to service the business. The training coordinator may also, for the life of an aid project, act as a project coordinator (see below), responsible for liaising with aid donors and integrating projects.

For an individual, a training plan will be a list of his or her training needs, ordered by the date at which they are to be met. Ideally the ordering of attendance will be such that training is supplied in sufficient time for the individual to be appropriately skilled to carry out his or her work. The branch or division manager will hold a branch/division training plan, which will be a table comprising each of the several ordered lists of his or her individual staff. The manager's priority will be to ensure that every member of staff is sufficiently trained to service the various projects that constitute his or her business. Divisional and corporate plans will be correspondingly more complicated.

The simplest solution to meeting such needs in a timely way is for the branch/division manager to hold branch training funds and spend those in a timely way to service the business of the branch. Few organizations are sufficiently wealthy to

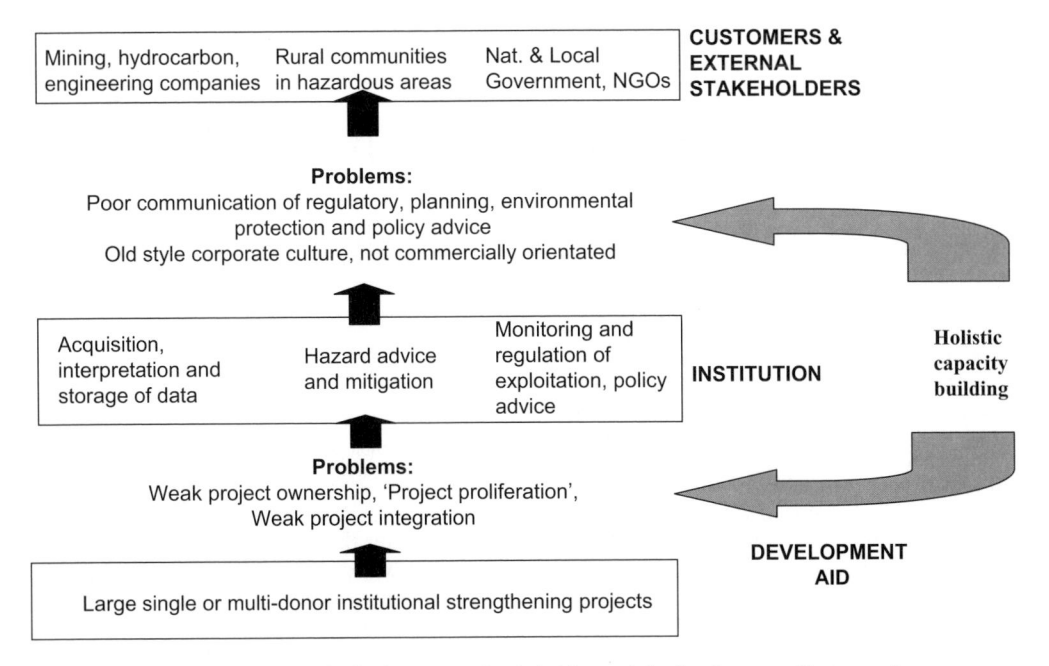

Fig. 1. Relationships between the institution, external stakeholders and the development aid community.

maintain this model of training delivery since it inevitably leads to duplicate costs where the same training is required across several branches. Neither is it generally cost-effective in organizations much larger than branch size, since the benefit of the reduction in unit training costs that arise when several staff require the same training from a supplier cannot be a readily realized unit. The rise of the importance of transferable IT skills, the rise of multidisciplinary team working requiring team members from different branches, and the rise of short-term contracts leading to staff redeployment (with consequent retraining) have all combined to lead to the corporate centralization of training planning and delivery. As a result, the training plans held by individuals and managers are generally subsets of the corporate plan. This has the benefit of leaving the manager free to concentrate on identifying need and attending to staff performance, but may erode the timeliness of training delivery since corporate business need (which may be crudely financial) may take precedence over individual or branch need.

Thus, any organizational training plan needs to be dynamic and, because of the variables involved in internal project planning and the imponderables deriving from the activities of external training suppliers, it cannot be planned in individual detail too far ahead. In fact, experience has shown that, for a large organization (>500 employees) it will not be possible to plan comprehensively for all individuals much more than three or four months ahead. It is possible, however, to set out critical milestones governing training delivery to dates much further ahead than this, leaving the detail to be completed as the various milestones approach.

The way aid is given

A local aid project coordinator is required to ensure that development projects are designed and circumscribed properly and so that their implementation can be monitored. This will lead to 'home-grown' capacity building, by fostering local ownership of projects. A project coordinator might have the following roles and responsibilities:

(1) Understand the theoretical need, scope, duration aims, outcomes and deliverables for each project;
(2) Be capable of modifying 'terms of reference' for projects, with agreement from the aid donor;
(3) Meet regularly with project managers and project personnel responsible for training, procurement and liaison.

It has been stated that donor organizations define the nature and scope of capacity-building projects, but that preparatory analytical work is often not done 'in country' or in partnership with other donors and, most importantly, with institutions of the recipient country. Similarly, the aims, outcomes and scope of projects may not be known to middle managerial and junior staff within a recipient organization, because corporate 'vertical communication' between senior managers and staff is poor. Thus, senior management should appraise staff members of the need for the proposed project, and the benefits of the project to the organization, the country as a whole, and to the individual employee. They might also arrange to have meetings at which staff are invited to comment on proposed projects with a view to improving them.

The aim of sustainability in the context of capacity building is to allow the organization to build itself, and become independent of the 'aid machine'. In practical terms this means that the organization must be able to monitor and change with business requirements without significant help from outside. To do this, an organization must commit to the concept that all business activities must be undertaken as part of a cycle of action, review and modification (Fig. 2). An example of how this concept may be put into practice is through the operation of a training system (Fig. 3), although this methodology is applicable to project and business management of all types. In all organizations, training is continuously required, since business continually changes. In the training system shown in Figure 3, the 'nucleus' comprises the employee requiring training and his/her branch/division manager, who are in continuous dialogue over the employee's training and development needs. The employee and branch/division manager are

Fig. 3. An example of how the business cycle can apply to training provision.

responsible for the conception, design and preparation of a bid for training, and must be sure, primarily, that a good business case is made before the bid is passed on to a training committee. Such a training proposal might be a course in database construction, hydrological well testing or an MSc in geophysics. The employee and the branch manager are expected to use their in-depth knowledge of the job, its business aim, and its context within the branch to construct a business case that makes clear that the training is indispensable. The branch manager will also have formed a view as to the suitability for training of the particular applicant based on the prospective trainee's experience, training record and ability. By virtue of his position in the hierarchy, the branch manager will also know the mind of the director of the organization as to the context of

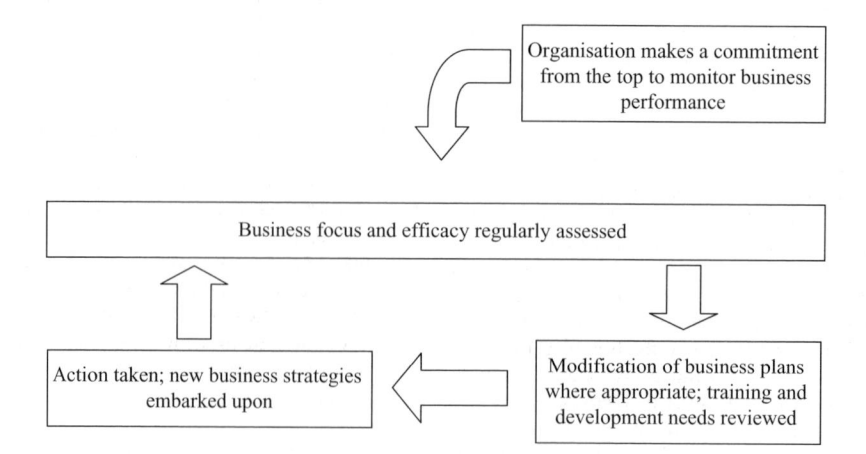

Fig. 2. The business cycle. All business activities should be undertaken as part of a cycle of action, review and modification.

the training bid within the division's priorities. The training coordinator will advise in the process of training bid preparation and may also advise on the creation of training projects that are appropriate to the organization's requirement. Thus, a training bid is submitted to the training committee who will be the decision-making body, and have a key role in balancing the corporate requirement across the various divisions. Following approval, training will take place. Evaluation of the training closes the loop: commitment, planning, action, evaluation. It informs the parties whether the training and development has benefited the person and the organization. Such evaluation should consist of two parts: evaluation of (1) outcomes and (2) impact.

Training outcomes record what has actually resulted from an individual's participation in a training event. This is the most important stage of evaluation since it seeks to measure primarily the effect that the training has had on the member of staff and his or her work. Evaluation of outcomes is made some time after the training event and when the trainee has had time to put what has been learned into practice. This may take place six months or a year subsequent to the event and should be assessed together by trainee and manager. It may be found useful, however, for each to make a separate assessment before discussing the matter, since differences of assessment may uncover valuable differences of perspective of the job performance. Several such individuals' evaluations showing the effect of the same training event upon staff performance are a powerful tool with which to judge the significance of the training. Actual outcomes may be many and varied. Usually an improvement in performance is anticipated; on the other hand the evaluation may reveal that further training is required, that a person is unsuited to a particular form of work or that the training was inappropriate or substandard.

Evaluation of training impact is a function of management and is more difficult. It seeks to relate the value to the business of particular training activities. It derives from the business plan of the particular manager who will want to show that the performance of his business has improved as a result of the training of his or her staff that are delivering the business products or services. At its most convincing, such an evaluation will show that income or profit has been generated at a cost outstripping that of the cost of the training. To do so requires the manager to cost the effect of staff 'down time' while staff are engaged in training, against the improved service, performance or product resulting from that training.

In conclusion, the premise of evaluation, whether it be of training, operation of individual projects or of general business, lies in bringing outside assessment to bear on its efficacy, measured by its effect on the employee or the department's business. Essentially, improvements in the organization's ability to do business come about because of shortcomings identified in evaluation. A system that continually monitors the efficacy of an institution's work and allows it to change with changing business, can help an institution to adapt to change. Such a 'sustainable business system' with strong local ownership should allow independence from the 'aid machine' and encourage 'home-grown' capacity building.

Conclusions

A comprehensive view of how well an institution interacts with its internal and external stakeholders, its capacity in terms of human and physical resources, and how to proceed to holistic capacity building, is gained from the methodology outlined in this paper. Among the crucial shortcomings of geological surveys and mines/energy departments are poor communication of regulatory, planning, environmental protection and policy advice and an old-fashioned corporate culture, which is not commercially orientated. Among the difficulties identified in the way that aid is given (or loaned) are weak project ownership, 'project proliferation', and poor project integration.

Holistic capacity building through training needs analysis forms a training plan that will solve present skills shortage problems in line with business need. This will focus as much on transferable skills such as information technology and writing/presentation as on science. Part of the holistic capacity-building programme is to install a local project coordinator who will be responsible for project formulation, in consultation with aid donors, and for coordinating projects once they start. Finally a 'sustainable business system' will be installed with strong local ownership, which will ensure that all business activities are undertaken as part of a cycle of action, review and modification to allow speedy independence from the 'aid machine'.

M. H. Stephenson and I. E. Penn publish with the permission of the Director, British Geological Survey. The authors would like to acknowledge the help of the

World Bank, the Islamic Transitional Government of Afghanistan, the Afghan Assistance Co-ordination Authority and the Department of Mines of the Government of Papua New Guinea during the consultancy work that formed the basis for this paper.

References

CARLSSON, J., SOMOLEKAE, G. & VAN DE VALLE, H. (eds) 1998. *Foreign Aid in Africa: Learning from Country Experiences*. Nordiska Afrikainstitutet.

DFID 2001. *Making government work for poor people*. DFID Strategy Paper, September 2001, 39 pp, http://www.dfid.gov.uk/Pubs/files/tsp_government.pdf.

FINDLAY, C. 1997. National geological surveys and the winds of change. *Nature and Resources*, **33**, 18–25.

KOGBE, A. A. 1991. Training of geoscientists in Africa. *In*: STOW, D. A. V. & LAMING, D. J. C. (eds) *Geosciences in Development*. AGID/Balkema, Rotterdam, 75–79.

O'CONNOR, E. A. 1996. One hundred years of geological partnership: British Egyptian geoscientific collaboration 1896–1996. British Geological Survey, Keyworth, 19 pp.

POWER, A. P. & HAGEN, P. C. 1996. The escalation of landowner benefits in the Kutubu Petroleum Development Project. *In*: BUCHANAN, P. G. (ed) *Petroleum Exploration, Development and Production in Papua New Guinea: Proceedings of the Third PNG Petroleum Convention*, Port Moresby, 9–11 September 1996, 751–772.

REEDMAN, A. J., CALOW, R. C. & MORTIMER, C. 1996. *Geological Surveys in Developing Countries: Strategies for Assistance*. British Geological Survey, Overseas Geology Series, Technical Report **WC/96/20**.

STEPHENSON, M. H. 2003. Reconstructing the Afghan Survey. *DFID Earthworks Magazine*, **16**, May 2003, 6.

STEPHENSON, M. H., PENN, I. E., GOWEN, J. & PEARSE, J. 2002. Training Needs Analysis for geoscience organizations. *DFID Earthworks Magazine*, **15**, October 2002, 3.

STEPHENSON, M. H. & PENN, I. E. 2003. Dynamism and optimism of Afghanistan's geologists will drive reconstruction of Afghanistan Geological Survey and the natural resources sector. *Geoscientist*, **13**, 16–17.

UK GOVERNMENT 1997. *Eliminating world poverty: a challenge for the 21st century, white paper*, November 1997, http://www.globalisation.gov.uk/.

WORLD BANK PARTNERSHIPS GROUP 1998. *Partnership for Development: Proposed Actions for the World Bank: A Discussion Paper*. World Bank, Washington DC, 20 May 1998.

WORLD BANK PUBLIC SECTOR GROUP 2000. *Reforming Public Institutions and Strengthening Governance: A World Bank Strategy Implementation Update*, World Bank, Washington DC, May 2003.

Sustainable development and minerals: measuring mining's contribution to society

D. J. SHIELDS[1] & S. V. ŠOLAR[2]

[1]*USDA Forest Service – Research and Development, 2150 Centre Ave, Bldg A, Fort Collins, CO 80526, USA (e-mail:dshields@fs.fed.us)*
[2]*Geological Survey of Slovenia, Dimičeva 14, 1000 Ljubljana, Slovenia*

Abstract: Traditional development theory focuses on two goals, income growth and poverty alleviation, and many mineral-rich developing countries have depended on resource exploitation to achieve those goals. In reality, mineral-driven economies have often experienced less growth than mineral-deficient economies. Conversely, the potential positive contributions of mineral development are often overlooked in countries with limited mineral endowment. Such under- or over-emphasis on the minerals sector can be lessened by addressing mineral development within the context of sustainable development. Each country identifies sustainability goals, with respect to social equity, environmental health, and economic growth that are appropriate to its circumstances. The contribution of mineral resources to the achievement of those goals will be similarly context dependent. In this paper we discuss the concept of sustainable mineral resource management. We then describe how indicators of sustainability can be used to measure the contribution of the minerals sector to a country's economic development and track progress toward its overall sustainability goals. We present an example of sustainable mineral management policy and related mineral indicators from a country with a transitional economy and limited mineral resources.

There are tremendous differences between the incomes, standards of living, and basic human well-being of citizens in rich, developed nations and those in the poor countries of the world. In response, government leaders have committed to the Millennium Development Goals (UNDP 2000), which include eradicating extreme poverty and ensuring environmental sustainability by 2015. Both the developed and developing nations will need to act if these goals are to be met. Developed nations will need to supply aid, debt relief and investment (Baird & Shetty 2003). Developing nations will need to improve policies and governance, and in addition make better use of development assistance (Shields & Šolar 2000*a*). This latter issue is of crucial importance. Despite numerous significant investments in developing nations, poverty remains a serious problem. Furthermore, even successful development projects sometimes generate enormous environmental and social costs.

We are coming to understand that many of the issues faced by both developed and developing nations alike are highly uncertain, urgent, complex, and interconnected. Human activities are having impacts that exceed the Earth's carrying capacity on global, and in many places, regional and local scales. We can no longer afford to address individual problems in a convenient isolation of their context, or their spatial or temporal scale (Funtowicz & Ravetz 2001).

The sustainability paradigm provides a framework for addressing complex types of problems because it is both comprehensive and flexible (Shields *et al.* 2002*b*). Sustainable development strives to improve the economy, environment, and society for the current generation, without compromising the ability of future generations to meet their needs. The interconnectedness of social, economic and environmental systems is explicitly recognized. The overarching goals of economic prosperity, environmental health and social equity are simple and flexible enough to allow for multiple interpretations and are applicable in a variety of circumstances. Further, economic growth and technological advancement are deemed to be essential, but need to be achieved in an environmentally sensitive and distributionally fair manner (Cordes 2000). Most definitions of sustainability also take into account system limitations and human needs. Physical, societal, and temporal limitations dictate that not everything is possible, desired or achievable in either a single time frame or at a single location (Bossel 1999). Universalism requires that we protect future generations, but

From: MARKER, B. R., PETTERSON, M. G., MCEVOY, F. & STEPHENSON, M. H. (eds) 2005. *Sustainable Minerals Operations in the Developing World.* Geological Society, London, Special Publications, **250**, 195–212. 0305-8719/05/$15.00 © The Geological Society of London 2005.

also address the pressing claims of the less privileged today (Anand & Sen 2000).

Minerals have the potential to contribute to a sustainable future. Mineral resources are fundamental to human well-being. They provide essential services to societies – fuel for transportation, mineral materials that are the basis of the built environment, and metals that are widely used in consumer products. Neither present nor future societies can be expected to forego the stream of benefits coming from the use of mineral resource products, and by extension from mining. Further, as the Extractive Industries Review pointed out, the extractive industries have the potential to contribute to poverty alleviation (World Bank 2003).

However, the history of mining as a development tool is mixed at best (Auty & Mikesell 1998). Mineral exploration, extraction, use and disposal pose environmental and social risks and yet developing nations have often waited for incomes to rise before curbing environmental damage and the wasteful use of resources (Auty 2003). Moreover, mineral wealth has sometimes been squandered through misuse and expropriation. If mining is to be used as a development tool, the implementation of new policies and practices will be essential. Developing nations will need to identify the contribution they want mining to make to their society *a priori*, and then track progress toward achievement of that contribution so that policies can be adapted if outcomes are not what was expected or desired.

A sustainable minerals paradigm is about creation of a viable dynamic equilibrium between industry, government, environment, community and other stakeholders (Findlay 1997). In this paper, we argue that strategic documents, such as mineral policies, and operational documents, such as mineral resource management plans, must be sustainability based. We offer the Sustainable Mineral Resource Management (SMRM) model as a tool for institutionalizing the sustainability paradigm, and operationalizing sustainable mineral policy. We also demonstrate how the indicators associated with an SMRM plan facilitate tracking progress and informing stakeholders.

In the next section of this paper, we briefly review the basic concepts of sustainability and sustainable development. We discuss the application of sustainability principles to minerals and then discuss aggregates and the role of construction materials in development. We next introduce the concept of sustainable mineral resource management and place particular emphasis on the need for sustainable aggregates management in developing nations. The use of indicators of sustainability for measuring the contribution of the minerals sector to a country's economic development and tracking progress toward overall sustainability goals is discussed. Last, an example of a sustainable mineral management policy and related mineral indicator is presented. We conclude with a few remarks about the transportability of indicator sets.

Sustainable development and natural resources

Sustainable development has four overarching goals: economic prosperity, environmental health, social equity for the present generation, and equal opportunities for future generations. The need to transition to a sustainable development path has been agreed to by the world's nations, first at the Earth Summit held in Rio de Janeiro in 1992 (UN 1992a, b), and again in 2002 at the World Summit on Sustainable Development held in Johannesburg (UN 2003). What has not been agreed upon is how to enhance human well-being while staying within the biosphere's capacities and limits, or the path countries should take to reach a sustainable future.

Disagreement stems from the fact that sustainability embodies values about the kind of world we want to live in and leave for future generations. Human values are not fixed or independent of social, economic, and ecological context, and as a result, there are multiple perspectives on what sustainability means, and how it should be achieved. Moreover, sustainability requires judgement about the state of the world. Inherent therein is valuation of those tangibles people believe should persist in space and over time (USDA Forest Service Inventorying and Monitoring Institute 2003). Sustainability is about choices regarding what to sustain, how, when, where, and for whom. Debates about sustainability reflect people's opinions about the appropriate answers to these questions and conclusions vary among countries because of differences in culture, values, and circumstances. There is no one, correct view of what sustainability means or how its principles should be implemented. In the next section we briefly review several alternative perspectives and identify the most well-know approaches to describing sustainability.

Sustainable development concepts

The Brundtland definition of sustainable development exhorts us to meet the needs of the present generation without compromising the ability of future generations to meet their own needs (WCED 1987). This general statement leads to a variety of different, and in some cases even conflicting, understandings of sustainable development. One of the most significant differences is between the eco-centric versus techno-centric perspectives. From the eco-centric perspective, the Brundtland definition focuses too much on material well-being; the development aspect is over emphasized relative to the sustainability aspect. Eco-centrism argues that humankind is just one part of the larger, global ecosystem and therefore humans must consider and respect the Earth's biological and physical constraints. The environment is placed at a different, more significant level than either the economy or social well-being because it is the source of both of these human necessities (Dawe & Ryan 2003). Eco-centric approaches can be criticized as inadequately sensitive to human needs.

The alternative, techno-centric perspective has greater faith in the ability of humans to address and to control the environmental issues with technological solutions and free market mechanisms. Neoclassical economic growth theory is often at the core of techno-centric models, which tend to acknowledge the importance of environmental protection and intergenerational equity, but emphasize the usefulness of conventional economic efficiency criteria (Solow 1993). The core of many such models is maximization of the present value of generational welfare, subject to a set of social, market, and environmental constraints (Toman et al. 1995). Techno-centric models are criticized as having unrealistic expectations about the ability of technology to solve our problems. They are further criticized on the grounds that free trade and a reliance on markets assumes that self-interested choice will lead to societal well-being, which may not be the case (Rammel & van den Bergh 2003), and that a present value-optimal path is not necessarily synonymous with a sustainability path (Islam et al. 2003).

One way to clarify the differences between these opposing perspectives is to think in terms of capital, or the related terms endowments and wealth (Costanza & Daly 1992; Toman 1994). The types of capital are natural capital (traditional natural resources), human-made capital (physical, produced assets and the built environment), human capital (the health and well-being of individuals) and social capital (the complex of social relations, norms and institutions). The eco-centric approach advocates strong sustainability, which requires that the amount of each type of capital be preserved independently through time. Capital of different types can be complements, but not substitutes, for each other, and discounting between time periods is not allowed. At the extreme, strong sustainability precludes mining.

The techno-centric approach is based on weak sustainability, which is distinguished from strong in terms of capital maintenance and augmentation, that is, by the degree to which alternative types of capital are deemed substitutable for one another (Pearce & Atkinson 1993; Faucheux et al. 1997). Weak sustainability preserves the net amount of capital, but not necessarily each of the four kinds of capital, so different types of capital, accruing in different time periods, are viewed as substitutable. At the extreme, weak sustainability allows mining virtually anywhere, in whatever manner and rate is profitable, so long as economic rent generated by extraction is reinvested in some alternative form of capital.

Models that embody a strong sustainability perspective include the ecological footprint (Wackernagel & Rees 1996), the pressure–state–(impact)–response model developed by the Organization for Economic Cooperation and Development and the United Nations (OECD 1994), and the sustainability barometer developed by the International Union for the Conservation of Nature (Prescott-Allen 2001). Two examples of approaches with a weak sustainability orientation are the genuine savings model developed by the World Bank (Hamilton & Clemens 1999) and Munasinghe's sustainomics (Munasinghe 2002).

Mineral resources are normally discussed in terms of weak sustainability, due to the inevitable transformations of natural capital that result from mineral extraction and use (Shields & Šolar 2000b). The transition from natural capital to human-made capital, and further to human and social capital, can be described as moving from primary means to ultimate ends, that is, well-being (Meadows 1998). However, not all mining generates economic rent that can be reinvested in other forms of capital (Dobra 2002) and in some cases the rent that is generated is not reinvested in any form of capital (Auty & Mikesell 1998). It is also important to keep in mind that measures of economic capital such as GDP do not adequately describe human well-being (Anand & Sen 2000). We recommend reporting measures of well-being, including the freedoms people enjoy and the life they lead, in addition to the materials they consume and the capital they accumulate.

Sustainable development applied to mineral resources

Natural resources are an important part of any country's sustainability considerations. Discussions about the role of natural resources in sustainability tend to focus on the need to sustain ecosystems and maintain biodiversity. For example, sustainable forest management requires that the capacity of forests to maintain their health, productivity, diversity, and overall integrity be protected, in the long run, in the context of human activity (USDA Forest Service 2004). The fundamental goal is sustaining the ecosystem. Minerals are as essential to a sustainable future as are ecosystems, but it is counterintuitive to speak of them as being sustainable in the same way. Individual deposits are finite in size and quantity. On a broader, global scale, minerals are seldom truly exhausted in any case, but rather redistributed from their location in deposits to products and waste materials.

There are several streams of thought on how minerals fit in sustainable development. One perspective focuses on mineral development as a source of wealth creation and by extension its value as tool for the eradication of poverty. Another focuses on present and future needs for minerals and fuels and points out that developed societies need a steady supply of material inputs. Yet another perspective focuses on the environmental and social consequences of mineral development, use and disposal, and argues that a reduction in the per capita use of materials will be essential to the achievement of sustainability. All these concerns can be captured by viewing the goal of sustainable mineral management as sustaining the stream of benefits that are provided to society by minerals and doing so in a manner such that those benefits are a net positive over the mine and product life cycle (IIED 2002a).

Unfortunately, mining's contribution to society has too often not been a net positive (IIED 2002b). Extraction, use and disposal of mineral resources have negatively impacted societies and the environment. Low environmental protection standards and limited implementation capacity have led to several preventable environmental accidents, for example, Anzalcóllar in Spain and Baia Mare in Romania. There are also problems with inadequate health and safety protection, as demonstrated by the number of underground mining accidents with casualities, and with fair benefit distribution to all stakeholders, and to local communities in particular.

The Minerals, Mining and Sustainable Development project (IIED 2002b) identified the range of stakeholders' concerns regarding the mining sector:

- viability of the minerals industry;
- the control, use, and management of land;
- minerals and economic development;
- local communities and mines;
- mining, minerals, and the environment;
- an integrated approach to using minerals;
- access to information;
- artisanal and small-scale mining; and
- sector governance: roles, responsibilities, and instruments for change.

In addition, although mineral-rich developing countries have often depended on resource exploitation to achieve their development goals, there are numerous examples of mineral-driven economies that have experienced less growth than mineral-deficient economies. There are many causes for the outcome, one of which is a tendency for the extractive sector to dominate the entire economy (Auty & Mikesell 1998). In addition, governments have either ignored or been unaware of the fact that income growth is necessary for human development, but not sufficient (Srinivasan 1994). Broader social issues related to well-being must be addressed as well.

The WSSD Plan of Implementation (UN 2003) took the foregoing issues into account when drafting Paragraph 46, which deals with minerals and mining. The text recognizes that minerals are important to the economic and social development of many countries and that they are essential for modern living, giving the wording a traditional development theory tone. However, the text went further, calling for enhancement of the contribution of minerals to sustainable development through actions at all levels to: (a) support efforts to address environmental, economic, health and social impacts and benefits, (b) enhance the participation of stakeholders, ... and (c) foster sustainable mining practices...

Sustainable development principles will need to be applied to minerals management if the goals of Paragraph 46 are to be achieved and unfair distributions of the costs and benefits from mineral resource extraction and use corrected. Each stakeholder will have a role. Government is one of the major stakeholders in society. Ascher and Healy (1990) demonstrated that development can be enhanced or retarded by public institutions. Lack of adequate governance, ineffective bureaucracies, and misguided policies limit nations' ability to develop. For the remainder of the paper we focus on how governments can design, implement, and track policies and management plans to sustainably manage mineral resources.

The governmental role in sustainability is to create an enabling economic environment that aligns a country's investments with its underlying comparative advantage, so as to improve the use of scarce capital and human resources (Auty 2003). They need to (Auty 2003, p. 3),

- maintain fiscal discipline by broadly matching public expenditure to a healthy diversified tax base;
- direct public expenditure away from administration, defence, ad hoc subsidies and white elephant mega-projects towards neglected areas offering high returns, notably those that improve asset distribution such as education, health and social infrastructure;
- secure property rights without excessive costs, including rights for the informal sector;
- maintain a competitive exchange rate and remove domestic price distortions; including subsidies that encourage wasteful overconsumption; and
- promote competitive markets, including efficient financial markets that facilitate entry by domestic and foreign firms and enhance the efficient allocation of investment.

With respect to minerals they should also:

- provide sufficient information on a country's mineral resources; and
- create, implement and maintain a proper, fair, and enforceable legal and regulatory framework.

These and other issues should be clearly expressed in the country's national mineral policy, which should include policy scope, sovereignty, economics, quality of life, legislative framework, and regulatory agencies (Otto 1997). The mineral policy clearly defines types of acceptable mineral activity and types of minerals that can be exploited (Otto 2001). Some countries have additional mineral policy guidelines embedded in other policies. This is often the case for construction materials, and aggregates in particular, due to the fact that their extraction and use affects many sectors of the economy and society. However, regardless of whether aggregates are specifically included in the mineral policy or some other policy, they should be treated in a manner that reflects their specific characteristics, an issue that will be discussed at greater length in the next section of this paper.

It is also essential that a sustainable mineral policy be correlated and consistent with other governmental policies (Shields *et al.* 2002*a*). Sustainable development is a policy concept in and of itself (Davis 1997), but it also makes demands on other policies (Dovers 1996). Operationalizing sustainable development principles takes place within geopolitical boundaries where policies are promulgated, codified in law, and implemented. Because of the interconnectedness among systems noted at the beginning of this paper, policy consistency across sectors is essential. A lack of harmonization within the legal framework will make the realization of national sustainability goals difficult (van der Straaten 1998).

Finally, it should be noted that in recent decades, some countries have begun to shift to more restrictive extraction policies, when other development opportunities are available. The main reasons for imposing restrictions are (a) reduced territory potential with regard to minerals, (b) decreased domestic demand by the industry due to economic restructuring, (c) increased supply of minerals from existing operations due to improved technologies and recycling, (d) liberalized importation due to globalization processes, (e) increased public concern for resource conservation and for negative environmental and social impacts during the whole mine cycle (Šolar 2003), and (f) an increasing societal attitude, particularly in developed countries, that a minimum standard of life quality is a 'right', combined with a failure to connect material benefits with the processes by which these benefits are realized; for example, everyone wants a car, but no one wants to live near a mine. The desire of governments to limit extraction opportunities may diminish in the presence of effective sustainable mineral management policies.

In the following section we will discuss aggregates and sustainable aggregate resource management.

Construction materials and aggregates

There has been an inappropriate tendency to focus exclusively on metals when minerals are used as an engine of development. As a result, the potential positive contributions of mineral development are often overlooked in countries with limited mineral endowment. Under- or overemphasis on the minerals sector can be lessened by addressing mineral development within the context of sustainable development. Each country identifies sustainability goals, with respect to social equity, environmental health, and economic growth that are appropriate to its circumstances. The contribution of mineral resources to the achievement of those goals will be similarly context dependent.

As previously noted, natural aggregate is an essential commodity in modern society. Developing nations need stable, adequate and secure supplies of construction materials to build the infrastructure needed to achieve the

Millennium Development Goals (Baird & Shetty 2003). This includes highways, roads, bridges, railroads, airports, seaports, water and waste treatment facilities, and energy generation facilities. Construction materials are also essential to the provision of sustainable housing and the expansion of industrial capacity (CIB and UNEP 2002). These large-volume materials will need to be provided in a rational integrated manner that maximizes their societal contributions and minimizes environmental impacts (Lungu & Price 2000).

Definition

Mineral resources are classified in many ways, most commonly into non-metals, metals and (solid or fluid) energy resources. Non-metals consist of two major groups: industrial minerals, and rocks and construction materials. Within the construction materials group, aggregates prevail over commodities (such as clay, dimension stone, etc.) in terms of volume produced (Šolar et al. 2004).

Natural aggregate is a material composed of rock fragments that may be used in their natural state or after mechanical processing such as crushing, washing, and sizing (Langer & Tucker 2003). There are two categories of aggregates, sand and gravel, and crushed stone. Sand and gravel generally is considered to be material whose particles are about 2.0 to 1024 mm in diameter. Crushed stone is of the same size range, but is artificially crushed rock, boulders, or large cobbles. Most or all of the surfaces of crushed stone are produced by the act of crushing, and the edges tend to be sharp and angular. Natural aggregate has hundreds of uses, from chicken grit to the granules on roofing shingles. However, most aggregate is used in road construction (the largest use), as well as in cement concrete, asphalt, and for other construction purposes. The average per capita consumption of aggregate generally ranges from 5 to 15 tons per year (Langer & Šolar 2002).

While aggregate is a non-renewable resource, supplies are nearly inexhaustible on a global scale. They have characteristics that differentiate them from most other mineral commodities:

- a high number of potential extraction sites;
- a high volume to value ratio, significantly different set of potential environmental impacts; and
- regional importance combined with a narrow economic transportation radius.

In general, demand for aggregate is met by local or in-country suppliers. And, in most cases, aggregate transportation is carried out on roads because transportation short distances by road is economically viable for a high volume-to-value ratio product, especially if the roads already exist. Constructing new, and maintaining existing, rail or channel (canal) networks is expensive, although in the long run transport by rail or canal/sea can be more energy efficient.

In many countries, aggregates have simplified legal frameworks (local level competence, licensing, taxation, control) compared to other minerals due to their characteristics. In some countries aggregates are the landowner's property, even if most other minerals are state owned.

Aggregates and sustainability

Aggregates are very often overlooked in minerals sustainability debates due to the fact that they are seldom export products of national importance like metals or energy resources. Moreover, trade is limited by transportation costs, particularly in the case of road transportation. Some export of aggregates is carried out by ships, especially where large extraction operations are close to water or sea transportation facilities. What export there is of aggregates tends to be carried out between suppliers situated close to a country's border and to foreign customers.

Aggregates typically do become part of the sustainability debate in countries with organized environmental protection groups that have active individual members. These are mostly developed countries where mining is a declining economic activity and most other types of mineral extraction have ended. In developing countries, the negative and positive effects of quarrying are not important issues in the development debate. The negative impacts of aggregate extraction that are passed on as burdens for future generations cannot be compared with the present desire for faster economic development and poverty alleviation. The fact that the costs of remediating negative social and environmental impacts of aggregate quarrying and use will be higher in the future than they are today is neither a priority nor a point of discussion in most developing countries.

Nonetheless, aggregates should be an integral part of any country's overall sustainability plans. Geological settings, and economic and social conditions, are influential factors in determining how aggregates are supplied. However, because the manner in which aggregates are upplied affects the ability of developing nations to achieve a sustainable future, it is important that a country's strategic and operational policy guidelines are based on sustainable development principles.

Construction materials and aggregates present a very clear example of the transition from natural to human-made (manufactured/physical) capital. In order to optimize this transition, the positive impacts of quarrying should be maximized and negative ones minimized. On the positive side of the equation, aggregates facilitate everyday life of humankind by providing shelter (housing), easing communication (traffic) and societal infrastructure (for economy, education, health, art and science). Negative impacts are linked mostly to the manner in which construction materials are extracted, transported, and disposed.

Lack of understanding about the links among different types of impacts of quarrying is a source of time-consuming disagreements between stakeholder groups, including the general public, industry, environmental, social and expert groups on local and national levels. One of the most effective ways to identify the full range of positive and negative impacts, as well as system interactions, is to examine the entire quarry and product life cycle. Societal, value-based, objectives expressed in policies emphasize certain parts of the life cycle and bring those issues to the attention of stakeholders.

Developing countries have at least as great a need for aggregate supply as do developed ones, but less resources to implement sustainable supply policies. One of the tools that can be used to ensure that mineral resources (aggregates) are provided in a manner that contributes to sustainability is 'sustainable mineral resource managements'.

Sustainable mineral resource management

A hierarchical model of resource management reflects two basic assumptions: (1) people's objectives are a reflection of a contextual application of their held value sets, and (2) management goals make sense only within the context of the human social system (Fig. 1). Consistent with that view, terminal and instrumental held values are placed at the top of the hierarchy. Those values influence, and are influenced by, the cultural, social, institutional, and economic framework within which that individual lives, and through that process become an ordered value set. Placed in context and mediated by experience and circumstance, that set is assumed to be the primary factor influencing an individual's selection and ordering of objectives. We assume that, in theory at least, decisions are made and actions taken with the intent of achieving stated objectives. Those actions have impacts on social, economic, and environmental systems that can be identified and measured. The purpose of the feedback loop in the hierarchical model is to pass back information about the state of the world and the consequences of fulfilling professed objectives (Shields & Mitchell 1997).

The purpose of policy is to codify objectives in a manner consistent with the social and political system, and also representative of the public's social value set. A policy is a connecting of desired ends with practical means; it is a course of action calculated to achieve a desired objective (Fenna 1998). Policies can be stand-alone documents or exist as part of other mineral

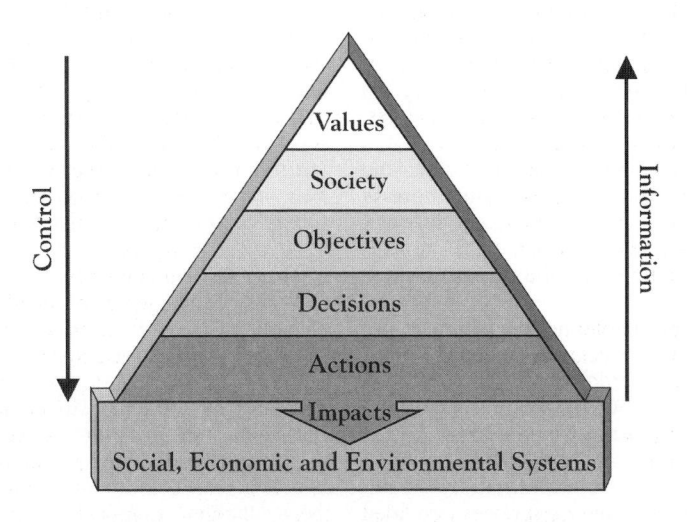

Fig. 1. Control and information flow: hierarchical model of resource management.

resources-related policies (such as a land use planning policy).

Policy is encoded in laws and acts. Every country has a mining law that regulates mineral resource extraction. In response to global changes in markets in the last few decades, many countries have changed their mining acts. The main change has been an effort to attract more foreign investment for mineral development, particularly in developing nations. This is a continuation of an historic tendency for traditional mining acts to be responsive to the desires of mining companies. Traditional mineral management decision making was hierarchical. There was a strong link between industry and government. Traditional policy incorporated internationally accepted principles regarding granting and transfer of title, technical and environmental monitoring, and fiscal requirements including bonding. However, environmental and social issues were seldom addressed in a democratic manner.

Once a policy is adopted and laws passed, the operational stage of designing and implementing regulatory framework begins (Šolar & Shields 2000a). In general, mineral management, including regulatory aspects, is the practical application of strategic planning, as laid out in policy documents and law. Mineral resource management includes operational measures on how to regulate all mineral resources related activities on a national level, which impacts also on the regional and local (mine site) level.

Sustainable mineral resource management (SMRM) differs from more traditional management approaches in several ways. The fundamental premise of SMRM is that decisions and actions should be consistent with the social, economic, and environmental principles of sustainable development (Šolar et al. 2002). Economic prosperity would include ensuring an adequate and secure supply of mineral resources, increases in competitiveness and employment, and mineral resource protection, including land access, exploration and characterizing mineral potential. The main feature of environmental health is minimizing environmental impacts over the whole mine cycle, the operating stage in particular. Social equity has intra- and inter-generational components, necessitating an effective legal framework, stakeholders' communication and participation in decision making, and analysis of the lasting benefits to communities and society of mining.

The overarching goal of SMRM is not to sustain a given deposit or industrial sector, but rather to sustain the flow of services provided by mineral commodities and to do so in a manner such that, over the commodity's life cycle, the net contribution to society is positive. Exploration and exploitation should be managed so that the natural capital embodied in mineral resources is transformed into built physical, economic or social capital of equal or greater value according to the weak sustainability paradigm. SMRM incorporates strict environmental protection and includes interaction with, and potentially permissions and concordances by, different stakeholders' groups over the mine life cycle. Moreover, SMRM provides a framework for addressing the trade-offs that society needs to make in terms of land use and resource management and can incorporate all relevant spatial scale hierarchies. It is often assumed by participants in sustainability debates that everything people value can be sustained. In reality, objectives are sometimes incompatible and only one can be achieved in a specific location.

Sustainable mineral resource management (SMRM) is an iterative, adaptive process. We start from the premise that people's objectives are value-based and context-dependent. Therefore the SMRM process begins with the identification of stakeholders, their value sets and related objectives for resource management. By recognizing the link between environmental and social issues, SMRM insists that mining should be conducted in a way that is acceptable to the public, so that firms do not lose the social licence to mine. Alternative management approaches are developed that reflect those objectives. Social and environmental impacts are predicted for each alternative, technical aspects are considered and costs estimated. Technically or economically unfeasible, or unsustainable, alternatives are revised or rejected. Feasible alternatives that support sustainable outcomes are then presented to the public for debate and negotiation with the goal of choosing an alternative that is acceptable to the public. Once an acceptable management alternative has been identified, it is implemented, monitored, evaluated, and revised as needed. The process of revision once again requires public participation and the cycle is repeated (Solar et al. 2002).

The SMRM framework reflects good science, public preferences, and financial and social constraints. Public preferences expressed in a multistakeholders' decision process can be properly implemented if public involvement is supported (assisted) by an access to sufficient, accurate, and understandable information. Such information is provided by scientists. In SMRM, the role of scientists is twofold. They participate in the evaluation of alternatives, but also provide information needed to support discussions that will lead to informed public judgement regarding resource management. Thus, a description of the state of the world, predictions about impacts, and measures of progress toward objectives will be needed (Šolar et al. 2002).

Mineral and mining information is gathered periodically (in most cases annually) by state institutions such as statistical offices, or may be collected randomly within different mineral sectors as part of research done for different end users and for various purposes. For SMRM, a mix of existing and new data will be needed. Some of the new information will respond to stakeholder questions, and other data will be needed to ensure a solid decision-making base.

As shown in Figure 2, SMRM is an iterative, adaptive process (Šolar et al. 2002). We start from the premise that people's objectives are value-based and context-dependent. Therefore, the SMRM process begins with the identification of stakeholders, their value sets and related objectives for resource management. Alternative management approaches are developed that reflect those objectives. Social and environmental impacts are predicted for each alternative, technical aspects are considered, and costs estimated. Technically or economically infeasible, or unsustainable, alternatives are revised or rejected. Feasible alternatives that support sustainable outcomes are then presented to the public for debate and negotiation with the goal of choosing an alternative that is acceptable to the public. Once an acceptable management alternative has been identified, it is implemented, monitored, evaluated, and revised as needed. The process of revision once again requires public participation and the cycle is repeated.

The issue of monitoring deserves special consideration. The purpose of monitoring is to ascertain if the objectives laid out in the laws and regulations are being achieved. However, existing data and conventional reporting methods may not provide sufficient information (Boyle et al. 2001). Monitoring programmes and data reporting requirements are a function of laws and regulations

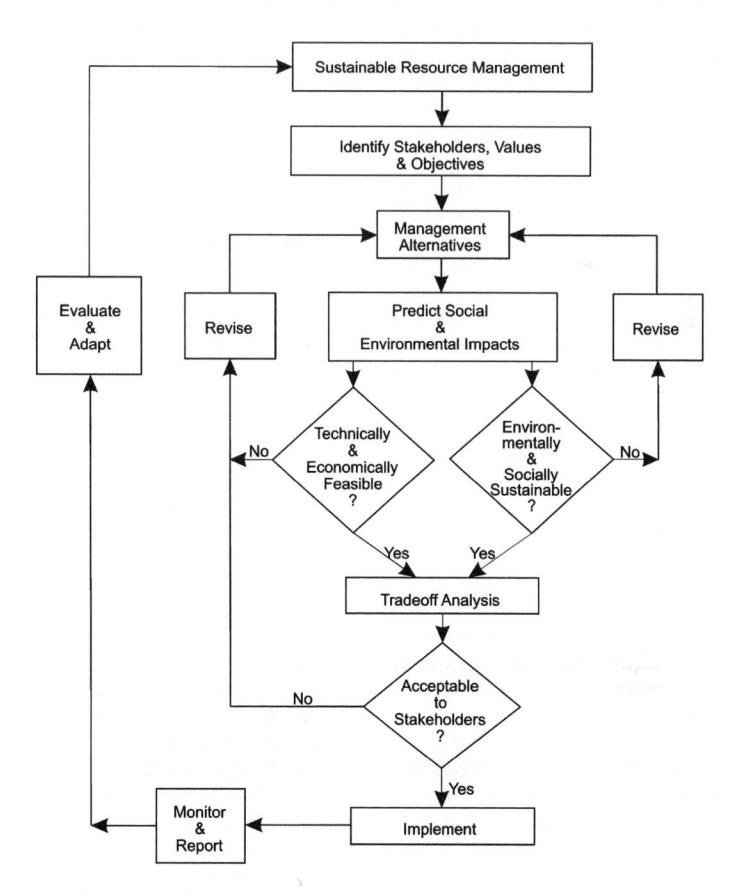

Fig. 2. The implementation process for sustainable mineral resource management.

that were promulgated in response to the policies in effect at the time, as well as the level of scientific understanding we had at the time the requirements were laid out, and what was thought to be important enough to measure. There will continue to be a need for data about the degree of regulatory compliance, but information about our progress in achieving the range of sustainability objectives embodied in policy will also be needed. One purpose of sustainability indicators is to provide these types of information.

Indicators and their use

Agenda 21 (UN 1992b) was a product of the World Conference on Environment and Development held in Rio de Janeiro in 1992. It laid out actions to forward the goal of sustainability. Chapter 40.4 of the Agenda called for the development of indicators of sustainable development (SDI) that could provide a basis for stages of the policy cycle, including decision making at all levels. Developing a set of SDI involves a series of hierarchical steps that was on its top an agreed-upon definition. This is a broad, overarching, vision statement that provides the rationale for policies, practice, and initiatives related to sustainable development. The ways in which the definition is carried out are defined by principles. A principle is a fundamental truth or law as the basis of reasoning or action. Principles are general, but can also have spatial, temporal and other limitations in order to make the definition more operational. Principles can be divided into three dimensions of sustainability.

Principles are in turn supported by criteria. Criteria describe what it means to be sustainable. They serve as basis for evaluation, comparison or assessment, and achievement is judged against relevant indicator(s). An indicator is a parameter (a property that is measured or observed), or value derived from a parameter, which provides information about the state of a phenomenon, environment, or area, with a significance extending beyond that directly associated with a parameter value. Indicators describe, display, or predict the status or trend of some aspect of sustainable development. An index is a set of aggregated or weighted parameters or indicators.

There are three basic functions of indicators: simplification, quantification, and communication. Ideally, an indicator should meet the following criteria: (a) be representative and scientifically valid; (b) be simple and easy to interpret; (c) show trends over time; (d) give early warning about irreversible trends where possible; (e) be sensitive to the changes in the environment or the economy it is meant to describe; (f) be based on readily available data or be available at reasonable cost; (g) be based on data adequately documented and of known quality; (h) be capable of being updated at regular intervals; and (i) have a target level or guideline against which to compare it (DETR 2000). In addition, special attention should be given to scale issue, because most misunderstandings are caused by improper scale interpretations.

The set of indicators and the process for creating them are equally important. Process democracy is one of the most important cornerstones of sustainability. There are many possible processes for defining indicators for various sectors on different scales, but recommendations and even requirements for the group defining the indicator set are similar. The conditions are (a) shared ownership of process, (b) fair decision-making processes, (c) transparency and accountability, (d) adequate participation and representation, (e) a mechanism for future revision, (f) clear grievance procedure, (g) clear structure, and (h) auditability (Scrase & Lindhe 2001). Last, but not least, the same rules that apply when creating the indicators should be also be enforced when creating documents (policy, management plan, and so on).

Indicators are used to track progress toward the goals and objectives set in policies (Šolar & Shields 2000b). Indicators describe policy success or failure. If policy is changed over the course of time, the set of indicators should be reviewed and changed as needed. However, to maintain continuity, the majority of indicators should be those previously utilized. If indicators are to describe the contribution of minerals to sustainable future, then not only must the policy itself be sustainability-based, a clear link must exist between policy and the information supporting or monitoring it.

Indicators and indices package complex mineral information into understandable forms for stakeholders, decision makers and public use. These mineral indicators must be useful as analytical, explanatory, communication, planning and performance assessment tools. Indicators help people understand the complexities associated with mineral resource management policy decisions, such as the interconnectedness of physical and environmental systems and the inevitability of making trade-offs among conflicting management policy objectives. Thus, the information contained in indicators can contribute to public understanding of the state of the world and the potential consequences of fulfilling various objectives, but only if the information contained therein is communicated in a manner that is appropriate to the audience for which it is intended. The basic rules of communication should be followed: (a) regularity/correctness, (b) justice/equity, (c) sincerity/honesty.

Mineral resource development, extraction, use and disposal are complex activities that can be described in many ways. It follows that there are multiple ways to organize mineral indicators. One method is to organize indicators according to the three dimensions of sustainability: economy, environment and society (as used in this paper case). An alternative is to use a life-cycle approach, organizing information based on rock, mine or product life cycle, or some interlinked combination of those cycles. A third approach is to organize issues according to topical hierarchies and then develop indicators applicable to one or more of the scales in each hierarchy.

Finally, indicators can be organized into environmental and human subsystems, each of which has state and process components (ISG 2004). As shown in Figure 3, the organizing framework distinguishes between the states or conditions of interest, and the processes through which changes in those states occur. Within the states of both subsystems the framework distinguishes between current conditions and the enduring capacities that humans rely upon to satisfy their needs. The enduring capacities are called social capacity, economic capital and natural resource capital.

A key feature of this approach is that the human and environmental subsystems are treated in an inclusive, even-handed and logically consistent manner. Tier 1 also shows processes that occur within both subsystems and, most importantly, the interactions that occur between them. The horizontal arrow is a place-holder for indicators of interactions between the human and environmental subsystems. The column running down the right side of this model represents the value- and goal-oriented bases for interpretation of the meaning of indicator sets and evaluation of system sustainability.

Case study

As previously noted, sustainability policies reflect the values and objectives of the people involved in their development. That is true for sustainable mineral policies as well. The objectives of a sustainable minerals policy or management plan and the form it takes differ between regions and countries due to the interplay of differing value sets, goals and objectives. This leads, in turn, to the development and implementation of context-specific indicators, which means that while indicator sets may overlap among countries, each set will have some unique members.

In Slovenia, minerals policy does not exist in the form of a stand-alone document. Some policy guidelines are set within the Mining Act, others reside in policies and laws that relate to other sectors, and policy components have yet to be defined. This will be done in the National

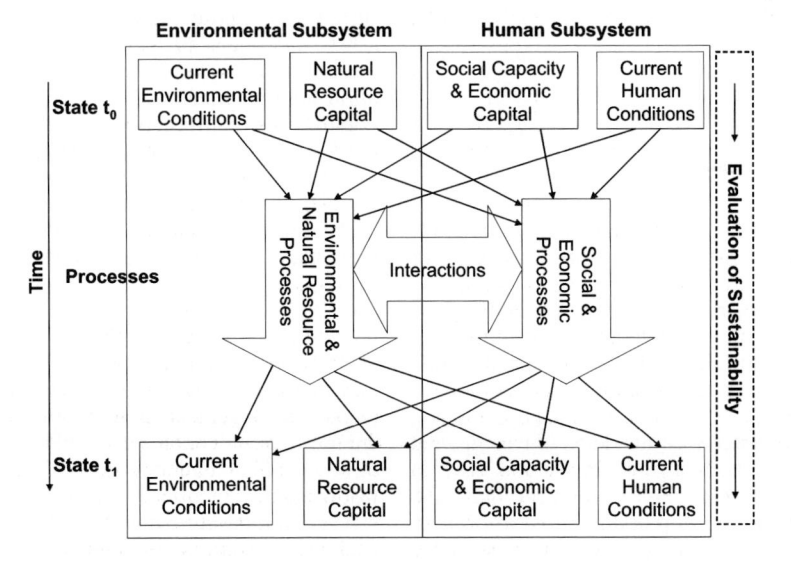

Fig. 3. A conceptual organizing framework for indicators.

Mineral Resource Management Programme (NMRMP), which is required by the Mining Act and is in preparation. Goals for aggregates in NMRMP draft are to secure the aggregate supply, eliminate illegal aggregate extraction, and reduce the overall number of aggregate extraction sites. The latter goal is strongly supported by public opinion on both the local and national scales, as well as by the land use planning and nature conservation authorities. Elimination of illegal quarrying has two aspects: (a) stricter law and regulation implementation to stop illegal activity, and (b) legalization of those quarries that fulfil land use and mining legal requirements. The latter is in conflict with the objective to reduce the number of aggregate extraction sites.

The mission of the NMRMP draft is stated to be ensuring minerals supply and land access by following sustainable development principles. One of the most important of those principles is stakeholder involvement; the plan to secure aggregate supply must reflect the objectives of various stakeholders. Following this principle, the objectives of three major stakeholders were determined (Šolar et al. 2002):

(1) Industry: a stable operating environment, including sufficient sales, production to support sales, and adequate reserves and resources.
(2) Government (Spatial Planning Department): reduction of environmental degradation (to be accomplished through a reduction in the number of quarry sites), regional supply of aggregates, and zoning areas of aggregate resources.
(3) Local Community: minimizing negative effects, including visual intrusions, and ensuring that quarry operations deal with environmental impacts, site closure, and reclamation.

The desired outcome (included in the NMRMP draft) is for a high percentage of legal quarry sites to have what are termed acceptable production and enough reserves/resources. For Slovenia, a 'proper' quarry would have (acceptable) production annually between 50 000 and 500 000 tonnes, and (enough) reserves for between 10 and 50 years of average production. These levels were chosen so as to address the competing objectives of the stakeholders listed above. In Slovenia, there is so far only one 'final' indicator, which resides at the top of a pyramid of primary (raw) and secondary (analysed) data: 'Percentage of 'proper' quarry sites by administrative unit, across spatial scales, that is, from municipality to country' (Langer et al. 2003).

Annual data for this indicator need to include the number of sites, their production, reserves, and resources of aggregates. These data are transformed into an indicator by combining them with land area (on 1000 km^2, or on administrative, statistical units) and average production per capita as a proxy for demand. Figure 4 presents the main indicator plus a series of auxiliary indicators.

The main indicator incorporates an upper limit on production that has been set in response to the significant negative environmental and social impacts of larger operations. An upper limit on reserves (resources) reflects the fact that for large reserves (resources) stock, larger areas need to be exclusively reserved for extraction. That may increase the possibility for potential land use conflict. The lower limit on production and low reserves/resources was included because small and short-term operations are not desired by the public.

These limits are arbitrary to some degree, especially upper limits. For that reason, the conditions were altered in the auxiliary indicators. Some of the auxiliary indicators remove the upper limit of production and reserves, or use only reserve estimates. Other auxiliary indicators use both reserves and resources (probable stock of mineral resources). The main indictor is more easily interpreted when contrasted with the auxiliary indicators.

The main and auxiliary indicators are national in scale. The main indicator's trend is negative; the number of improper quarry sites increased in the period 1983–2001. Two major discontinuities occur because data became more accurate in 1998, and then the revised Mining Act was passed in 1999. From 1999 on, control over operation licences transferred from local to national control. Starting in 2001, all operations were obligated to pay a mandatory royalty.

In 2001, the number of sites with insufficient production and reserves (and resources) was higher then desired by stakeholders and the government. Conversely, there are only a very few locations that have production above the upper limit. This can be seen clearly by examining the auxiliary indicators that do not have an upper limit on either production or reserves (and resources) or use only reserves (and resources). Approximately 5% of all locations are larger than desired, but over 70% of locations have insufficient production and reserves (and resources).

The national-level data are also available for each of Slovenia's 12 regions. These regions are grouped into three main regions to create

Indicators for Aggregates

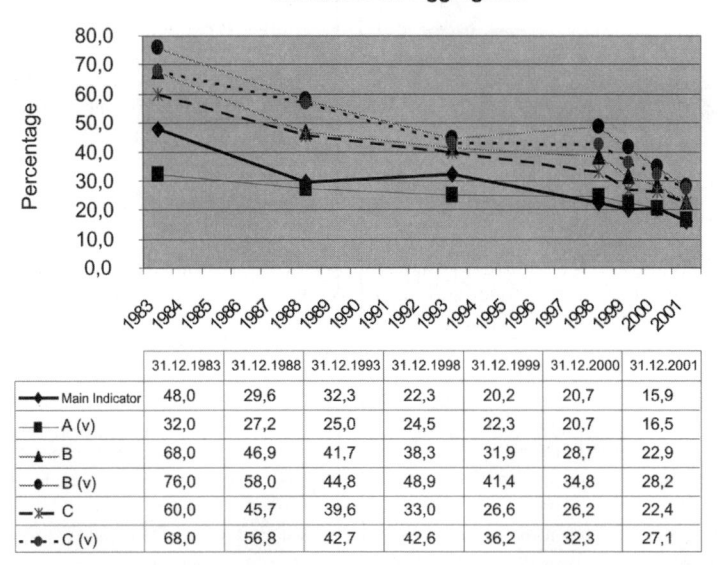

	31.12.1983	31.12.1988	31.12.1993	31.12.1998	31.12.1999	31.12.2000	31.12.2001
Main Indicator	48,0	29,6	32,3	22,3	20,2	20,7	15,9
A (v)	32,0	27,2	25,0	24,5	22,3	20,7	16,5
B	68,0	46,9	41,7	38,3	31,9	28,7	22,9
B (v)	76,0	58,0	44,8	48,9	41,4	34,8	28,2
C	60,0	45,7	39,6	33,0	26,6	26,2	22,4
C (v)	68,0	56,8	42,7	42,6	36,2	32,3	27,1

Fig. 4. Indicators for aggregates for years 1983–2001: A (v), percentage of legal quarry sites that have (acceptable) production annually between 50 000 and 500 000 tonnes, and (enough) reserves and resources for between 10 and 50 years of average production (based on last 5 years); B, percentage of legal quarry sites that have (acceptable) production annually above 50 000 tonnes, and (enough) reserves for more than 10 years of average production (based on last 5 years); B (v), percentage of legal quarry sites that have (acceptable) production annually above 50 000 tonnes, and (enough) reserves and resources for more than 10 years of average production (based on last 5 years); C, percentage of legal quarry sites that have (acceptable) production annually between 50 000 and 500 000 tonnes, and (enough) reserves for more than 10 years of average production (based on last 5 years); C (v), percentage of legal quarry sites that have (acceptable) production annually between 50 000 and 500 000 tonnes, and (enough) reserves and resources for more than 10 years of average production (based on last 5 years).

sets of data and indicators for the country, as well as for the three and 12 regions. Comparison at the most disaggregated scale clearly demonstrates those parts of the country in which the value of the indicator is below or above the national value (Fig. 5).

Additional insight can be gained by considering the frame of extraction activity; overall number of sites, production, reserves and resources and their ratio to population and, area (Table 1). Aggregate data are also divided into groups of crushed stone, and sand and gravel. The crushed stone group is further divided into limestone, dolomite and other rocks. Three and seven production and reserve (and resources) size classes were created. The three main indicator classes are for production (below 50 000 tonnes of annual production, between 50 000 and 500 000 tonnes, and above 500 000 tonnes) and for reserves (resources) (below 10 years, between 10 and 50 years, and above 50 years). The three classes were also broken into seven subclasses. All is information is part of the indicator information pyramid, which helps explain the details of the main indicator and also exposes other related information that may be of interest to many stakeholders at the regional and national level.

In the end, the process of creating the Slovenian NMRMP was as important as the programme itself because of the need for public acceptance. This was true also for the choice of indicators. An information pyramid was used to demonstrate how data were concentrated. The aggregation of data made complex and comprehensive information understandable. Data on reserves, resources, and production of mineral resources were captured within a simple, flexible database system and reported as a single indicator. The data manipulation tools will also provide opportunities for an inquiry of individual data as well as analysis of indicator information on different spatial and temporal scales (Šolar 2003).

Inevitably, many practical problems have to be solved during the process of indicator creation. In the Slovenian case, the indicator selection process had to deal with unrealistically high

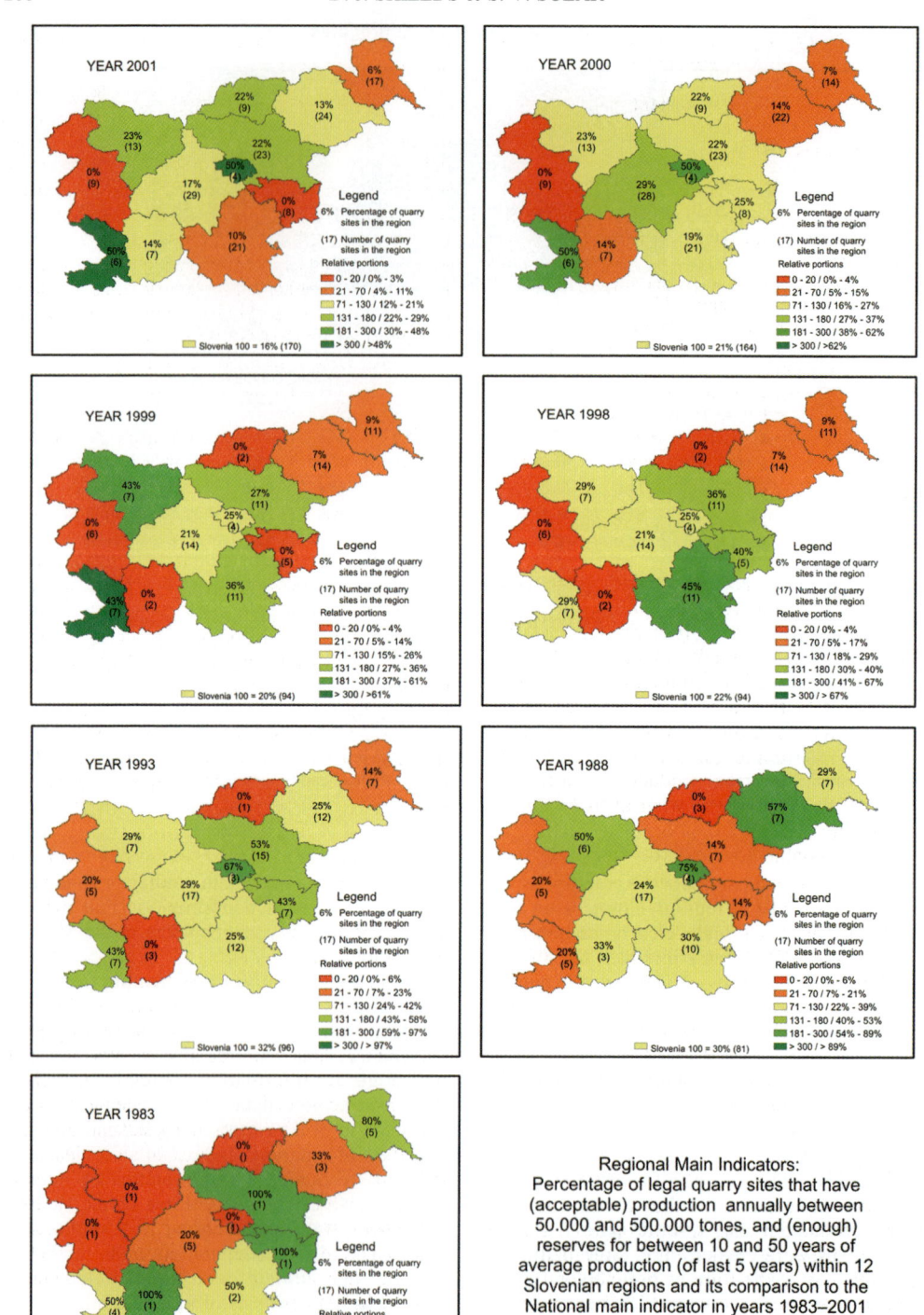

Fig. 5. Regional indicators.

Table 1. *Data for aggregates for year 2001*

Number of sites 170
- Annual production 15 500 000 t
- Reserves: 590 million t
- Reserves and resources: 1155 million t
- Per site
  - Production per site: 91 000 t
  - Reserves per site: 3.5 million t
  - Reserves and resources per site: 6.8 million t
- Per capita
  - Number of inhabitants per site: 11 718
  - Production per capita: 7.7 t
  - Reserves per capita: 295.76 t
  - Reserves and resources per capita: 580.24 t
- Per area (1000 km^2)
  - Number of sites per area: 8.39
  - Production per area: 763 662 t
  - Reserves per area: 29.1 million t
  - Reserves and resources per area: 57.0 million t

stakeholder expectations that complete information would be incorporated into a single indicator, when in reality only a limited mount of information was actually incorporated in the selected indicator. That indicator was based on available data that had been collected over a long period of time. During the reporting period, data accuracy was inconsistent; in the 1980s, data on location, production and reserves were collected only from what at that time was termed large aggregate sites. Data accuracy is also questionable, due to the fact that many quarry operators were not filling out the annual questionnaires in the same manner. There were particular problems related to reporting on reserves and production cubic metres. No clear distinction was made between cubic metres of intact mineral resources (in site) and dispersed cubic metres of mineral resources on tracks (of production). As a result, data were calculated into tonnes under the assumption that all cubic metres were dispersed.

Conclusions

Nations have used their mineral endowment as an engine of development throughout history, and part of the legacy of doing so remains with us to this day in the form of polluted and unreclaimed mine sites and altered landscapes. There is also a long history of mineral development that was approved because it was expected to create wealth that could be used to better people's lives, but that actually enriched only the mining company and select elites. As a result, the mining industry is in danger of losing the social licence to operate. This does not need to happen.

Societies, both developed and developing, need a stream of material inputs. And developing nations need to generate wealth to alleviate poverty and fund programmes that increase the well-being of their citizens. Domestic mineral development can provide both materials and new wealth. Moreover, creation of a value-added sector for minerals can reduce the need for imported materials, while allowing the domestic economy to capture the economic benefits (profits, employment, tax revenues) that would otherwise accrue in another country. The question is how to do so in a manner that most benefits the nation.

We believe that sustainable mineral resource management offers a structure within which nations can maximize the benefits, and recognize and control the costs, of mineral development. For this to occur, developing nations must determine *a priori* what their development goals are and how mineral resources can and should contribute to the achievement of those goals. They must then carefully track and regulate the behaviour of the extractive sector to ensure that their contributions to society are net positive.

That tracking is accomplished with mineral indicators of sustainability. A selected set of mineral indicators should express a need for balance: (a) among stakeholders; (b) between the process of defining indicators and the set of chosen indicators; and (c) among dimensions of sustainability. Mineral indicators of sustainability should be used: (a) as tools for knowledge and, information transfer; (b) as integral parts of other initiatives and sets of indicators; and (c) as a solid base for decision making.

In this paper, we have described how Slovenia identified its goals with respect to aggregate resources and then developed an indicator that provided information about the degree to which that goal was being achieved. Creating the indicator required input from many stakeholders and the collection and aggregation of data. The main indicator for aggregates in Slovenia clearly points out that policy goals are not being met, but rather just the opposite. Declared stakeholders' objectives are in greater opposition to the actual state of the world with each passing year. This is a clear sign of a need for policy action.

Slovenia's indicator is not necessarily useful to other countries. Each country must identify for itself the information needed for decision making. Moreover, the indicator presented here cannot stand alone. It needs to be placed in the context of a complete set of indicators that describe the economic, environmental, and social situation vis à vis mining.

References

ANAND, S. & SEN, A. 2000. Human development and economic sustainability. *World Development*, **48**, 2029–2049.

ASCHER, W. & HEALY, R. 1990. *Natural Resource Policymaking in Developing Countries*. Duke University Press, Durham, NC.

AUTY, R. 2003. *Natural resources, development models and sustainable development*. IISD Environmental Economics Programme, Discussion Paper 03–01, http://www.iied.org/eep.

AUTY, R. & MIKESELL, R. 1998. *Sustainable Development in Mineral Economies*. Oxford University Press, Oxford.

BAIRD, M. & SHETTY, S. 2003. Getting there. *Finance and Development*, **40**(4), 14–19.

BOSSEL, H. 1999. *Indicators for Sustainable Development: Theory, Method, Applications, A Report to the Balaton Group*. International Institute of Sustainable Development, Winnipeg, Manitoba, Canada.

BOYLE, M., KAY, J. J. & POND, B. 2001. Monitoring in support of policy: an adaptive ecosystem approach. *In*: MUNN, T. (ed) *Encyclopaedia of Global Environmental Change*. Wiley & Son, New York.

CIB UN ENVIRONMENT PROGRAM 2002. *Agenda 21 for Sustainable Construction in Developing Countries*. CSIR – Council for Scientific and Industrial Research (CIB – International Council for Research and Innovation in Building and Construction), Pretoria.

CORDES, J. 2000. Introduction. *In*: OTTO, J. & CORDES, J. (eds) *Sustainable Development and the Future of Mineral Development*. UNEP, Paris.

COSTANZA, R. & DALY, H. E. 1992. Natural capital and sustainable development. *Conservation Biology*, **6**(1), 1–37.

DAVIS, S. K. 1997. Fighting over public lands: interest groups, states and the federal government. *In*: DAVIS, C. (ed) *Western Public Lands and Environmental Politics*. Westview Press, Boulder, CO, 11–31.

DAWE, N. & RYAN, K. 2003. The faulty three-legged stool model of sustainable development. *Conservation Biology*, **17**(5), 1458–1460.

DEPARTMENT OF THE ENVIRONMENT, TRANSPORT, AND THE REGIONS (DETR) 2000. *Planning for the Supply of Aggregates in England*. DETR, United Kingdom, http://www.odpm.gov.uk/stellent/groups/odpm_planning/documents/pdf/odpm_plan_pdf_605804.pdf

DOBRA, J. 2002. *The US Gold Industry 2001*. Nevada Bureau of Mines and Geology Special Publication 32.

DOVERS, S. 1996. Sustainability: demands on policy. *Journal of Public Policy*, **16**(3), 303–318.

FAUCHEUX, S., MUIR, E. & O'CONNOR, M. 1997. Neoclassical natural capital theory and 'weak' indicators for sustainability. *Land Economics*, **73**(4), 528–552.

FENNA, A. 1998. *Introduction to Australian Public Policy*. Addison Wesley Longman, Melbourne.

FINDLAY, C. 1997. National geological surveys and the winds of change. *Nature and Resources*, **33**(1), 18–25.

FUNTOWICZ, S. & RAVETZ, J. 2001. Post-Normal Science: environmental policy under conditions of complexity, http://www.jvds.nl/pns/pns.htm

HAMILTON, K. & CLEMENS, M. 1999. Genuine savings rates in developing countries. *World Bank Economic Review*, **13**(2), 333–356.

INTEGRATION AND SYNTHESIS GROUP (ISG) 2004. *Progress Report of the Roundtable Network Integration and Synthesis Group*. Available from D. Shields, USDA Forest Service, Rocky Mountain Research Station, 2150A Centre Ave, Ft. Collins, CO 80526 USA.

INTERNATIONAL INSTITUTE FOR SUSTAINABLE DEVELOPMENT (IIED) 2002a. *Seven Questions to Sustainability*. IIED, Winnipeg.

INTERNATIONAL INSTITUTE FOR SUSTAINABLE DEVELOPMENT (IIED) AND THE WORLD BUSINESS COUNCIL FOR SUSTAINABLE DEVELOPMENT (WBCSD) 2002b. *Breaking New Ground – The Report of the Mining, Minerals and Sustainable Development Project*, First Edition. Earthscan Publication Ltd., London, UK.

ISLAM, S., MUNASINGHE, M. & CLARK, M. 2003. Making long-term growth more sustainable: evaluating the costs and benefits. *Ecological Economics*, **47**, 149–166.

LAMBERT, I. B. 2001. Mining and sustainable development: Considerations for minerals supply. *Natural Resources Forum*, **25**(4), 1–19.

LANGER, W. H. & ŠOLAR, S. V. 2002. Natural aggregate resources – environmental issues and resource managements report of Working Group 5. *In*: FABBRI, A. G., GAAL, G. & MCCAMMON, R. B. (eds) *Deposit and Geoenvironmental Models for Resource Exploitation and Environmental Security*. NATO Sciences Series, Series 2, Environmental Security, Vol. 80. Kluwer, Dordrecht, Netherlands, 325–332.

LANGER, W. H. & TUCKER, M. L. 2003. *Specification Aggregate Quarry Expansion – A Case Study Demonstrating Sustainable Management of Natural Aggregate Resources*. Open-File Report 03–121, U.S. Department of the Interior, U.S. Geological Survey, http://pubs.usgs.gov/of/2003/ofr-03-121/OFR-03-121-508.pdf)

LANGER, W. H., ŠOLAR, S. V., SHIELDS, D. J. & GIUSTI, C. 2003: Sustainability indicators for aggregates. *In*: AGIOUTANTIS, Z. (ed) *Sustainable Development Indicators in the Mineral Industries*. Milos Conference Center – George Eliopoulos, Greece, 251–257.

LUNGU, S. & PRICE, A. 2000. Sustainability Considerations and Funding Criteria for Infrastructure Projects in Developing Countries. *1st International Conference: Creating a Sustainable Construction Industry in Developing Countries*, 7, http://buildnet.csir.co.za/cdcproc/docs/3rd/lungu_price.pdf

MEADOWS, D. H. 1998. *Indicators and Information Systems for Sustainable Development – A Report* to the Balaton Group. The Sustainability Institute, Hartland Four Corners, VT.

MUNASINGHE, M. 2002. The sustainomics transdisciplinary meta-framework for making development more sustainable: applications to energy issues. *International Journal of Sustainable Development*, **5**(1/2), 125–182.

ORGANIZATION FOR ECONOMIC COOPERATION AND DEVELOPMENT (OECD) 1994. *Environmental Indicators*. OECD, Paris.

OTTO, J. 1997. A national mineral policy as a regulatory tool. *Resources Policy*, **23**(1/2), 1–7.

OTTO, J. M. 2001. *Mineral Policy, Legislation and Regulation/advanced copy/*. United Nations Conference on Trade and Development (UNCTAD), http://www.naturalresources.org/minerals/generalforum/docs/pdfs

PEARCE, D. W. & ATKINSON, G. 1993. Capital theory and the measurement of sustainable development: an indicator of weak sustainability. *Ecological Economics*, **8**, 103–108.

PRESCOTT-ALLEN, R. 2001. *The Well-being of Nations*. Island Press, Washington, DC.

RAMMEL, C. & VAN DEN BERGH, J. 2003. Evolutionary policies for sustainable development: adaptive flexibility and risk minimizing. *Ecological Economics*, **47**, 121–133.

SCRASE, H. & LINDHE, A. 2001. *Developing Forest Stewardship Standards – A Survival Guide*, Taiga Rescue Network, Jokkmokk, Sweden, http://www.taigarescue.org

SHIELDS, D. J. & MITCHELL, J. E. 1997. *A Hierarchical Systems Model of Ecosystem Management*, Working Paper. USDA Forest Service, Rocky Mountain Research Station, Fort Collins, CO.

SHIELDS, D. & ŠOLAR, S. V. 2000a. Challenges to sustainable development in the mining sector. *Industry and Environment*, Special Issue, 16–19.

SHIELDS, D. J. & ŠOLAR, S. V. 2000b. Alternative approaches to sustainable development: implications for mineral resource management. *31st International Geological Congress*, Rio de Janeiro, Brazil, 6–17 August 2000, abstracts volume, 1 p. (+poster), available from the authors.

SHIELDS, D., ŠOLAR, S. V., & MARTIN, W. 2002a. The role of values and objectives in communicating indicators of sustainability. *Ecological Indicators*, **2**, 149–160.

SHIELDS, D. J., ŠOLAR, S. V., MARTIN, W. & MARTIN, I. 2002b. Conflicting objectives for mineral resource management: the case of aggregate quarries in Slovenia. *In*: CICCU, R. (ed) *Proceedings of the Seventh International Symposium on Environmental Issues and Waste Management in Energy and Mineral Production*, Cagliari, Italy, 7–10 October, 2002.

ŠOLAR, S. V. 2003. *Indicators of Sustainable Development for a Mineral Resource Management Plan: The Case of Open Pits*. PhD thesis, University of Ljubljana, 182 p., 3 app.

ŠOLAR, S. V. & SHIELDS, D. 2000a. The need for geology in sustainable development policies. *European Geology*, **10**, 134–138.

ŠOLAR, S. V. & SHIELDS, D. 2000b. Mineral indicators of sustainability: review and systemization. *In*: *31st International Geological Congress*, Rio de Janeiro, Brazil, 6–17 August 2000, abstracts volume, 1 p. (+poster), available from the authors.

ŠOLAR, S. V., SHIELDS, D., MARTIN, W. & MARTIN, I. 2002. Balancing the costs and benefits of aggregate extraction in Slovenia: are existing data adequate? Raw materials planning in europe change of conditions! New perspectives? *3rd European Conference on Mineral Planning*, Krefeld, Germany, Geological Survey of North Rhine-Westphalia, 73–78.

ŠOLAR, S. V., SHIELDS, D. & LANGER, W. H. 2004. Important features of sustainable aggregate resource management. *Geologia*, **47**, 99–108.

SOLOW, R. 1993. An almost practical step toward sustainability. *Resources Policy*, **19**(30), 162–172.

SRINIVASAN, T. 1994. Human development: a new paradigm or reinvention of the wheel? *American Economic Review*, **84**(2), 238–243.

TOMAN, M. A. 1994. Economics and 'sustainability': balancing trade-offs and imperatives. *Land Economics*, **70**(4), 399–413.

TOMAN, M., PESSEY, J. & KRAUTKREMER, J. 1995. Neoclassical economic growth theory and 'sustainability'. *In*: BROMLY, D. (ed) *Handbook of Environmental Economics*. Blackwell, Oxford, 139–165.

UNITED NATIONS 1992a. *Rio Declaration on Environment and Development*, 13 June 1992, U.N. Doc./CONF.151/5/Rev.1.

UNITED NATIONS 1992b. *Agenda 21 – Report of the United Nations Conference on Environment and Development*, Rio de Janeiro, 3–14 June 1992. United Nations, New York.

UNITED NATIONS 2003. *Report of the World Summit on Sustainable Development*. UN Doc. A/Conf.199/20. United Nations, New York, http://www.johannesburgsummit.org/html/documents/documents.html

UNITED NATIONS DEVELOPMENT PROGRAM (UNDP) 2000. *Millennium Development Goals*, //http://www.un.org/millenniumgoals/

USDA FOREST SERVICE 2004. *National Report on Sustainable Forests – 2003*, FS-766. Government Printing Office, Washington, DC.

USDA FOREST SERVICE, INVENTORYING AND MONITORING INSTITUTE 2003. *Monitoring for Sustainability*, http://www.fs.fed.us/institute/monitoring/ sustainability_monitoring.htm

VAN DER STRAATEN, J. 1998. Sustainable development and public policy. *In*: FAUCHEUX, S., O'CONNOR, M. & VAN DER STRAATEN, J. (eds) *Sustainable Development: Concepts, Rationalities and Strategies*. Kluwer, Dordrecht, 69–83.

WACKERNAGEL, M. & REES, W. 1996. *Our Ecological Footprint: Reducing Human Impact on the Earth*. New Society Press, Gabriola Island.

WORLD COMMISSION ON ENVIRONMENT AND DEVELOPMENT (WCED) 1987. *Our Common Future*. Oxford University Press, Oxford.

WORLD BANK. 2003. *Extractive Industries Review*: Volume I, *Striking a Better Balance*, available at http://www.eireview.org

System dynamics modelling: a more effective tool for assessing the impact of sustainable development policies on the mining industry

B. O'REGAN & R. MOLES

Centre for Environmental Research, University of Limerick, Limerick, Ireland
(e-mail: bernadette.oregan@ul.ie)

Abstract: The high mobility of mining investment is frequently cited in the literature. Consequently, the concept of relative attractiveness is particularly important. This paper describes a detailed computer simulation 'feedback' model. The model provides a means of examining the effects of varied environmental, fiscal and corporate policies on the flow of investment funds and mineral resources between a number of simulated mining firms and competing countries. Through a quantitative analysis of existing data, the model exposes, within the context of sustainable development, the underlying assumptions used as a basis for corporate decisions. Through the compression of time, the model provides a means of taking these assumptions to their logical conclusions. Exposing assumptions in this way leaves less room for misinterpretation and provides a solid basis for enhancing the understanding of system structure. It is by better understanding system structure that more effective sustainable development policies may be designed and implemented. An outline of the system dynamics method is also presented.

The most recent Irish Government statement pertaining to the mining industry, the *National Minerals Policy Review Group (NMPRG) Report*, begins by stressing the need to 'recognize the complexity and interdependence of the factors which have influenced investment in the minerals sector in Ireland over the past three decades' (NMPRG 1995). In the introduction to this report, the authors expose their mental models of the dynamics of the mining industry in Ireland, as can be seen from the following extracts:

- The relative attractiveness of Ireland as a location for minerals exploration and development is not determined solely by geology and national domestic policy, but must also be seen in the context of very competitive international markets and the aggressive marketing strategies pursued by other countries competing for very mobile international investment funds.
- The Minerals Industry has not been successful in attracting sufficient investment interest from overseas to enable it to reach its true potential.

To enable a minerals industry to 'reach its true potential', it is necessary to increase significantly the level of exploration activity while protecting the environment. Given that most of the exploration licences in Ireland in 2003 are held by multinational firms, this means attracting an increasing share of these firms' exploration budgets at the expense of other competing countries. The high mobility of mining investment is often cited in the literature and, in this respect, the concept of relative attractiveness is particularly appropriate. In a free market, a mining firm will concentrate its exploration budget in countries that present the most attractive investment opportunities. This creates a positive feedback loop towards further exploration activity, until, eventually, some limit is reached, thus reducing the particular country's relative attractiveness. Also, in times of shrinking minerals markets firms have shifted their investment to more stable regions, reflecting the fact that, in their opinion, these countries have become relatively more attractive as the firms themselves become more risk averse. This is an important point, as it shows that relative attractiveness is a dynamic concept. A government can control the minerals policy within its own boundaries, but, in order to remain competitive, policies must constantly evolve to be at least equally attractive as the policies of other countries. This situation is particularly relevant today, with some developing countries offering less stringent environmental restrictions and increased land access in order to attract mineral investment funds. In order to move towards a more sustainable development, all mineral-producing countries must recognize and identify the trade-offs concerned.

Modelling complexity

Base metals are traded on the international commodity markets. These markets, which can be considered exogenous to both individual firm behaviour and government policy, are subject to fluctuations that impact on the performance of mining firms and the relative attractiveness of mineral-producing countries. A hypothetical mining firm might decide to invest a proportion of its exploration budget in a particular country, on the basis of the prevailing investment climate in that country. It is more likely that it will spread its investment among a number of countries as a function of their relative attractiveness. Furthermore, the firm's decision mechanisms do not exist in isolation, but are dependent on the activities of its competitors, reflected in the behaviour of the international minerals market. In this respect organizational boundaries become crucial when the implications of policy decisions are examined. Decisions taken in one organization, for example, government, have a direct effect on other organizations, mining firms, and many problems may be caused by the lack of integration in multiorganizational systems (Fig. 1).

[Government] Policy makers who understand the specific factors which influence private sector mining company investment decision-making are in a better position to evolve appropriate mining investment policy, laws and agreements to meet their national objectives. (Otto 1994a)

A computer simulation model, developed by the authors in the system dynamics tradition, provides a powerful means of exposing system complexity and, thus, increasing understanding. The model shows how exploration spending is a function of the relative attractiveness of individual deposits in competing countries. This relative attractiveness, as defined by the expected net value of exploration, changes over time as determined by the interrelationships between many other factors such as the host government environmental regulatory and planning requirements, the level of taxation and the availability of accurate geological information. The exact nature of these complex interrelationships, as assumed by the model, is made explicit through the variable definitions. These assumptions (definitions) can be modified and the resulting changes in behaviour patterns examined.

The greatest advantage in adopting system dynamics as an analytical tool is that it exposes the many interrelationships (structure) that influence the behaviour of a complex system. In a

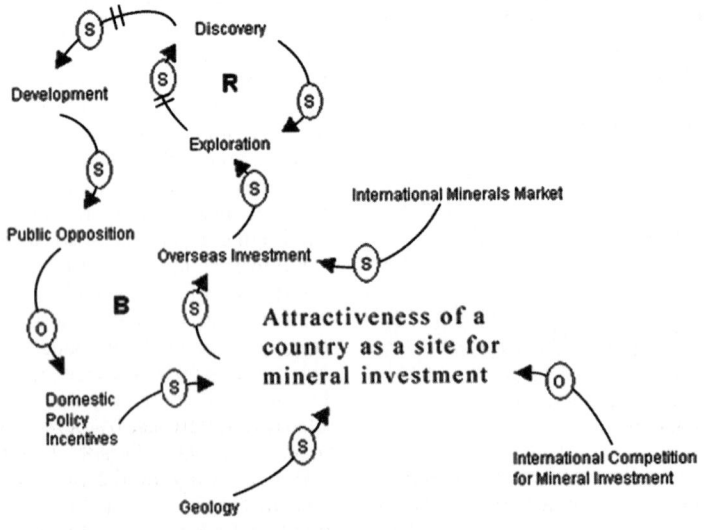

S: Change in the (S)ame direction
O: Change in the (O)pposite direction
B: (B)alancing loop R: (R)einforcing loop

Fig. 1. Attractiveness of a country as a site for mineral investment.

complex system, such as the flow of mineral investment funds, the same change to environmental or fiscal policy does not always have the same effect. Instead, the effect is dependent on the 'state' of the system at a particular point in time. Through its effectiveness at capturing and exposing the state of the system, this model improves on more conventional methods for evaluating the effectiveness of environmental and socio-economic policies.

> To manage complex social systems effectively, policy makers must bring together a variety of mental models, both ecological and technological; translate them into a common language; and determine simultaneously all their important implications. That process of synthesis requires formal models, that is, models whose assumptions are stated explicitly so that they can be widely examined and discussed. (Meadows *et al.* 1974)

Modelling policies and decisions

Critics of computer simulation modelling sometimes argue that, as yet, not enough is known to construct models of social systems. Yet, by necessity, governments and institutions continue to redesign them, usually through legislation and 'improvement programmes'. Because social systems are complex, their behaviour patterns are often difficult to understand and predict. The reason for this is that, intuitively, people are more used to dealing with simple systems, where the cause and effect are closely related in space and time. These systems usually take the form of first-order, negative-feedback loops containing a single important system-level variable. The resulting intuition and judgement do not hold true for complex systems that exhibit counterintuitive behaviour. One of the reasons for this is that complex systems often present an apparent cause that is closely related in space and time to the observed symptoms.

> No plea about the inadequacy of our understanding of the decision-making processes can excuse us from estimating decision-making criteria. To omit a decision point is to deny its presence – a mistake of far greater magnitude than any errors in our best estimate of the process. (Forrester 1994)

Complex systems are insensitive to parameter changes, particularly external policy changes (Richardson 1991). When a policy is introduced with the intention of modifying system behaviour, the non-linear nature of the interconnected feedback loops often causes a shift in dominance whereby the system relaxes its internal pressures, thus compensating for the external intervention.

For this reason it is always better to concentrate effort on making system modifications that change the internal incentives of the system. Consider the example depicted in Figure 2.

System dynamics fosters a feedback view of management as a process that converts information into action. This process is, in essence, a decision process and success depends on selecting the right information and using it effectively. From this perspective, a policy is a guiding rule, an aid to decision making. The decision process is complicated by the fact that information about the outcome of actions taken is never immediately available. Therefore, it becomes necessary to distinguish between perceived and actual information (e.g. perceived/actual geology, expected/actual price, estimated/actual mine life).

Through a quantitative analysis of existing data, the model exposes, within the context of the problem area, the underlying assumptions used as a basis for policy formulation and corporate decisions (Fig. 3). Without criteria for evaluating the comparative advantages and disadvantages of various policy options it is difficult to select, on rational grounds, an appropriate policy response for a particular problem. Perhaps most critical to policy makers is the difficulty of comparing alternative action plans and their long- and short-term consequences (Legasto & Maciariello 1980).

Furthermore, through the compression of time, the model provides a means of taking these assumptions to their logical conclusions. Exposing assumptions in this way leaves less room for misinterpretation and provides a solid basis for enhancing the understanding of system structure and improving the effectiveness of sustainable development policies.

The model underpinning this paper emphasizes the role of environmental policy within the socio-economic constraints of mining firms. Environmental policy has evolved over time as a reaction to increasing exposure to a series of limiting conditions in the natural environment. Insofar as its formulation is documented, it typically comprises a set of written descriptions encapsulating the collective knowledge of some of the more experienced practitioners. These verbal descriptions, by definition, are linear in nature. Therefore, they are unable to expose the true complexity of the problems that they are created to address, and they do not allow full prediction of the consequences of their implementation. In creating the simulation model described here, the intention is to enrich existing verbal and written models of the factors impacting on the flow of international mineral investment. These factors include host

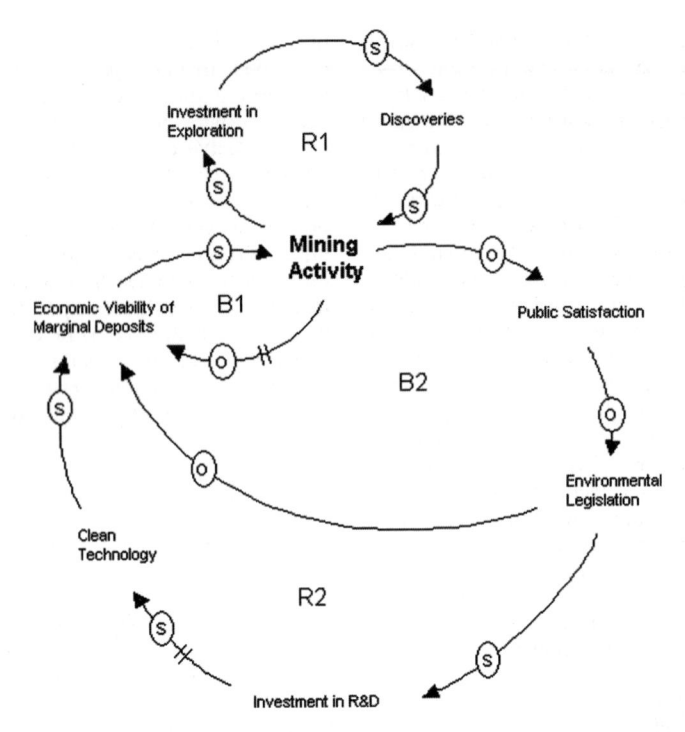

Fig. 2. Systems resist change. One of the principles of complex systems is that they are insensitive to parameter changes, particularly in the form of external policies. For example, these more stringent environmental regulations will, in the long term, force companies to find ways to deal more effectively with their waste. This is achieved through increased investment in R&D, which, in turn, often results in improved engineering and processing technologies, thus again increasing the economic viability of marginal deposits R2. This is an example of a shift in dominance between the feedback loops in a system. These shifts in dominance are one of the principal reasons why complex systems exhibit counterintuitive behaviour.

government environmental and socio-economic policy. This policy mix is in effect the sustainable development policy of the host country as it covers the areas that are necessary to sustainability. Others relate to the various exploration and development decisions of individual mining firms in response to these policies. As such, the model may be thought of as the synthesis of existing documentation to form a more transparent and dynamic tool for policy analysis.

Model variables and relationships

The model contains over 8000 individual model objects (array elements and scalars). Almost all of these objects are dependent variables; that is, their value at any particular time is determined mathematically based on the current 'state' of the system. System state is defined by the collective values of the level variables in the system, of which there are 31. These level variables, together with the 24 model constants, act as initial conditions for the model and can be changed at the beginning of, or during, the simulation run to reflect particular circumstances such as changes in corporate or government policy. Many of the levels are derived

Fig. 3. Modelling actions and decisions.

variables (as opposed to fundamental variables) in that their values are directly dependent on the values of other levels. Some of the most significant level variables and constants are presented in Table 1.

Zinc, the most commonly mined mineral in Ireland, was chosen as a basis for parameter values in the construction of the model, but any base metal can readily be substituted with relevant parameter changes. The structure of the model is captured through the relationships between the variables. Many of these relationships are simple (proportional) in nature and therefore easy to define. Non-proportional relationships are modelled through the use of multipliers or graph functions. The main use of a multiplier in a system dynamics model is to act as a changing (dynamic) pressure on decision making. This is in contrast to the static, normal, value, which takes effect when the system is in equilibrium. For example, the decision of how much to invest in exploration, represented by the Exploration Budget, is defined as the product of static and dynamic pressures as follows:

Exploration_Budget =
Exploration_Budget_Normal*
Expected_Price_to_Exploration_Budget_ Multiplier *
Market_Share_to_Exploration_Budget_ Multiplier   (1)

Exploration_Budget_Normal represents the typical exploration budget, all other things being equal. The size of the exploration budget is affected by changes in expected price. This price pressure, which cannot be ignored, is modelled through the use of an expected price to exploration budget multiplier. When the expected future price equates to the long-term median price ($1235/tonne), then expected price neither has a positive or negative pressure on the size of the exploration budget, point 1 on the y-axis (Fig. 4). However, as the expected price of a tonne of the mineral exceeds $1235/tonne, then there is an incentive (pressure) to increase the size of the exploration budget to maximize gains from expected improvements in market conditions. Similarly, when price is expected to fall, there is an incentive to reduce the exploration budget as it is perceived that there will be less opportunity for profit.

Perceived risk

As exploration and development take such a long time, there are risks associated with any possible investment location. During this time, many important factors such as the mineral policies, the government itself and the public attitude to mining within the country may alter significantly. These risks must be factored in to decrease the expected value of the exploration investment. As the firm is attempting to predict events over a very long time period, the estimates that are made at this stage of the mining process are very uncertain (Astakhov et al. 1988).

For example, clear and transparent terms under which tenure rights are granted and terminated at all stages in the exploration to development process are very important to mining firms. Johnson (1990) identified the right to mine any deposit as the most important non-negotiable

Table 1. *Sample of model variables*

Actual_Geology	Available_Resources
Average_Ore_Grade	Book_Value_of_Mine
Cash	Cumulative_Income_Reserves
Debt	Discovery_Delay
E_Cost_of_Exploration_Effort	Equity
Expected_Demand	Expected_Price
Exploration_Spending_by_Country	Explored_Resources
Extracted_Ore	LME_Inventory
Metal_Recovery	Mineable_Reserves
Paid_in_Capital	Perceived_Geological_Potential
Perceived_Ore_Grade	Proven_Reserves
Refined_Mineral	Regulatory_and_Planning_ Requirements
Retained_Earnings	Waste
Exploration_Budget_Allocation	Construction_Delay_Normal
Costs_of_Local_Inputs	Interest_Rate
Perceived_Political_Stability	Perceived_Security_of_Tenure
Percentage_of_Profits_Reinvested_in_New_Mines	Planning_Delay_Normal
Pollution_Tax_Rate	Price_Normal
Profits_Reinvisted_in_Exploration	Taxation_Percentage

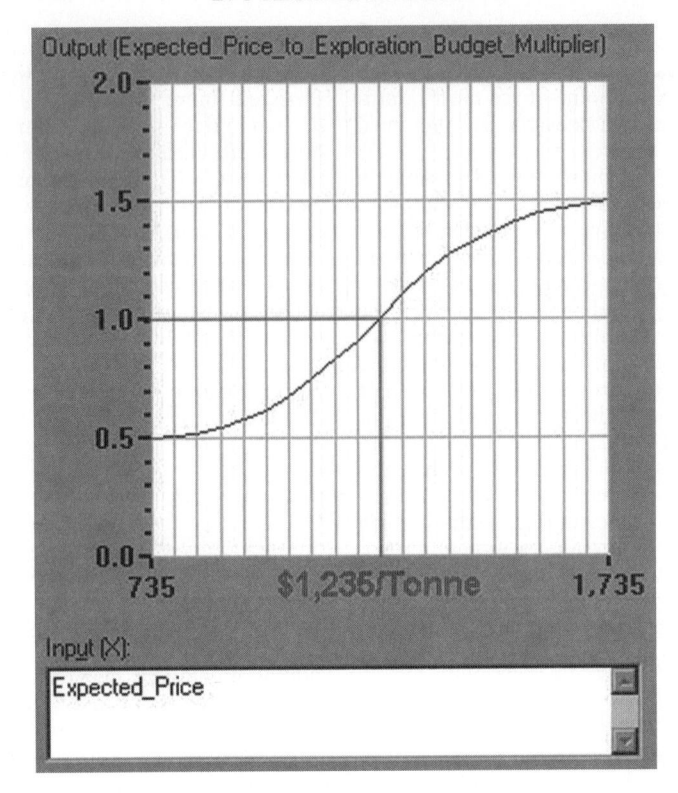

Fig. 4. The effect of Expected Price on Exploration Budget modelled as a graph function or multiplier.

factor before a mining firm will consider investing in a country. The economic argument for security of tenure is that exploration risk should only be assumed if there is an expectation of obtaining subsequent mining rights. In some developing countries, due to political problems, security of tenure has been a contentious issue. However, it is important to note that tenure issues also arise in developed countries. For example, in Ireland during the 1990s, a number of potentially economically viable deposits failed to be developed. In particular, the Irish Government refused permission to Ennex International to develop a gold mine on Croagh Patrick, Co. Mayo, on environmental and religious grounds. Permission to develop the deposit was refused only after considerable investment in exploration of the site by Ennex. This resulted in an increased risk associated with investing in Ireland.

Figure 5 shows the output for a deposit from two simulation runs, one where construction commences as planned in simulation year 2008 and the other where the mining licence is refused.

Figure 6 shows how this refusal not only affects the development of the deposit in question, but also has negative feedback on the future expected net value of exploration in the country in question, in the short term. More specifically, the perceived security of tenure is adversely affected immediately on refusal to grant the mining licence. This, in turn, impacts the perceived risk, for all mining firms, of carrying out further exploration in that country.

However, a firm's perception of security of tenure (and other risk factors) will return to normal over time if there are no other unexpected policy changes. For this reason, the expected net value of exploration is only affected for a short period, until approximately 2015. Nevertheless, if there is a second refusal of a mining licence, as happens in 2064, then the effect on perceived security of tenure is compounded and so it takes longer for the perception of risk to return to normal. This underlines the need for consistency over time in government policy if the inflow of minerals investment is to be maintained.

Firms tend to invest in countries where there has been prior successful investment by multinational mining firms. For example, most zinc exploration is focused around existing zinc mines. This is seen as an indication that there is a workable

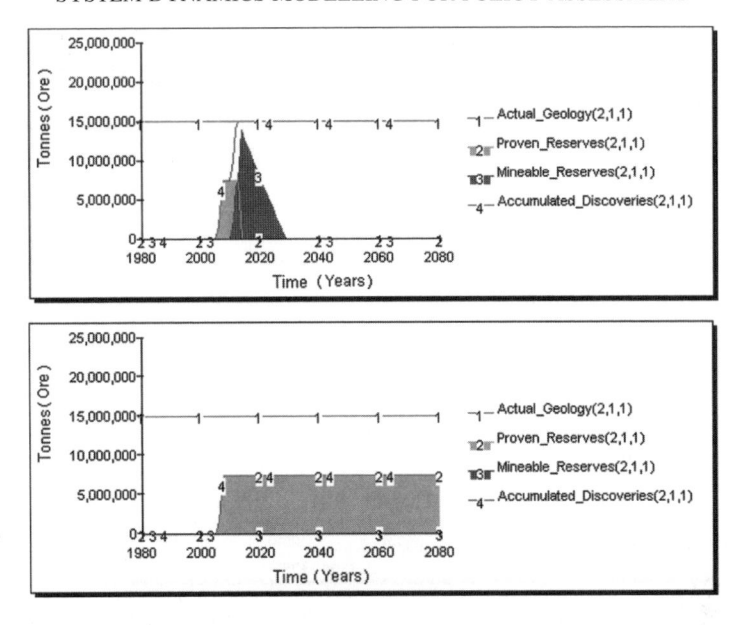

Fig. 5. Standard run with construction commencing in 2008 (top) and separate run where mining licence is refused for the same deposit (bottom).

minerals policy, support industries and suitable infrastructure as well as geological potential.

Results of a survey of the investment policies of 39 multinational mining firms indicate that 40% of these firms will not invest in exploration or development in countries where other similar firms have not invested (Otto 1994a, table 4). Mining firms also tend to be conservative in their investment decisions and often reinvest in countries in which they have already made discoveries and developed mines (Otto 1992). This is because the perceived geological potential of the country increases as discoveries are made, the firm is familiar with the mineral policies, and it may already have offices and employees *in situ*.

Political risks affect the profitability of mining investment. Although some developing countries remain politically unstable, the more developed countries have growing problems of access to land and more stringent environmental legislation coupled with, at least in theory, a desire to move towards a more sustainable development. Political changes that might affect the operation or profitability of a mining or exploration investment may occur even in a country widely regarded by investors and political observers to be politically stable. All countries have some element of risk as well as some potential to produce profits. However, the difference is in the degree of risk that is present.

Also, there is the fear that as the developing countries begin to industrialize there may be a problem with exporting minerals, as they will be needed to supply development (Hargreaves *et al.* 1994). These difficulties increase the uncertainty associated with exploration and subsequent development of any mine, and consequently alter the risk–reward balance (Anon 1995). It must be recognized that while political risk cannot be avoided completely, it may be managed. The methods of managing and reducing political risk, besides avoiding high-risk countries altogether, are by having a fair and flexible contract that contains a reference to international law and arbitration, by insuring the investment either with a public or private investment insurance scheme, risk spreading through joint ventures involving international institutions, and investing in more than one country (Kolo 1997).

Information regarding the risks associated with a particular country is available from international mining agencies (such as MEG). This is used by mining firms to assess the risks associated with each potential investment location. A mining firm will typically assess the risks associated with each country and attach a weighting to these risk factors. This weighting is determined by the firm's attitude to risk and its strategic objectives (Otto 1992, p. 3). Therefore, each potential investment location may be rated differently by individual mining firms.

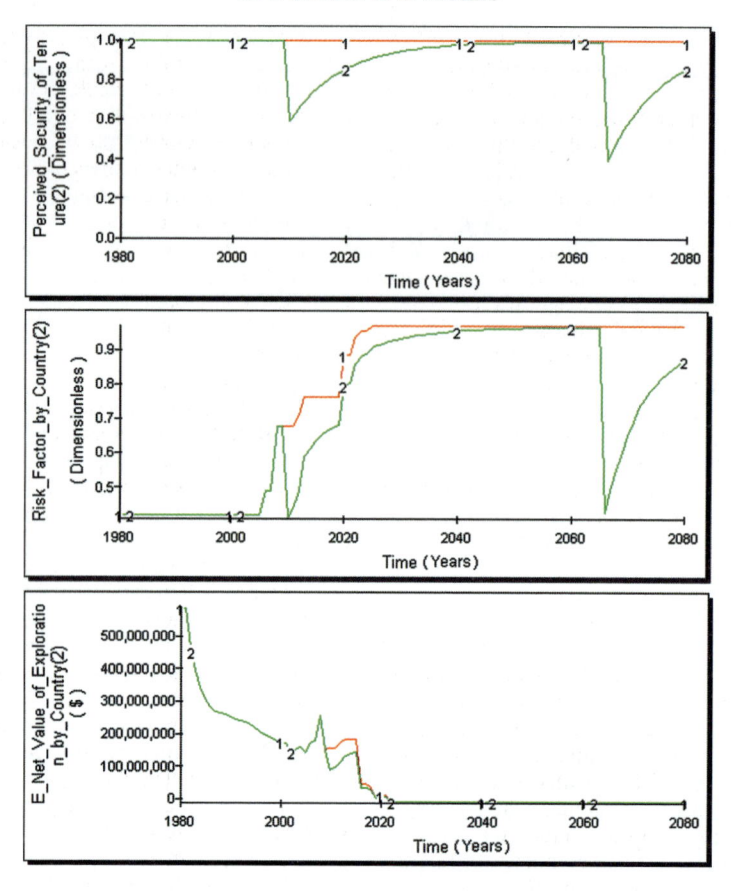

Fig. 6. Security of tenure, perceived risk and expected value of exploration. These graphs show the superimposed output from the two simulation runs presented in Figure 5. The red line (line 1) plots the output from the first run, where the mining licence is granted. The green line (line 2) plots the output from the run where the licence is refused. Where the output is the same, as it is in all cases before 2008, then the green line shows through as it represents the second simulation. The moving average of the expected net value of exploration curve declines over time because the most favourable deposits are worked first. By 2025, all available sites are explored and so the average expected net value declines to zero. The second refusal of a mining licence, in 2064, is included here to show the compounding effect it has on perceived security of tenure.

Expected value of exploration

The geological potential of a country is obviously an important factor in determining the attractiveness of that country to mineral exploration and investment. Worldwide, governments are the main source of geological information. In 1994, 120 countries had Geological Survey Organizations whose purpose is to collect, organize and disseminate information on the geological potential of the country (Otto 1994b).

When ranking possible exploration sites, mining firms will make an estimate of the probable size and type of a deposit in each site. This estimate is affected by the number and size of any discoveries already made in that region and, if there are no discoveries, then the firm will use the concept of a model deposit. This information is used, along with expectations about costs, to make an estimate of the expected value of a discovery in each country. The costs of finding a deposit are termed the exploration costs. These are broken down into the cost of the exploration licence and the cost of physically carrying out the exploration, that is, the labour, drilling and equipment costs.

To arrive at these estimates the firm draws on its own experiences and those of other mining firms, as well as industry-wide forecasts. To estimate the value of a deposit, the firm must first estimate its likely physical characteristics.

This is based on discoveries already made by the firm in the particular exploration site or, if the firm is deciding whether or not to invest in a country for the first time, on any discoveries made by other firms. Based on this estimate and on commonly used industry rules, as well as its own experience and knowledge about the country, the firm estimates the planning, capital and operating costs for that deposit. The expected revenue from the deposit is the product of the estimated tonnage of recoverable mineral and the expected price for that mineral. The costs are subtracted from the revenue to give the expected value of the deposit (MacKenzie & Woodall 1988).

It is recognized that, at least in the short run, the environmental protection requirements of the country also have a significant impact on costs. Increasing environmental regulatory and planning requirements reduce the relative attractiveness of potential deposits. This is because increased planning and regulatory costs reduce the expected net value of exploration, thus shifting exploration funds elsewhere in the short term. This can be seen in Figure 7, in which relatively high planning and regulatory requirements in Country 4 increased the environmentally related proportion of planning, capital and operating costs by 50%. This results in a short-term reduction in exploration spending in Country 4, as it reduces the expected return on investing there. However, after the more favourable (highest expected values) deposits of Country 2 and Country 3 are explored, Country 4 again becomes a target for exploration funds (in the late 1990s of the simulation run).

Initially, scarce exploration funds, in the form of paid-in capital, are distributed between potential countries, favouring those that show higher expected net value of exploration. Later, when the more favourable locations are explored, profits from initial exploration/development efforts can be reinvested in other less favourable locations or countries. By the year 2040 of the simulation, Country 1 (developing country) has secured almost twice the exploration funding of each of the other countries, but this is because the cost of exploration is higher in the developing country (Country 1), thus explaining why it is left until last to explore.

In effect, tightening environmental regulations, as in the case of Country 4 in this example, results in shifting the country's potential deposits down the rank. Indeed, this effect is not specific to environmental parameters. Increasing or decreasing one parameter while holding all others constant will have an obvious effect on exploration expenditure in the country in question. The use of a simulation model is not necessary to expose this behaviour.

In practice, however, the situation is more complex. Regulatory requirements change over time, both within and between countries, shifting the relative attractiveness up and down in a dynamic fashion. For example, tightening environmental regulations need not necessarily reduce the relative attractiveness of exploration sites in a particular country if it is combined with changes to other important policies. This is where the recognition of the trade-offs between environmental and socio-economic policies is necessary and can help in the move towards a more sustainable development.

Figure 8 shows the effect on exploration spending of increasing worldwide regulatory and planning requirements to the level of the developed country (Country 4). In this instance, Countries 2 and 3 secure most exploration funds

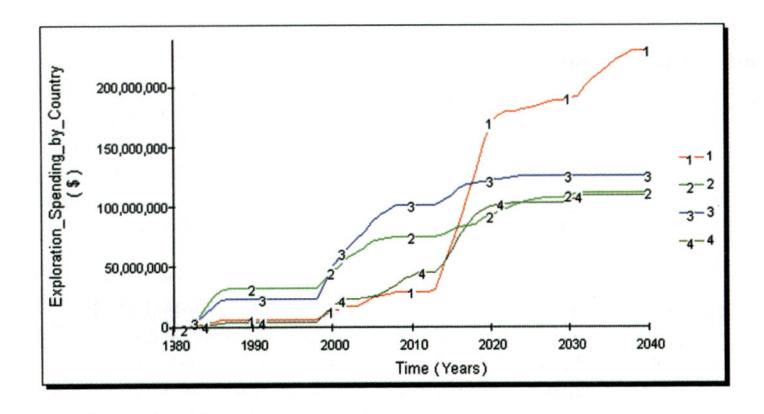

Fig. 7. Exploration spending by country (1 to 4 in the legend).

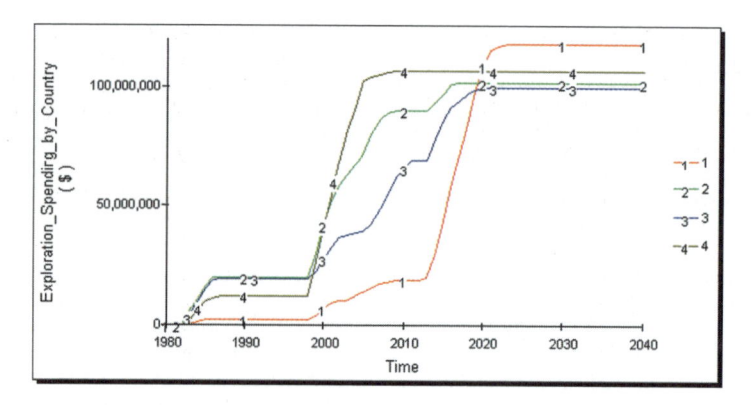

Fig. 8. The effect on exploration spending of worldwide coincidence of environmental regulatory and planning.

initially, as Country 4 still has higher local input costs and Country 1 has higher exploration costs. There is then no exploration activity during the next 15 years as exploration funds have been exhausted. However, once the initial development phase becomes profitable, further exploration funds become available and are directed primarily towards Country 4 in the first instance, reflecting its increased perceived geological potential resulting from the initial round of exploration. Country 4 is initialized with a more favourable geology along with more stringent regulatory and planning requirements and higher cost of local inputs. Knowledge of the more favourable geology is not available to firms in advance and so the model deposit size is used as a basis for determining the net value of exploration, until more accurate information becomes available. The effect of making this knowledge available from the outset is reflected in Figure 8.

Figure 8 shows the importance of providing firms with accurate geological information as a means of securing exploration funding. The expectation of increased tonnage has a positive effect on expected net value of exploration calculations, thus increasing the relative attractiveness of Country 4 for exploration investment. Therefore, improving geological information facilities is an example of how a country with a favourable geology may compensate for the effect of applying more stringent environmental regulatory and planning requirements.

Concluding comments

Within the concept of sustainable development, it is necessary to develop a framework for assessing the environmental and socio-economic as well as organizational impacts of changes to policies.

The model described here facilitates this through the inclusion of the socio-economic impacts of environmental policies as implemented through host government decisions.

Through a quantitative analysis of existing data, a simulation model, such as outlined here, exposes – within the context of the problem area – the underlying assumptions used as a basis for policy formulation and corporate decisions. Through the compression of time, it provides a means of taking these assumptions to their logical conclusions. Exposing assumptions in this way leaves less room for misinterpretation and provides a solid basis for enhancing the understanding of system structure, vital to the success of sustainable development policies.

References

ANONYMOUS 1995. Global mining trends – the hotspots. *Mining Journal*, March 3, 165.

ASTAKHOV, A. S., DENSION, M. N. & PAVLOV, V. K. 1988. Prospecting and exploration in the Soviet Union. *In*: TILTON, J. E., EGGERT, R. G., LANDSBERG, H. H., (eds) *World Mineral Exploration – Trends and Economic Issues*. Resources for the Future Publications, New York, USA, 199–226.

FORRESTER, J. 1994. Policies, decisions and information sources for modeling. *In*: STERMAN, J. & MORECROFT, J. (eds) *Modeling for Learning Organisations*. Productivity Press, New York, USA, 51–84.

HARGREAVES, D., EDEN-GREEN, M., & DEVANEY, J. 1994. *World Index of Resources and Population*. Dartmouth Publishing Company, Hants, UK.

JOHNSON, J. 1990. Ranking countries for mineral exploration. *Natural Resources Forum*, August Edition, 178–186.

KOLO, A. 1997. *Managing Political Risks in Transnational Investment Contracts*. Centre for

Petroleum and Mineral Law and Policy, University of Dundee, Scotland.

LEGASTO, A. A. & MACIARIELLO, J. 1980. Modelling policies and decisions. *In*: LEGASTO, A. A., FORRESTER, J. W. & LYNEIS, J. M. (eds) *System Dynamics: A Critical Review.* North-Holland Publishing Company, Amsterdam.

MACKENZIE, B. & WOODALL, R. 1988. Economic productivity of base metal exploration in Australia and Canada. *In*: TILTON, J. E., EGGERT, R. G. & LANDSBERG, H. H. (eds) *World Mineral Exploration – Trends and Economic Issues.* Resources for the Future Publications, USA. 363–418.

MEADOWS, D. L., WILLIAM, W. *ET AL.* 1974. *Dynamics of Growth in a Finite World.* Wright Allen Press, Inc., Cambridge, MA, USA.

NATIONAL MINERALS POLICY REVIEW GROUP 1995. *A New Minerals Policy.* The Government Stationery Office, Dublin.

OTTO, J. 1994*a*. *The International Competition for Mineral Investment: Implications for Asia-Pacific*, Seminar Paper SP15. Centre for Petroleum and Mineral Law and Policy, University of Dundee, Scotland.

OTTO, J. 1994*b*. *Tabulated Results of a Global Survey of Geological Survey Organisations*, Professional Paper PP10. Centre for Petroleum and Mineral Law and Policy, University of Dundee, Scotland.

OTTO, J. 1992. *Criteria and Methodology for Assessing Mineral Investment Conditions*, Seminar Paper No. SP6. Centre for Petroleum and Mineral Law and Policy, University of Dundee, Scotland.

RICHARDSON, G. 1991. *Feedback Thought in Social Science and Systems Theory*, University of Pennsylvania Press, USA.

Sustainable mineral development: possibilities and pitfalls illustrated by the rise and fall of Dutch mineral planning guidance

MICHIEL J. VAN DER MEULEN

TNO, Geological Survey of the Netherlands, PO Box 80015, NL-3508 TA Utrecht, The Netherlands (e-mail: michiel.vandermeulen@tno.nl)

Abstract: The Netherlands has major resources of sand, gravel and clay, exploited mainly for construction works and the building materials industry. As in most Western countries, mineral extraction meets with considerable societal resistance. To this end, Dutch minerals policy aims to prevent extraction by promoting economical use of materials, and the use of alternative (secondary or renewable) materials. Until recently, it also included a system of production planning to sustain supplies of regionally scarce materials. Dutch policy development is reviewed and discussed in terms of pitfalls and possibilities for mineral planning in general. Promoting secondary substitution has been quite successful, and presents an example. In contrast, the production planning system has been controversial from the start and ineffective as a result, mainly because it attempted to solve supply problems without properly addressing the underlying resistance. For this reason the system is in the process of being abandoned.

Building, construction and several process industries require bulk amounts of industrial minerals, which are largely supplied by the extractive industry. In the developed world, minerals are used to sustain and expand established economies. Quality of life is paramount, which presents the paradox that people, although dependent on mineral supplies, have come to reject their extraction when visible. To some extent, this trend has been acknowledged in mineral planning policies in Europe and Northern America. Typically, more effort is made in balancing the needs and requirements of various stakeholders, using the concept of sustainable development as a guideline.

Under-developed countries show a low consumption of minerals, sustaining largely subsistence lifestyles. Minerals are primarily extracted for exports, but as this is mostly undertaken by foreign companies, and value is added in the importing countries, revenues are limited. Building and industrialization in rapidly developing countries, such as China, require large amounts of minerals. These are either obtained from domestic extraction, or on the world market; exports of bulk materials are currently limited. The exploitation of mineral resources in the developing world often does not comply with any definition of sustainability and presents concerns to the global community.

This paper reviews a full cycle of mineral planning development in the Netherlands, from the first outlines of a national regulatory system in the early 1980s to its recent abandonment. The starting point of this cycle hardly compares with the current situation in the developing world. Nonetheless, the Dutch case presents some possibilities and pitfalls for sustainable mineral development with general relevance.

Dutch resources

The Dutch shallow subsurface consists predominantly of unconsolidated Quaternary clastic deposits (Fig. 1). Clay and fine sand are nationally abundant. Coarse sand occurs, roughly, in the southeastern half of the country. Exploitable gravel resources are limited to the extreme south; the Netherlands is dependent on imports for its coarse aggregates supplies. The occurrence of bedrock at or near the surface, mostly Mesozoic carbonates, is limited to the southernmost and central eastern parts of the country. Except for silica sand, virtually all Dutch non-energy minerals resources are exploited for construction works or the building materials industry. For a full account of Dutch surface mineral resources see Van der Meulen *et al.* (2005).

Policy building

The 1960s and 1970s showed a gradual increase in the Dutch demand and production of industrial minerals. A combination of spatial planning

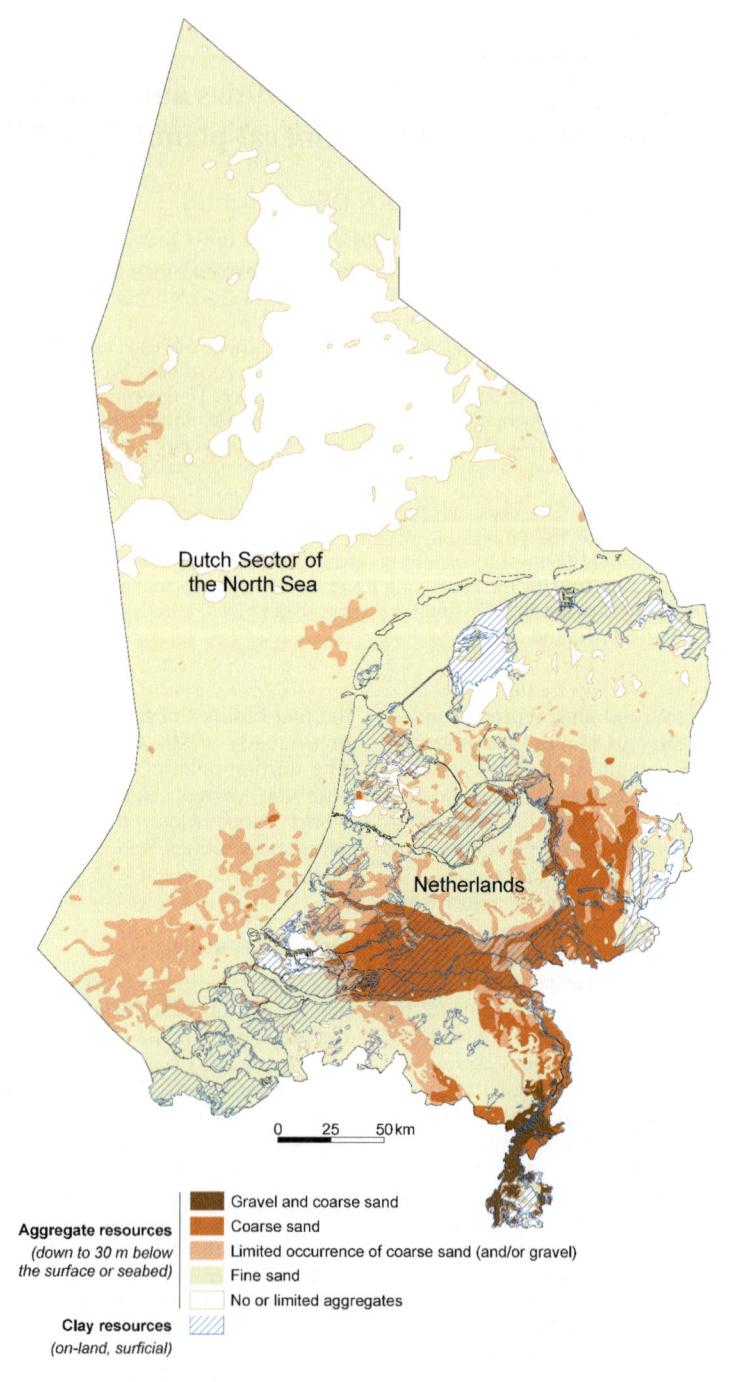

Fig. 1. Resources of aggregates and clay in the Netherlands. Occurrences of other materials (i.e. carbonates and silica sand) cannot be displayed on this scale.

problems and a growing environmental awareness resulted in resistance against their extraction. Provincial administrations became increasingly reluctant to grant extraction permits, which eventually put the business continuity of the quarrying industry at risk. At that time, mineral planning was a task for the subnational permit-issuing authorities, that is, the provinces for land-based extractions and the Directorate-General of Public Works and Water Management ('Rijkswaterstaat') for the state waters. The Mineral Extraction Law (Anonymous 1965) did not then provide for intervention or coordination by the national government.

By the mid-1970s, supplies of gravel, concrete and mortar sand, and clay for the structural ceramics industry, became a matter of national concern (Ike 2000). (Concrete and mortar sand is translated from the Dutch term for a range of industrially produced coarse sands. About 75% is used for concrete production and about 15% in masonry mortars. The remaining $\sim 10\%$ is used for a wide range of purposes such as the production of asphalt and bricks, and for unconsolidated applications such as drains and filters). The supply problem was taken up by the Minister of Transport, Public Works and Water Management. He installed the LCCO, the National Commission for the Co-ordination of Mineral Planning Policy, consisting of representatives of the permit-issuing authorities. In 1980, the LCCO produced a series of reports (Anonymous 1980a–d) that prepared for the start of national mineral planning guidance: a set of agreements, reached in 1981 between the national government and provincial administrations on the extraction of gravel, concrete and mortar sand, and clay until 1989.

A policy document issued in 1983 outlined problems and set objectives for the Dutch building materials supplies over the long term, that is, after 1989 (Anonymous 1983). The ensuing long-term national mineral policy plan stated that the exploitation of Dutch surface mineral resources required planning and coordination, and that mineral planning should be integrated in other policy fields, especially spatial planning (Anonymous 1989). The policy from then on, even though its wording and legal status changed in later issues (Anonymous 1994a–b, 1995, 1996, 2001), aimed at a sustainable exploitation of surface mineral resources to meet the demand for construction and building materials, at an economical use of materials, and maximum use of renewable, secondary and sea-won materials. The Mineral Extraction Law was changed accordingly: the 1994 amendment arranged for national coordination and for the embedding of mineral planning into spatial planning.

Sustainable mineral development

The extent to which materials use is sustainable is a matter of ample scientific debate. However, it is often the public perception rather than scientific definitions that defines sustainability in policy. Resources of building raw materials such as aggregates or clay are, except when considered on local scales, virtually indepletable. Hence, resource depletion is not a primary challenge for sustainable building materials provision. The key issue seems to be the impact of extraction, which can bring about irreversible land-use changes and loss of function if effective rehabilitation is not undertaken.

Either the natural or the human environment prevails in the perception of impacts. It is not surprising that in the densely populated Netherlands, it is primarily the human environment has appeared unable to sustain minerals extraction. The primary manifestation of this was 'NIMBY' behaviour, with the complication that mineral extraction in the Netherlands is always in somebody's 'backyard'. Minerals policy has been labelled sustainable, but it was basically a pragmatic response to this NIMBY problem. Whether or not individual policy measures would qualify as sustainable will be discussed by case.

Recycling and economical use of materials ('prevention')

Measures and effects

The use of secondary materials has been stimulated by means of product quality control protocols, the taxation and banning of landfilling with recyclable materials, and various promotional activities. The national government took on a pioneering role in the use of secondary materials. This has had a fairly large impact as it commissions, about 20% of the infrastructure works, and its construction specifications are adopted by most subnational authorities.

Altogether the secondary substitution policy has been quite successful (Broers *et al.* 2002; Van der Meulen *et al.* 2003, 2005). Coarse and medium-grained secondary aggregates (such as granulated construction and demolition waste, steel slag and phosphorous slag, asphalt waste, colliery spoil and municipal solid waste incineration bottom ashes), have come to be widely used as fill or foundation material in road and hydraulic engineering. Fine-grained secondary materials, such as flue-gas desulphurization gypsum, and coal combustion and waste incineration fly ashes have become well-accepted raw materials in the building materials industry. The use of secondary materials rose from ~7 Mt a^{-1} in the early 1980s to 33 Mt in 2000; their share in the total provision rose from 6 to 15% (Fig. 2). While the total industrial minerals use per capita rose by approx. 17% between 1980 and 2000, the consumption of primary materials remained more or less stable (Fig. 3). The use of industrial minerals per unit of economic activity shows a dematerialization (Herman et al. 1989) of some 25% (Fig. 3). The latter trend suggests successes in economical use of materials but, other than for secondary substitution, there is no straightforward link with specific mineral policy measures.

Stimulating the use of renewable materials in building has not been very successful. Other than for earthy and stony secondary materials in construction, this requires adapted building techniques and styles. With a few exceptions, both the Dutch building industry and its clients appear unprepared to divert from traditions.

Sustainability

Aiming for recycling and an economical use of materials, in order to prevent extraction, is consistent with sustainable development in the widely used definition 'development which meets the needs of the present without compromising the ability of future generations to meet their own needs' (Brundtland 1987). On the operational level, however, waste processing and application presented some problems. Construction and demolition waste had been recycled, for instance, for about two decades, before proper arrangements were made to isolate asbestos from this waste stream in the late 1990s. As a result, a very large amount of granulated construction and demolition waste containing asbestos has been produced and used, mainly as road foundation material. Even

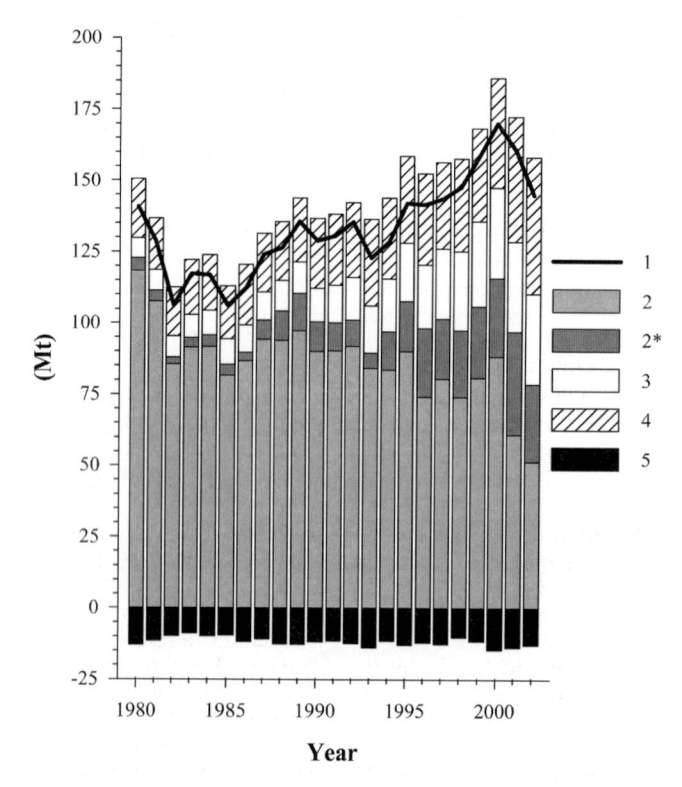

Fig. 2. The demand for aggregates, clay and carbonates in the Netherlands from 1980 to 2002 (1), and its provision by home production (2; 2* is the share of North Sea extraction), secondary (recycled or waste) materials (3) and imports (4). The growing demand was met with increasing amounts of sea-won, imported and secondary materials, mainly aggregates. Exports (5, shown as negative values) have remained more or less constant (Van der Meulen et al. 2005).

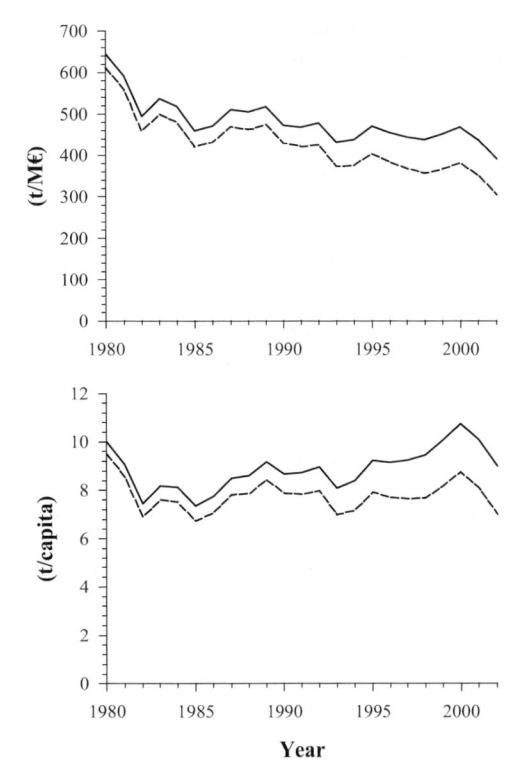

Fig. 3. Lower panel: industrial minerals use per capita in the Netherlands from 1980 to 2002 (minerals worked in the Netherlands only): for primary and secondary materials (solid line), and for primary materials only (dashed line). Upper panel: industrial minerals use (t) per M€ of GDP (price level of 1995), lines same as for lower panel.

Fig. 4. The share of sea-won sand in the Dutch filling sand provision.

though asbestos concentrations are generally low, recycling has presented an unintended challenge to future generations.

Site selection

Site impacts are evaluated in two ways. Mineral extraction and related planning and/or permitting policies are subjected to EU-defined environmental impact assessments (Anonymous 1985). Less straightforwardly, the site selection process should be compliant with the policy favouring extraction combined with, or embedded in other activities (Van der Meulen et al. 2004). The strategy has worked well for clay, which became coupled with nature development, and filling sand, of which significant amounts are obtained from navigation channel and seaway maintenance. Policy favouring seabed extraction of sand altogether, that is, both embedded and non-embedded, has resulted in a gradual increase of the share of sea-won sand in the filling sand provision from 1980 to date (Fig. 4). In fact, hardly any land-won filling sand is currently used in the coastal provinces.

Water engineering works are probably the best types of projects in the Netherlands for embedded aggregates extraction. The combination of sand and gravel extraction and widening of the Meuse River, with huge potential yields, has suffered from a difficult cooperation between river managers and the extractive industry in the planning stage. It is as yet unclear what the aggregate yields will amount to. Probably as a result of this, the embedding concept has become less popular, and it will probably not be incorporated to a significant extent in the currently drafted plans for the widening of the Rhine.

Production planning

Dutch minerals policy included production planning. To this end, three categories of commodities are distinguished. The production of *nationally abundant* materials, such as (fine) filling sand and clay, is not considered to require national planning and coordination. The involvement of the national government is limited to monitoring, aimed at an early identification of supply problems. The extraction of *nationally scarce* materials, is either being reduced to a level of regional self-supply (gravel), phased out eventually (carbonates), or maintained at low levels (silica sand). Concrete and mortar sand is a *regionally scarce* commodity. In order to meet the national demand, assignments ('taakstellingen'), that is, amounts of sands for which permits are to be granted, are negotiated between national and provincial rulers. Up to 1998, the assignments for concrete and mortar sand added up to the approximate

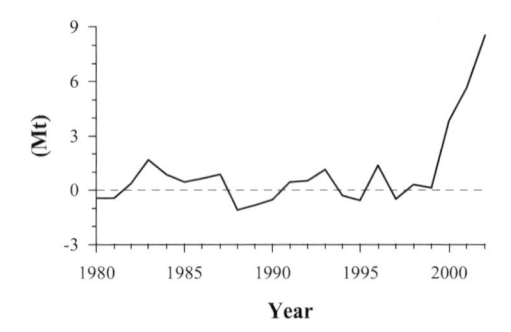

Fig. 5. Net imports of concrete and mortar sand, most of which originates from the German federal state of North Rhine-Westphalia. The sharp increase in 2000 is related to reduced home production levels.

level of the national demand. For 1999–2008, the authorities agreed on an underproduction of about 20%, in order to stimulate the use of secondary alternatives (Anonymous 2001).

In spite of the assignments, concrete and mortar sand has remained a problematic commodity, from the perspectives of both policy makers and the industry. Some provinces have not accepted the outcomes of assignment negotiations. In fact, the negotiations were usually concluded halfway rather than before the ten-year assignment periods. The government presented the production restrictions for 1999–2008 as sustainable, as they aimed at higher levels of secondary substitution. They were qualified by the industry as simply another expression of resistance against extraction. In any case, restrictions seem to have brought about increased imports rather than secondary substitution (Fig. 5; Van der Meulen et al. 2003, 2005). This effect, apparently not taken into account a priori by policy makers, would render restrictions at least partially counterproductive on the issue of sustainability.

The size of extraction sites has become the main problem associated with the home production of concrete and mortar sand. The production planning system distinguished between extraction for regional and national supplies. The latter component was the excess production in provinces with sufficient resources for the supplies of provinces lacking resources. Extraction for national supplies has been concentrated in large sites, which met with severe objections, both during the permit procedures and the production stage. In one notorious case, permit procedures took about 18 years before the project was cancelled.

Current developments: the abandonment of mineral planning

Because of budget cuts, the Ministry of Transport, Public Works and Water Management has recently decided to take on a lesser role in mineral planning. Regulatory policy elements, especially the production planning for concrete and mortar sand, are in the process of being abandoned. A skeleton version of the most recent mineral plan (Anonymous 2001) has been included in the National Spatial Strategy, a policy document on spatial planning (Anonymous 2004). It is unclear what the effects of deregulation will be, but they will undoubtedly be the largest for the provision of concrete and mortar sand. The industry initially reacted negatively, and predicted supply problems. In a later analysis, the industry seems to have accepted the fact that production guarantees would become history, and presented a strategy in which the extraction of concrete and mortar sand is to become embedded in various types of land reconstruction projects (Braakhekke et al. 2003). Commitment of other stakeholders is a prerequisite for this strategy; if it can be raised, the effects of deregulation could be positive in the long term. If not, it would probably only result in increased import levels. The latter possibility is expected to raise objections in the neighbouring German federal state of North Rhine-Westphalia: its government had already expressed its concerns on the aforementioned Dutch production restrictions agreed for 1999–2008 (Anonymous 2002).

Although the initial reason for abandoning mineral planning guidance was budgetary, the full arguments included the statement that production planning for concrete and mortar sand was ineffective (Schultz van Haegen-Maas Geesteranus 2003). When adding this remark to the aforementioned objections of the provinces, it can be concluded that the production planning system had fully lost its support after about two decades of being in operation.

Discussion and conclusions

The most unambiguous successes of Dutch mineral planning have been achieved on the consumption side of the Dutch mineral economy. Developing economies having a high minerals consumption may benefit from the experience in the processing and application of secondary materials, and from the experience in their promotion. It also indirectly supports the point often made that significant responsibility for sustainable mineral extraction in underdeveloped countries rests with the importing countries (Atkinson & Hamilton 2002).

Production planning has been developed to keep up Dutch aggregates supplies, which would otherwise probably have fallen. In doing so, the system added to the cause for this tendency, i.e. societal resistance. In the end, the system was unsuccessful and is being dismantled, while the reasons for its development still prevail. Slovenia, in the process of designing a mineral planning policy, presents another approach to counter objections against minerals extraction (Solar et al. 2003). The objectives are (1) to maintain access to adequate reserves and receive permission to mine in acceptable locations; (2) to eliminate small, unregulated quarries; (3) to disallow super-quarries that cause significant disruption; and (4) to close or disallow quarries that are so distant from their markets that commodity transport becomes disruptive to communities. The combination of arranging for access to resources and setting targets for the site size population may proof to be more effective and acceptable than imposing production targets.

Brian Marker (Office of the Deputy Prime Minister, London), Ian Thomas (National Stone Centre, Wirksworth) and Joris Broers (Ministry of Transport, Public Works and Water Management, The Hague, Netherlands) are kindly thanked for helpful suggestions.

References

ANONYMOUS 1965. *Ontgrondingenwet* (in Dutch). Staatsblad van het Koninkrijk der Nederlanden 1965, **509**.

ANONYMOUS 1980a. *Rapportage van de Interprovinciale Werkgroepen Beton – en Metselzand, Grind en Klei – Algemene Inleiding* (in Dutch). LCCO, The Hague, Netherlands.

ANONYMOUS 1980b. *Rapportage van de Interprovinciale Werkgroepen Beton – en Metselzand, Grind en Klei – Rapportage Beton – en Metselzand* (in Dutch). LCCO, The Hague, Netherlands.

ANONYMOUS 1980c. *Rapportage van de Interprovinciale Werkgroepen Beton – en Metselzand, Grind en Klei – Rapportage Grind* (in Dutch). LCCO, The Hague, Netherlands.

ANONYMOUS 1980d. *Rapportage van de Interprovinciale Werkgroepen Beton – en Metselzand, Grind en Klei – Rapportage Klei* (in Dutch). LCCO, The Hague, Netherlands.

ANONYMOUS 1983. *Nota Uitgangspunten, Probleemstelling en Doelstellingen over Oppervlaktedelfstoffenvoorziening op Lange Termijn* (in Dutch). Tweede Kamer, Zitting 1982–1983 (Parliamentary Records 1982–1983), **15436/6**.

ANONYMOUS 1985. Council Directive of 27 June 1985 on the assessment of the effects of certain public and private projects on the environment. *Official Journal of the European Communities*, **L 175**, 40–48.

ANONYMOUS 1989. *Gegrond Ontgronden* (in Dutch), Tweede Kamer, Zitting 1988–1989 (Parliamentary Records 1988–1989), **21199/1**.

ANONYMOUS 1994a. *Structuurschema Oppervlaktedelfstoffen, Deel 1, Ontwerp Planologische Kernbeslissing* (in Dutch). Ministry of Transport, Public Works and Water Management, The Hague, Netherlands.

ANONYMOUS 1994b. *Structuurschema Oppervlaktedelfstoffen, Deel 2, Reacties op de Ontwerp Planologische Kernbeslissing* (in Dutch). Ministry of Transport, Public Works and Water Management, The Hague, Netherlands.

ANONYMOUS 1995. *Structuurschema Oppervlaktedelfstoffen, Deel 3, Kabinetsstandpunt* (in Dutch). Ministry of Transport, Public Works and Water Management, The Hague, Netherlands.

ANONYMOUS 1996. *Structuurschema Oppervlaktedelfstoffen, Deel 4, Planologische kernbeslissing* (in Dutch). Ministry of Transport, Public Works and Water Management, The Hague, Netherlands.

ANONYMOUS 2001. *2e Structuurschema Oppervlaktedelfstoffen, Deel 1, Ontwerp Planologische Kernbeslissing* (in Dutch). Ministry of Transport, Public Works and Water Management, The Hague, Netherlands.

ANONYMOUS 2002. *2e Structuurschema Oppervlaktedelfstoffen, Deel 2, Reacties op de Ontwerp Planologische Kernbeslissing* (in Dutch). Ministry of Transport, Public Works and Water Management, The Hague, Netherlands.

ANONYMOUS 2004. *Vijfde Nota over de Ruimtelijke Ordening, PKB deel 4 Planologische Kernbeslissing* (in Dutch). Ministry of Spatial Planning, Housing and the Environment, The Hague, Netherlands.

ATKINSON, G. & HAMILTON, K. 2002. International trade and the 'ecological balance of payments'. *Resources Policy*, **28**, 27–37.

BRAAKHEKKE, W., LITJENS, G. & WINDEN, A. 2003. *Over Winnen – Zandwinning als hefboom* (in Dutch). IZGP (Dutch Association of Sand and Gravel Producers), Beuningen, Netherlands.

BRUNDTLAND, G. H. 1987. *Our Common Future*. Oxford University Press, Oxford, UK.

BROERS, J. W., PIETERSEN, H. S. & SMITS, R. G. 2002. Secondary raw materials in the Dutch building industry – an overview of policy, current research and practices. *Third European Conference on Mineral Planning (ECMP'02)* – Conference Transcript and Field Trip Guide. Geological Survey of North Rhine-Westphalia, Krefeld, Germany, 161–167.

HERMAN, R., ARDEKANI, S. A. & AUSUBEL, J. H. 1989. Dematerialization. *In*: AUSUBEL, J. H. & SLADOVICH, H.E. (eds) *Technology and Environment*. National Academy Press, Washington, DC, 50–69.

IKE, P. 2000. *De planning van ontgrondingen* (in Dutch). PhD thesis, Groningen University, Geo Pers, Groningen, Netherlands.

SCHULTZ VAN HAEGEN-MAAS GEESTERANUS, M. H. 2003. *Vaststelling van de Begrotingsstaat van het*

Ministerie van Verkeer en Waterstaat (XII) voor het Jaar 2003 (in Dutch). Tweede Kamer, Vergaderjaar 2002–2003, 28 600 XII (Parliamentary records 2002–2003, 28 600 XII), **114**.

SOLAR, S.V., SHIELDS, D. J., MARTIN, W. E. & MARTIN, I. M. 2003. Balancing the costs and benefits of aggregate extraction in Slovenia: are existing data adequate? *Third European Conference on Mineral Planning (ECMP'02)* – Conference Transcript and Field Trip Guide. Geological Survey of North Rhine-Westphalia, Krefeld, Germany, 73–78.

VAN DER MEULEN, M. J., KOOPMANS, T. P. F. & PIETERSEN, H. S. 2003. Construction raw materials policy and supply practices in Northwestern Europe. *In*: ELSEN, J. & DEGRYSE, P. (eds) *Industrial Minerals – Resources, Characteristics and Applications*. Aardkundige Mededelingen **13**, 19–30.

VAN DER MEULEN, M. J., DE KLEINE, M. P. E., VELDKAMP, J. G., DUBELAAR, C. W. & PIETERSEN, H. S. 2004. The sand extraction potential of embedded land surface lowering in the Netherlands. *Netherlands Journal of Geosciences*, **83**, 147–151.

VAN DER MEULEN, M. J., BROERS, J. W., HAKSTEGE, A. L., VAN HEIJST, M. W. I. M., KOOPMANS, T. P. F. & PIETERSEN, H. S. 2005. Surface mineral resources. *In*: WONG, TH. E., BATJES, D. A. J. & DE JAGER, J. (eds) *The Geology of the Netherlands*. Netherlands Institute of Applied Geoscience TNO, Utrecht, Netherlands, in press.

The emperor's new clothes: sustainable mining?

ANDY WHITMORE

Mines and Community Network, 41 Thornhill Square, London N1 1BE, UK
(e-mail: web@minesandcommunities.org)

Abstract: Over the last few years, the idea of 'sustainable mining' has, thanks to industry sponsorship, been working its way into the agenda of many international processes. There is now a push in many countries to invite in multinational mining companies with the idea that there is a 'new, sustainable mining' that is different from the old, bad practices of the past. Yet what has actually changed in the industry to match this shift in rhetoric? From the perspective of mine-affected communities, nothing seems to have changed. Their land is still being taken from them without giving their free, prior and informed consent, and they are suffering the same ill effects on their ways of life, health and environment. This paper will illustrate, using case studies from the Philippines and West Papua, how under this rhetoric, the mining industry 'emperor' has the same old naked ambitions. This paper intends to look at how 'sustainable mining' is perceived from the viewpoint of mines-affected communities and their supporters. Ideally a representative of such a community should be writing this, but as this was not possible, I am writing this as a member of the editorial board of Mines and Communities (MAC). (MAC is a network of organizations across the world seeking to empower mining-affected communities in their struggles against damaging proposals and projects. More information on Mines and Communities, including the members of its editorial board, can be viewed at http://www.minesandcommunities.org.) Many of the communities MAC works with are made up of indigenous (or first, aboriginal) peoples, who have been unfairly disadvantaged by mineral development on or near their land. The paper will therefore concentrate to some extent on the issues of indigenous communities.

Industry initiatives

Over the last few years, the idea of 'sustainable mining' has, thanks to mining industry sponsorship, been working its way onto the agenda of many international processes. The idea was effectively born under the banner of the Global Mining Initiative (GMI) in 1998, leading up to the questionable insertion of the term 'sustainable mining' into the post-Johannesburg plan of action at the World Summit on Sustainable Development (WSSD) in 2002. The industry's preparation for, and indeed integration with the processes of, the WSSD was partly a wake-up call from its relative exclusion at the 1992 predecessor, the Earth Summit (MMSD 2002; Moody 2002).

The GMI was initiated in a meeting of industry leaders and was based in London at Rio Tinto's head office in St James Square. The founders of GMI included Rio Tinto, Anglo American, BHPBilliton, Freeport McMoRan, Newmont and WMC. These are among the largest companies in the industry, and each has been frequently criticized over their own problematic environmental and social records (Moody 2002).

The GMI established the two-year research and dialogue initiative entitled the Mines Minerals and Sustainable Development (MMSD) project. This was followed by a global conference and the formation of the International Council of Mining and Metals (ICMM) – which is also based in London and features many of the same companies – to better represent the industry to its critics (MMSD 2002).

Despite an engagement with some, particularly Northern-based, NGOs, MMSD predictably was widely criticized and boycotted by indigenous peoples' organizations, NGOs with expertise on the issue and mine-affected communities, as can be seen in the joint letter by the Mineral Policy Centre and others dated 2 August 2000 (Nettleton *et al.* 2004). This is because, while it projected itself as being consultative and inclusive, its framework, objectives and structure were all unilaterally predetermined by the corporate sector. Even the members of the nominally independent Assurance Board (established to monitor rather than manage the process) were originally chosen unilaterally by industry, without the broad consultations and self-selection processes associated with a more credible multistakeholder process (Moody 2002).

The aims and outcomes of MMSD were therefore tailored to the industry's priority agenda of linking mining to sustainable development, and according to critics did not reflect those of communities. This meant that, as argued by Moody (2002) in his paper 'Sustainable Development Unsustained', MMSD did not gain broad acceptance or credibility as an independent body, and as a result the project failed to generate any meaningful dialogue between those most affected by mining and those most responsible. MMSD, in its own conclusions, noted lack of trust as an important qualifier in its own history and for future dialogue in the industry. This failure stunted its capacity in terms of its stated goals of engaging with critics (MMSD 2002; Moody 2002).

Even some within the industry, such as Philip Crowson in a letter published in the *Mining Journal*, 12 April 2002, called MMSD the 'flawed outcome of a flawed process'. He finds its introductory pages '... no more than a recycling of currently fashionable ideas and phrases, preceded by an alarmist sketch of the world today and laced with some crudely misleading economic history.' Crowson cannot stand the idea that the MMSD team has now 'dignified' NGOs as 'civil society'. He deplores the 'working assumption' that miners would need a 'license to operate' through 'multistakeholder engagement' before they can lift a sod of other peoples soil, (Moody 2002).

However, none of this has stopped the industry engaging with some large, Northern-based NGOs, such as Conservation International and CARE International, much to the frustration of mines-affected communities, as these organizations have no grassroots consent on which to enter into dialogues that will affect those communities (Moody 2001b; Choudry 2003). As a slight aside it is worth looking at Marc Chapin's examination of the tension between indigenous peoples and large conservation organizations in an excellent article that exposes just how environmental organizations frequently fall into the same arrogant traps as multinational companies in dealing with the expectations of local communities (Chapin 2004). Nor has the MMSD project prevented subsequent 'dialogue' processes (including that between ICMM and IUCN concerning the future of mining exploration within Protected Areas) from repeating the same exclusion of the indigenous peoples whose lands were under discussion (ICMM/IUCN 2002). It has also not stopped the industry, aided by the World Bank, promoting 'sustainable mining' within the liberalization of various country mining codes, such as the National Minerals Policy in the Philippines (Corpuz 2003).

London Declaration complaints

It is partly owing to the above industry initiatives that MAC was born. There was a meeting of community activists and their supporters in London in May 2001. London was chosen because it is increasingly seen as the centre of mining companies, mining financing (note the rapid growth of mining stocks on London's AIM market), and most importantly to the industry-sponsored initiatives mentioned above.

A joint declaration came out of that meeting, called the 'London Declaration', which forms the basis for the following critiques of 'sustainable mining' (MAC 2001).

The Declaration makes various demands upon the mining industry and civil society, but from the point of view of this paper its most relevant section is in how industry-sponsored initiatives promote at least four half-truths or myths. These are:

(1) the supposed need for more and more minerals from ever more mines;
(2) the claim that mining catalyses development;
(3) the belief that technical fixes can solve almost all problems; and
(4) the inference that those opposed to mining mainly comprise ignorant and 'anti-development' communities and NGOs.

I intend to look at each in turn to see how it exposes the fallacy of sustainable mining.

The supposed need for more and more minerals from ever more mines

This is perhaps the key question; how can the naked laws of supply and demand, in a growing world of consumption, result in a sustainable extractive industry? In 1999 some 9.6 billion tons of marketable materials were dug out of the earth, nearly double the amount for 1970. This only accounts for minerals that reach markets, and not the unused and/or waste products, or overburden. If you add this in, then the amounts removed are staggering. In 2000, mines around the world extracted some 900 million tons of metal – and left behind some 6 billion tons of waste ore (not including overburden). The Ok Tedi mine in Papua New Guinea generated 200 000 tons of waste a day on average – more than all the cities in Japan, Australia and Canada combined (MMSD 2002, p. 243; Sampat 2003, p. 112).

It can be argued that at least part of the increase in production is not just pure global economic demand, but is a result of subsidies provided by government policies and the promotion of mining via International Finance Institutions (IFIs), such as the World Bank. The subsidies that mining firms enjoy can come in the form of pro-mining laws, which charge very little in taxes or royalties, for example, the 1872 U.S. mining law, or any of a number of newly introduced codes in developing countries. Taxpayers, who pick up the bills for cleaning up costly mistakes, such as through the U.S. 'superfund', can also subsidize mining. IFIs have actively promoted mining in developing countries through loans, investment guarantees, insurance covers and pressure on law making. Between 1995 and 1999 the World Bank spent close to US$6 billion to fund mining projects around the world (Sampat 2003, pp. 123–124).

Based on figures from the late 1990s, on the negative side mining consumed close to 10% of world energy, is responsible for 13% of sulphur dioxide emissions and, it is estimated, threatens nearly 40% of the world's undeveloped tracts of forest. Yet it directly accounts for only 0.5% of employment and 0.9% of 'gross world product'. When considered as a 'complete balance sheet', it just does not look sustainable (Sampat 2003, p. 111).

Given these figures, there is definitely a logic and market for increased production of recycled materials (or even the use of current stockpiles – there is after all enough gold stockpiled to meet demand for 17 years). About half of the world's lead comes from recycled sources, as does a third of aluminium, steel and gold. Energy savings of up to 70% are available through recycling, not to mention the benefits from reduced toxic emissions, as well as to occupational health and safety.

Yet, partly thanks to subsidies, companies search for new greenfield sites when so much useful metal lies in landfills (although some pilots on mining landfills are happening in the United States). These subsidies can also be hidden ones, because it is very difficult to quantify any costs to air or water quality, the consumption of 'free natural resources' and the loss of alternative livelihoods. In the same way that the heavily subsidized arms industry is being asked to transfer skills towards more peaceful industries, so the corporate mining industry should be looking towards switching its skills and knowledge towards investing in recycling (Sampat 2003, pp. 123–124; Lehman Brothers Inc. 2000).

The claim that mining catalyses development

It is interesting that even in the 18th century, Adam Smith had already noted how capricious the benefits of mining were, when he said, 'of all those expensive and uncertain projects which bring bankruptcy upon the greater part of the people that engage in them, there is none more ruinous than the search for new silver and gold mines. It is perhaps the most disadvantageous lottery in the world' (Smith 1776).

There is now a well-published argument upon the issue of mining's contribution to developing countries; a proper exploration of which would require an article many times the length of this one. Aside from the 'boomtown' effect caused by short-term projects, often referred to around a set of symptoms known as 'Dutch disease', there are also the problems associated with extractive industries called the 'mining curse'. Mineral development, it is argued, has paradoxically been shown to have a negative effect on economic growth in minerals-dependent developing countries, or at the very least not to benefit the poor in that country. According to the United Nations, the proportion of people living on less than $1 a day in minerals-exporting countries rose from 61% in 1981 to 82% in 1999 (OXFAM 2003). A study from Britain's Lancaster University concluded that mineral-driven, resource-rich countries were among the poorest economic performers in a study from 1960 to 1993 (Auty 1998).

Many economists consider it 'intuitively obvious' that mining is the vehicle to drive sustained economic growth to lift countries out of poverty. A recent World Bank report makes this familiar case; '... natural resources-based activities can lead growth for long periods of time. This is patently evident in the development history of natural resource-rich developed countries, such as Australia, Finland, Sweden, and the United States. Mining was the main driver of growth and industrialization in Australia and the United States over more than a century...' (Power 2002, pp. 4–7).

However, a recent publication from OXFAM America, 'Digging for Development', debunks many of these standard myths. The author, Thomas Michael Power (2002) shows how the idea that mining was an engine for growth in countries such as Canada, the United States and Australia, and therefore can be so for other countries, is simply not true. He shows how relatively unimportant mining was for those countries, how their large geographical size and internal markets assisted a diversified mineral sector and how they were effectively already middle income countries anyway. Professor Power stresses how these positive factors just do not hold for most mineral-rich developing countries, with most lucky to have even one of these characteristics. More importantly he notes how in the last 20 years – with increased globalization leading to cheaper transportation prices and more volatile markets – mineral-dependent countries are faring even worse (Power 2002).

Although there has been an ongoing argument between Professor Power and the World Bank over the figures used and the conclusions drawn, it is obvious to many that even if a country can be seen to benefit, communities who are close to the operation seldom seem to in comparison to the costs that are imposed upon them. Research has shown that many of the poorest regions within countries are those associated with mining, from Bihar in India to the Appalachian coal mine areas of the United States. An excellent example is in Peru, where aside from a general proof of the above statement there is one province, Cajamarca, the site of Latin America's largest gold mine, Yanacocha, which has, after over ten years of mining, dropped from being the fourth poorest province in the 1980s to the second poorest. Also, certain previously disadvantaged groups, such as women or indigenous peoples, suffer the greatest negative effects, as confirmed by the extensive research undertaken for the World Bank's Extractive Industries Review (Moody & Flood 1997; Salim 2003; Glennie 2004, p. 11–12).

It is, of course, arguable why this should be, when claims are constantly made of the benefits of jobs and infrastructure that mining can bring to local communities. Although there is some obvious circumstantial evidence that mining tends to take place in already remote and often mountainous areas, there is an argument that notes the potential damage to the environment and loss of land that leads to the loss of livelihoods. Also, the point that taxes paid to some national governments for community benefit simply does not make it back to the community not only holds up to as much empirical inspection as the subject of corruption can, but also stresses the role that corruption, as well as conflict, seems to play in the 'resource curse' (Jaques 2003; Nettleton et al. 2004).

The belief that technical fixes can solve almost all problems

The corporate mining industry is exceedingly quick to herald beneficial improvements in technology, which are promoted as the products of, or associated with, large multinational companies. However, history has shown us that even if beneficial, these technologies are not always necessarily used, as costs can sometimes be prohibitive. For example, in the case of Ok Tedi, when confronted with problems of building a tailings dam in a high rainfall and geologically unstable area, BHPBilliton opted to dump waste straight into the local river system, which has been the source of ongoing legal action since. It seems that at no point when a reasonable technical solution could not be found did the simple solution not to mine in such an area occur to the company (Kirsch 2000).

There are still arguments over so-called new solutions. A good example is submarine tailings disposal, which in the developing world started in Papua New Guinea and at the urging of industry is spreading to other Asia-Pacific countries, despite serious civil society opposition. This is a relatively untested technology, and already objections and problems have been noted in the Lihir, Misima and Buyut Bay mines. In the case of the latter, the company responsible, Newmont, is at the time of writing the subject of civil and criminal proceedings, and had five executives detained in jail for questioning, for the alleged poisoning of residents of the eponymous bay via submarine tailings disposal (Moody 2001*a*).

If you look back over the period when sustainable mining has been on the agenda, the two outstanding technological developments have been the extraction of minerals from lower-grade ores and heap leaching. Lower-grade ore, open pit mining produces less ore for energy consumed and waste produced, compared to underground mining. It also employs fewer people and obviously is less attractive to communities, who stand to lose more of their otherwise productive land, and run a greater risk of airborne pollution (Sampat 2003). It is difficult to see how this has encouraged 'sustainability'.

Heap leaching, heap leach SX-EW and High Pressure Acid Leaching (HPAL) benefit industry, but not necessarily the environment or local communities. In fact it could be argued that some of these processes have opened up new regions for mining where it has not been previously practical. The MMSD report commends HPAL, as an example, for its lower capital and operating costs and superior metal recovery, which 'may have a significant effect on the location and nature of nickel mining in the future' (although in hindsight constant problems at mine sites have left the whole initial HPAL hype somewhat tarnished). There is absolutely no recognition that by opening up new areas of laterite (as dramatized via the Crew Mindoro project in the Philippines) this 'new location' factor may sound the death knell for traditional farming or fishing communities (MMSD 2002, p. 29).

It is difficult to see how these types of technology are really helping sustainable development. If technical fixes are the answer to everything, then we should have seen reductions in the number of tailings dam problems in the period that 'sustainable mining' has been in discussion. Recent research by Indigenous Peoples Links on tailings dams in the Philippines, and by the World Information Service on Energy in Australia on uranium mines, seems to suggest that this has not happened (Nettleton et al. 2004; Wise 2005).

As a good example of where even the largest and best-equipped companies have technical problems, the Kelian gold mine in Indonesia, which lasted for less than a decade, is closing before its scheduled decommissioning, and has left Rio Tinto with an enormous problem of rehabilitation (including not being able to return its mine site to productive agricultural use as undertaken in the EIA) (McDonald & Ross 2003).

Finally, if technology is really improving safety, then surely companies should be putting their dollars where the their company reports are, and agreeing to the 'polluter pays' and 'precautionary' principle. However, companies have been investing much time and energy lobbying against this principle, despite civil society worries that some fees for polluting can be so low as to actually make it affordable for the polluter to just 'pay and go' (Corpuz 2003; REHN 2003).

The inference that those opposed to mining mainly comprise ignorant and 'anti-development' communities and NGOs

It is important to stress that many of the groups that MAC works with are not 'anti-mining' per se, and many work with, or are, miners themselves (notably small-scale or artisanal miners). However, their dealings with the corporate mining industry to date have often made those same groups suspicious or hostile of the intentions of the industry.

As previously alluded to, there is a perception that the mining industry is using bodies of engagement, such as the MMSD, to seek out northern – often environmental – NGOs, with which it feels it can work. This tactic is frequently used to isolate those who are critical of larger scale mining, but the worst part of the situation is that those organizations seeking to engage are then taken to represent 'civil society' positions on mining and therefore to speak on behalf of communities who have never endorsed such a position. Understanding and dealing with the different groups within 'civil society', and as a result dealing directly with affected communities, is essential to creating real trust (Moody 2001b).

The key question is who has the right to make the decision over the future of communities: companies, governments, NGOs or the communities themselves? In any decision-making process there must be the concept of free, prior and information consent (FPIC). With indigenous peoples this must take account of collective decision making. Respect for FPIC, alongside a human rights-based view of development, was strongly recommended in the authoritative Extractive Industries Review (Salim 2003, pp. 21–24, 52–53, 61–64).

It is crucial that corporate mining representatives understand what this means. It means identifying and dealing fairly with all the affected communities from the very conception of a project. It also means accepting a 'no' to a project, if after fully and transparently explaining the costs and the benefits, that is what the community wishes.

Yet, so far there has been a history of the abuse of communities' right to consent in order to promote projects. In order to explore this further it is worth concentrating on two specific areas: the first is a study of the types of abuse of such consent, focussing on examples across the Philippines. The second case study is looking specifically at the Freeport mine in West Papua/Irian Jaya.

The abuse of FPIC in the Philippines

The Philippines is a good example to concentrate on with regard to the abuse of FPIC, especially with regard to indigenous peoples, for a number of reasons. Mining, mainly for gold, copper, chromite and nickel, is widespread in the country. It has, however, a negative legacy – including the famous 1996 Marcopper tailings dam disaster – which has led to widespread community opposition to ongoing or proposed mining projects. The country also has around 10 million indigenous peoples (approximately 15% of the population) who retain a close link with their land and traditions, and many of whom live in the more remote areas where mining is to take place (Nettleton *et al.* 2004).

This potential clash between mining and the wishes of indigenous peoples is played out in the field of law. First, there are pro-mining laws that have been enacted by the Philippine Government in order to encourage overseas investment, such as the 1995 Mining Act, 2003 National Minerals Policy and 2004 Mineral Action Plan. However, the Philippines also has arguably the most advanced law with regard to the rights of indigenous peoples. The 1997 Indigenous Peoples Rights Act (IPRA) is modelled on the provisions of the UN Draft Declaration of Indigenous Peoples Rights, and states that before exploration or extraction rights are granted, the whole community must be informed and agree to the decision (Corpuz 2003; Nettleton *et al.* 2004).

The two laws are often diametrically opposed, even to the point of having separate land uses, under both Acts, covering the same piece of ground. The courts, and subsequent legislation, are now trying to unravel the inconsistencies of the situation. At present the cash-strapped Philippine Government's stated priority is applying pressure in favour of mining, to the point where the operating regulations of IPRA are being re-written with a mind to 'fast-track' mineral operations.

However, it is not just in legal terms, but practically on the ground, that indigenous peoples are losing out despite the theoretical rights they hold under IPRA. There are numerous examples of breaches of good faith or duplicity, where the consent is either not freely given, not informed, not done prior to commencement or in some cases consent is not given at all (Nettleton *et al.* 2004). The following are just a small number of illustrations of such behaviour.

Western Mining Corporation (WMC) allegedly tried to persuade leaders of the B'laan people to endorse its Tampakan mine in Mindanao by giving them gifts, building tribal halls, employing relatives and taking them on trips where they received lavish entertainment. Despite this, some community elders refused to agree to the mine, with one leader withdrawing his consent after visiting an existing mine and seeing its environmental impact. On one occasion, as part of its negotiations with the B'laan, a joint team from WMC and the government presented a 75-page Memorandum of Agreement to the community in English. The Government lawyer described the content of the agreement to community leaders in rough oral translation and advised them to sign. The whole negotiation was completed in a one-day session (Muntz 2001).

On the subject of bribery, Councillor Peter Duyapat of Didipio, Nueva Vizcaya, is among several indigenous leaders who make claims of attempted bribery. In exchange for stopping his vocal opposition to the entry of the mining company Climax Arimco, he says he was 'offered by the company a 20-hectare farm in Bayombong, in Nueva Vizcaya, a sum of money that I could not spend in my lifetime, a house and lot, and a Pajero land cruiser' (Colchester 2003).

In Pagadian, on Zamboanga del Sur, British-based mining company Rio Tinto called a meeting to talk about its mining applications in the region. The consultation took place in the provincial capital, far from indigenous communities, and less than one week's notice was given. However, assisted by civil society groups, more than 300 Subanon did attend to register their strong and unanimous opposition to the plans. Despite this hostile response, the local Rio Tinto manager told the Philippine Government that the company had fulfilled its obligation to consult, and that the meeting had been 'very successful' (Nettleton *et al.* 2004).

Finally, another tactic is to undermine the indigenous leaders who oppose mining. In Mindoro, the Norwegian company Mindex – now part of UK-based Crew Development Corporation – cooperated with regional NCIP (the government body charged with indigenous rights) officials to establish a new indigenous organization, Kabilogan, within their claim area. There were, however, already two other indigenous organizations who held prior, officially registered, ancestral domain claims to the land covered by the concession claim of Mindex. Despite strong and repeated protest from local organizations and elected officials, the NCIP chose to recognize and deal only with the newly formed Kabilogan. The project was endorsed as having gained consensual free prior informed consent, and a licence was granted. At least in this case the licence was later revoked following an unprecedented level of local protest (Gariguez 1999).

The Grasberg mine in West Papua

Moving south across the Asia-Pacific region, the huge Freeport-McMoRan Grasberg mine was carved out of the sacred Graberg mountain in West Papua (otherwise known as the Indonesian-occupied province of Irian Jaya). Several features mark it out as worth concentrating on from the point of view of community resistance to projects. The first is its size, as by the turn of the century it was set to become the world's biggest single mine. Secondly, it produces more mine waste than any other mine on the planet – averaging over 200 000 tonnes a day deposited straight into the local river system (Evans *et al.* 2001; Moody 2005).

However, from the perspective of community rights, it is the extreme repression meted out by armed forces against the local indigenous peoples around the mine site – on whose land it was constructed – that is most shocking. When the American company Freeport arrived in the late 1960s, the Amungme and Komoro people were living subsistence lifestyles in a landscape full of spiritual significance. Aside from the impact of the pit, waste and technology, the mine has brought mass migration and Indonesian soldiers and police, provisioned and paid for by the company with a mandate to protect it. There was no 'good neighbour' scheme adopted, no attempt at consultation. Freeport simply arrived, reinforced by Indonesian military power, and imposed itself upon the occupied people nearby without their consent.

Since that time, the human rights abuses have been widespread and well documented. Peaceful protest gave way to resistance, which provoked 'counter insurgency' operations and an escalating conflict. The mine has become a lightning rod for the armed Free Papua Movement (OPM), and independence struggles of the West Papuan peoples. Yet the main people who suffer as a result are the unarmed, local community around the mine (Evans *et al.* 2001; Moody 2005).

It is interesting to note that the first contract of work signed between Indonesia and Freeport – to mine the associated Ertsberg copper deposit – took place two years before the United Nations recognized Indonesian sovereignty over West Papua. The contract was the first to be signed by a multinational corporation with the Suharto New Order Government, and was to set a pattern of American support for the regime, in protection of its business interests, despite the human rights violations and corruption of the Suharto dictatorship.

There were belated attempts to engage with the local communities. Seven years after their arrival, the company was forced into a three-way meeting with the Amungme community and the provincial government. This led to the 1974 January Agreement, wherein Freeport pledged to construct facilities, including a school and health clinic, if indigenous landowners gave their approval for the mine (Evans *et al.* 2001).

However, despite the agreement, for the next 20 years the land grabbing without consent, and forced resettlement, continued and expanded. It was in 1995 that community members understood for the first time that they had, according to government records, ceded all their ancestral rights in the Timika area (nearly one million hectares) to the government for transmigration settlers (migrants from Java) associated with the mining.

The local landowners have been struggling since that first agreement to have their concerns for compensation and an end to the environmental and human rights abuses effectively addressed. They have appealed to the Indonesian Government, the United States Government and courts, the United Nations and directly to the managers and shareholders of Freeport, and its partner Rio Tinto.

The company's response has been to increase the amount of money offered to local communities – namely through a 'one per cent fund' for local development. Although apparently a generous offer, it does not address the concerns of the people over their rights and the destruction being wrought – and therefore is generally seen as a belated bribe. Inevitably, the decision over whether to accept the offer, and even how it should be distributed, created more divisions in the community, leading to the deaths of 18 people in an inter-ethnic conflict over how the money would be distributed.

Eventually, for that reason, a local leader with some credibility, Tom Beanal, with huge reluctance decided to sit on PT Freeport's board of commissioners in June 1999 to attempt to mitigate the damage the money was doing in his community. What he says of this bitter decision sums up the feeling of powerlessness in so many communities faced with huge company pressure: 'What Freeport has done to me is to present me with a limited choice, prepared by the company, so that I was not able to choose freely, but was always obliged to choose what was desired by Freeport. People see me as working with Freeport now' (Evans *et al.* 2001, p. 88).

Statement from John Rumbiak

Given the history of the Grasberg mine, and the fact I am not a direct community representative, it would be useful to quote at some length John Rumbiak of ELSHAM, a human rights group based in West Papua. It is important to note what he has to say of the Grasberg mine, and its effect on the local Amungme people, but also his conclusions on the 'right to say no'.

> In its 1967 contract of work, [Freeport] was the first foreign investor to enter into a [newly

defined business] agreement with Suharto's New Order regime [in Indonesia]. The company gave itself broad powers to resettle local indigenous populations whose traditional lands the company seized free of charge for its operations. It also gave itself, in writing its own contract of work with the Indonesian government, the power to take, free of charge, whatever natural resources it required for its operations.

In the three decades that Freeport has operated in Papua, the company has single-handedly succeeded in establishing its own fiefdom. With the assistance of the Indonesian armed forces, paid by Freeport to safeguard its operations, the company decides who can enter the area surrounding its mine and who cannot. Although there are commercial flights in and out of Timika, the main town near the mine, visitors have been deported and blacklisted for attempting to be in the area without Freeport's permission.

In September 1995, Indonesia's National Commission on Human Rights concluded that clear and identifiable human rights violations had occurred in and around Freeport's project area, including indiscriminate killings, torture, and inhuman or degrading treatment, unlawful arrest and arbitrary detention, disappearance, excessive surveillance, and destruction of property. The commission noted that these violations 'are directly connected to [the Indonesian army] acting as protection for the mining business of PT Freeport Indonesia'.

To put an end to these dynamics of destruction and violence, the international community – particularly international investors – must, first and foremost, recognize indigenous communities' basic rights to chart their own development paths, to manage their own resources, to pursue their traditional livelihoods and cultures, and to say NO to multinational operations on their lands. The failure to respect communities' basic right to "just say no" exists at the heart of the nexus of human rights violations, environmental degradation and conflict (Rumbiak 2003).

Conclusion

The truth is that sustainability implies something quite different depending on which side of the bulldozer you are on. Attempts by the industry to green-wash itself as a new, improved, sustainable industry simply will not wash, as even some corporate mining cheerleaders, such as Philip Crowson as mentioned earlier, have pointed out. Those at the other side of the bulldozer can easily see that the emperor is naked, and the more he insists it is not so, the further away is an honest dialogue.

To emphasize this point, it is worth finishing with a quote from an Indian colleague who was recently questioning the use of the term 'stakeholder'. 'The meaning of "stakeholder" got ruined the day it got coined by Rio Tinto, a major mining multinational corporation, to give itself legitimacy and pose its demands of somebody else's land as reasonable... The stakeholder engagement process is purported to be an exchange of information and views between all parties concerned by one project. In fact, a "stakeholder engagement process" stands for communities being continually told of companies' plans and invited to modify them. But it does not mean that these communities are permitted to reject the projects per se. It does not mean that they are empowered to present their own development plans' (pers. comm. 2003).

References

AUTY, R. M. 1998. *Resource Abundance and Economic Development: Improving the Performance of Resource-Rich Countries*, UNU/WIDNER, Research for Action No. 44.

CHAPIN, M. 2004. A challenge to conservationists. *Worldwatch Magazine*, November/December 2004.

CHOUDRY, A. 2003. *Beware the Wolf at the Door*, Seedling (October 2003), GRAIN, Barcelona, Spain.

COLCHESTER, M. 2003. *Extracting Promises: Indigenous Peoples, Extractive Industries and the World Bank*. Tebtebba Foundation, Baguio, Philippines, and Forest Peoples Programme, Moreton-in-the Marsh, UK.

CORPUZ, C. 2003. *Comments and Recommendations on the National Minerals Policy Framework*. Tebtebba Foundation, Baguio, Philippines.

EVANS, G., GOODMAN, J. & LANSBURY, N. 2001. *Moving Mountains – Communities Confront Mining and Globalisation*. MPI, Sydney, Australia & Oxford Press, Oxford, UK.

GARIGUEZ, E. A. 1999. The Mangyans and the Mindoro Nickel Project of Mindex/Crew/Aglubang Mining Corporation: a case of deception and manipulation in obtaining free, prior and informed consent. *Philippines International Review*, 2(1).

GLENNIE, J. 2004. *Unearthing the Truth – Mining in Peru*. Christian Aid, London, UK.

ICMM/IUCN 2002. *ICMM/IUCN Announcement*, 'Mining Industry and IUCN: The World Conservation Union Announce Partnership on Mining and Biodiversity, Johannesburg, 31 August 2002', http://www.minesandcommunities.org/Charter/iucn1.htm

JAQUES, A. 2003. *Fuelling Poverty – Oil, War and Corruption*. Christian Aid, London, UK.

KIRSCH, S. 2000. *Incompatible with our Environmental Values: What's Next for Ok Tedi?* Higher Values (February 2000) Minewatch, London, UK.

LEHMAN BROTHERS INC. 2000. *Reverse Alchemy: The Commoditization of Gold Accelerates*. Lehman Brothers Inc., New York, USA.

MAC 2001. *London Declaration, Mines and Communities*, http://www.minesandcommunities.org/Charter/londondec.htm

McDONALD, I. & ROSS, B. 2003. *Oxfam CAA Mining Ombudsman Annual Report* 2003. OXFAM (Australia), Fitzroy, Australia.

MMSD 2002. *Mines, Minerals and Sustainable Development (MMSD) Project, Breaking New Ground*. Earthscan, London, UK.

MOODY, R. 2001a. *Into The Unknown Regions – The Hazards of STD*. Down to Earth, London.

MOODY, R. 2001b. *Sleepwalking with the Enemy – or Waking to the Truth?* Nostromo Research, http://www.minesandcommunities.org/Charter/sleepwalk1.htm

MOODY, R. 2002. *Sustainable Development Unsustained: A critique of the MMSD project*, Nostromo Research, http://www.minesandcommunities.org/Charter/mmsd1.htm

MOODY, R. 2005. The *Risks We Run – Mining, Communities and Political Risk Insurance*. International Books, Amsterdam, Netherlands.

MOODY, R & FLOOD, M. 1997. *Minerals Extraction and the Environment*. Powerful Information, Milton Keynes, UK.

MUNTZ, B. 2001. *Mining and Community Rights – A Case Study: WMC Resources Ltd. and the Tampakan Copper Project at South Cotabato, Mindanao in the Philippines 1991–1999*. Community Aid Abroad, Fitzroy, Australia.

NETTLETON, G., GLENNIE, J. & WHITMORE, A. 2004. *Breaking Promises, Making Profits – Mining in the Philippines*. Christian Aid, London, UK.

OXFAM AMERICA 2003. *Investing in Destruction*, OXFAM America briefing paper. OXFAM, Washington DC, USA.

POWER, T. M. 2002. *Digging to Development?* OXFAM America, Washington DC, USA.

REHN 2003. *Rachel's Environment & Health News #778*. Corporate Campaign Against Precaution, New Brunswick, USA.

RUMBIAK, J. 2003. *Globalization, Rights and Poverty* (speech). Columbia University's Center for the Study of Human Rights, New York, USA.

SALIM, E. 2003. *Conclusions and Recommendations of the Final Report of the Extractive Industries Review*. World Bank, New York, USA, **1**.

SAMPAT, P. 2003. *Scrapping Mining Dependence*. State of the World 2003, World Resources Institute, Washington DC, USA.

SMITH, A. 1776. An inquiry into the nature and causes of the wealth of nations. *In: Of Colonies*, Chapter 7, Edinburgh University, Edinburgh.

WISE 2005. *World Information Service on Energy Figures on Uranium Dam Disasters*. Amsterdam, Netherlands.

Index

Note: page numbers in *italics* indicate Figures, while those in **bold** denote Tables.

AAC *see* Anglo American Corporation
Abosso Goldfields Ltd (AGL) 68–69
accidents, small-scale mining 107–108
ADB *see* Asian Development Bank
added value 115–116
afforestation 145
Afghanistan, donor support 189, **190**
African Mining Partnership 22
aggregates
   china clay waste 54
   Jamaica 36–43
   Netherlands 225, *226*
   river mining 35–46
   slate waste 55–56
   Slovenia 207, **209**
   sustainability 199–201
   Timor Leste 177–180, *181*
AGL *see* Abosso Goldfields Ltd
agriculture
   coal mining impact 143–144
   lime 121–126
aid
   donors 188–189, **190**
   projects 189, 191–193
Aileu silt-rich deposits 170–172
alluvial deposits 63, 178–180
   see also river aggregate mining
Anglo American Corporation (AAC) 73, 78
armour stone 55
artisanal and small-scale mining (ASM) 95–120
   added value 115–116
   assistance schemes 101–102, 113, 118
   contribution 18–21, 30
   cooperatives 104–105
   economic impacts 99–101
   environmental impacts 88–90, 109–110
   finance and credit 116
   governance 102–103
   health and safety management 107–109
   Kenya 87–91
   key characteristics 95–96, *97–99*, **97**
   labour issues 105–107
   large-scale mining co-existence 112–114
   legislative framework 104
   obstacles and remedial measures 136–139
   Pakistan 135–140
   poverty 101, 103
   pro-poor policy 116–117
   reform goals 117–118
   sector value, Africa 96–99
   sustainability strategies 70–72
   technology and training 110–112
   Timor Leste 181
   trading and markets 114–115
   Zambia 82, 84
Ashanti Goldfields Company 68
Asian Development Bank (ADB) 189
ASM *see* artisanal and small-scale mining
assistance schemes 101–102, 113, 118
attractiveness for investment 213–214, 221
awards 146

Bangladesh 127–134
   mineral-based industries 132–133
   mineral resources 128–132, *129*, **131**
barite waste 56
base metals 6, 56, 214
Beheda limestones 175–177
Beheda marbles 173–175
benefits
   economic 26, 27, 236
   Gold Ridge Mine 157–158
   small-scale gold mining 65–66
   waste utilisation 48
bentonitic clays 169–170
best practice technology 111
BGL *see* Bogoso Gold Limited
BGS *see* British Geological Survey
BHP Billiton 30, 31
biodiversity, coal mining impact 144
blame 29
Bogoso Gold Limited (BGL) 69
BPD *see* Business Partners for Development
British Geological Survey (BGS) 186, 189
Bruntland definition 196–197
business cycle 192
Business Partners for Development (BPD) 14
business practices 31

capacity building
   business cycle 192
   public sector institutions 185–194
   training needs analysis 186–188, 190–193
capital 9, 84, 197
capitalism 29
carbonate resources 122, 125
CASM *see* Communities and Small-scale Mining initiative
ceramic raw materials 132, 165–173, 180
channel mining 35, 36, *39*

child labour 97, *100*, 106–107
china clay 54, 130
CIL *see* Coal India Ltd
clays
   china clay 54, 130
   Timor Leste deposits 165–170, 172–173, 180
   wastes 54
Coal India Ltd (CIL) 142, 144, 145
coal mining
   Bangladesh 128–130
   Ib valley, India 141–147
   Zambia 78, *80*
cobalt production 77, 78–82, **79**
Code of Practice, river mining 36, 43–44
coffee production 162, 164
comminution 64
communities
   impacts 233–242
   small-scale gold mining 71–72
   Solomon Islands 153–155
Communities and Small-scale Mining (CASM) initiative 20–21
concentration of gold 64–65
concrete and mortar sand 227, 230
conflicts
   financing 114–115
   post-conflict situations 185–194
   small-scale gold mining 67–70
   Timor Leste 162, 164, 180
construction raw materials
   see also aggregates
   Netherlands 225
   sustainability 199–201
   Timor Leste 161–184, **166–167**
consultation process 155–157
consumption 7, 28–30, 234–235
cooperatives 104–105
copper
   global production 6
   large-scale mining 78–82
   production, Zambia 73–75, 77, 78–82, **79**
Cornwall, UK 10, *11–12*, 54
corporate social responsibility 113–114
cost benefit analysis 124–125
costs
   agricultural lime production 123
   exploration 220–221
   mineral waste handling 48
craft markets 115–116, 188
credit 116
crop trials 123–124
crushed stone aggregates 178, 181, 200
cultural beliefs 111
curse of mining 16, 25, 235, 236

decentralization 103
decision making 215–216, 237, 238

deforestation 144
demand 5, 234–235
Democratic Republic of the Congo (DRC), governance 26–27
Department for International Development (DFID) (UK) 185, 189
deregulation 230
development
   aid 188–189, *191*
   India 141
   mining as catalyst 196, 209, 235–236
   Timor Leste 162–164
DFID *see* Department for International Development
diamonds 17
dimension stone waste 55
disasters, environmental 12–13
disposal of mining wastes **49–52**
diversification 84, 96, 123
dolomite rock 122, 125
DRC *see* Democratic Republic of the Congo
Dutch mineral planning guidance 225–232

earnings, transparency 114
eco-centric perspective 197
economics
   benefits 26, 27
   impacts 26, 99–101
   significance in development 127–134
ecosystem impacts 41–42, 144
education 110, 139
   see also training
EIAs *see* Environmental Impact Assessments
EIR *see* Extractive Industries Review
EITI *see* Extractive Industries Transparency Initiative
EMP *see* Environmental Management Plan
employment
   child labour 97, *100*, 106–107
   women 105–106
empowerment 113, 116–117
environment
   audit 157
   disasters 12–13
   impacts
   global 25–28, 29
   Indian coal mining 142–146
   mining waste usage 53
   Pakistan 139
   river mining 35, 40–41
   small-scale mining 88–90, *99*, 109
   management 13–14, 92, 109–110, 146
   monitoring and awareness 145–146
   policy 215–216
   protection 144–146, 202, 221
Environmental Impact Assessments (EIAs) 109–110

INDEX

Environmental Management Plan (EMP) 144–145
epithermal gold deposits 152
ethical trade 30–31
exploration
   budget 213, 214, 217, *218*
   costs 220–221
   expected value 220–222
   Solomon Islands 152
   spending 221–222
external stakeholder relationships 186–187
extraction issues 199, 227
Extractive Industries Review (EIR) 15–16, 18
Extractive Industries Transparency Initiative (EITI) 16–17, 114

fair prices 30–31
farming, small-scale 121–126
FarmLime project 121–126
feldspar waste 56–57
finance
   conflicts 114–115
   lack of assistance 138
   small-scale mining 116
fluorite waste 56
free, prior and information consent (FPIC) 237–238
Freeport 239, 240

galamsey operations 68
gas reserves 128, 137
gemstones
   exchange scheme 84
   Kenya 88
   Pakistan 135
   small-scale mining 97
   Zambia 76, 82, *83*, 84
gender mainstreaming 105–106
geochemical baseline *151*
geological survey departments 186, 187
geology
   information 85, 222
   Jamaica *37*, 38
   Kenya 87–88, *89*
   Timor Leste 161–162, *163*, 165–177
   Zambia 74, 75, 76
Ghana small-scale gold mining *18*, 61–72
   see also gold mining
glass sands, Bangladesh 130
global expertise 7–8
Global Mining Dialogue (GMD) 113
Global Mining Initiative (GMI) 233
Global Reporting Initiative (GRI) 17
globalization 32–33
GMD *see* Global Mining Dialogue
GMI *see* Global Mining Initiative

gold mining
   impacts 65–70
   mining efficiency 90–91
   mining methods 63
   processing 64–65
   production, Ghana **65**
   river mining 88, *90*
   small-scale *18*, 19–20, 61–72, 97, *154*
   Solomon Islands 149–160
Gold Ridge Mine, Guadalcanal 149–160
governance 3, 15–16, 17, 26–27, 102–103
governments
   departments 186
   inter-governmental forum 22
   small-scale gold mining 70
   sustainability strategies 83–85, 198–199
Grasberg mine, West Papua 238–240
gravel 35–46, 178–180, 200
   see also aggregates
green metals 31
green revenues 32
GRI *see* Global Reporting Initiative
Guadalcanal, Solomon Islands 149–160

hard rock 130
health and safety 107–109, 142–143
heavy minerals 53
hierarchical model resource management 201
High Pressure Acid Leaching (HPAL) 236–237
history
   mining industry 9–10, 18–19, 26
   small-scale mining 61, 96
   Zambia 73–75
holistic capacity building 189–193
HPAL *see* High Pressure Acid Leaching
human rights violations 26, 239, 240

Ib valley coalfield, India 41–47
ICMM *see* International Council on Mining and Metals
illegal small-scale mining 68–70, 112–113
illite 170
ILO *see* International Labour Organisation
impacts *see* environment, impacts; social impacts
India, Ib valley coalfield 141–147
indicators of sustainable development 204–209
Indigenous People's Rights Act (IPRA) (Philippines) 238
industrial minerals
   see also aggregates; construction raw materials
   Africa 82, 97
   low profile 137
   mining wastes **49–52**
   Netherlands 225, *228–229*

Timor Leste 164, **166–167**
information 90, 187, 203, 222
infrastructure damage and reconstruction 162, 164, 180
institutions 70–71, 102–103, 185–194
  see also governments
internal stakeholder relationships 187–188
International Council on Mining and Metals (ICMM) 15, 113, 233
International Labour Organisation (ILO) 105
intrusive rocks 178
investment
  attractiveness 213–214
  policies 218–219
  profitability 219
  Zambian mining 83–85
IPRA see Indigenous People's Rights Act
Ireland 213, 218

Jamaica
  alternative materials 39–40
  resource evaluation 36–39
  river mining impacts 40–42
jewellery sector 115–116

kaolin 172
Karoo Supergroup, Zambia 75
KCM see Konkola Copper Mines Limited
Kenya 87–94
  environmental management 92
  information and education 90–91
  legal framework 91–92
  mineral potential 87–88
  titanium mining 92–93
Kimberley Process 17, 115
Kisii soapstone 88
Konkola Copper Mines Limited (KCM) (Zambia) 78

labour see employment
land
  degradation 66–67
  disputes 67–70
  ownership 154–155, 156, 157, 158, 239
  reclamation 145
  tenure 112
large–scale mining
  Ghana 71
  precursor 66
  small-scale mining coexistence 112–114
  Zambia 78–82, 85, 86
leaching 236–237
legal framework
  Kenya 91–92
  old laws 136–137
  policy 202
  small-scale mining 61–62, 104, 109
  Zambian mining sector 85
licences **82**, 92, 218–219
lime for small-scale farming 121–126
limestones 36, 130–131, 175–180
London Declaration 234
low-cost lime 121–126
low-technology mining 95, **97**
  see also artisanal and small-scale mining

MAC see Mines and Communities
magnetite *80*
Mahandi Coalfield Ltd (MCL) 142, 144–146
manufactured sand 39–40
marbles 173–175
marine sand and gravel 39
marketing facilities 137–138
markets
  base metals 214
  craft industries 115–116, 188
  mining wastes 47–59, **49–52**
  small-scale mining 114–115
MCL see Mahandi Coalfield Ltd
MCM see Mopani Copper Mines Limited
MDGs see Millennium Development Goals
mercury pollution 108
mica waste 56–57
Millennium Development Goals (MDGs) **17**, 19, 22, 118, 195, 200
mine workers
  conditions 137, 139
  cooperatives 104–105
  education 110
  health 142–143
mineral-driven economies 195–212
mineral planning
  abandonment 230
  Netherlands 225–232
  production planning 229–230
  river mining 43
  site selection 229
mineral resources
  Bangladesh 128–132, *129*, **131**
  management *132*
  Netherlands 225, *226*
  Pakistan 135–140, **136**
  sustainable development 197–199
  Timor Leste strategy 164–165
  Zambia 75–78
mineral sands 131–132
minerals policies 199, 213
  Netherlands 227
  Philippines 238
  river mining 43
  Slovenia 205–206
Mines and Communities (MAC) 234, 237

INDEX

Mining Act of Kenya (1940) 87, 91
mining industry
   see also artisanal and small-scale mining; large-scale mining
   catalyst for development 196, 209, 235–236
   community impacts 233–242
   contribution to society 195–212
   global impacts 25–28
   law 202
   rights 218
   'rushes' 96, 101, 102
   sustainability 9–24
   wealth creation 198, 209
Mining, Minerals and Sustainable Development (MMSD) project 15, 116, 198, 233–234
Mining Sector Diversification Programme (MSDP) 84
Mining University, Ghana 71
mining wastes 47–59
   benefits of utilization 47–48
   composition and utilization 48–53
   profitability scenarios 53–57
   secondary aggregates 40
MMSD *see* Minerals, Mining and Sustainable Development project
modelling 197, 213–224
Mopani Copper Mines Limited (MCM) (Zambia) 80–81
MSDP *see* Mining Sector Diversification Programme

National Environmental and Coordination Act (2000) (Kenya) 91–92
National Mineral Resource Management Programme (NMRMP) (Slovenia) 206, 207
Netherlands, policy building 225–227
NMRMP *see* National Mineral Resource Management Programme
non-governmental organizations (NGOs) 71
non-renewable resources 9

oil 128, 137
open cut gold mining 152
Orissa, Ib valley coalfield 141–147
ornamental stones 173–177, 180

Pakistan, small-scale mining 135–140
participatory approach 91
   see also stakeholders
payments labour 29–30, 107
peat 130
pegmatite mining waste 56–57
perceived risk 217–220
petroleum 128, 137

Philippines, FPIC abuse 237–238
planning guidance 42–43, 225–232
plant hire scheme 84
policies
   see also minerals policies
   environment 215–216
   investment 218–219
   legal framework 202
   Netherlands 225–227
   pro-poor 16–17
political risks 219
pollution 67, 108, 141–143, 145
population
   displacement 144
   gold mining economy 155
   growth 28–29, 67
post-conflict situations 185–194
poverty
   alleviation 164, 196, 198, 209
   pro-poor policy 116–117
   small-scale mining 101, 103, 113, 118
Poverty Reduction Strategy Paper (PRSP) (Zambia) 123
prices
   copper and cobalt, Zambia *80*
   fair 30–31
   small-scale mining 114
private sector mining 78–82, 112–114
   see also large-scale mining
pro-poor policy framework 116–117
processing 53–54, 64–65
production planning 229–230
productivity 28–29
profitability 53–57, 219
profits 30–32, 114–115
PRSP *see* Poverty Reduction Strategy Paper
public sector institutions 185–194
   external stakeholders 186–187
   internal stakeholders 187–188
pyrite production 78, *80*

quarrying 201, 206, *208*

rare mineral waste 57
rates of return 31–32
re-use of mining wastes 48–53
recycling 227–229, 235
reform goals 117–118
regional indicators, Slovenia *208*
regulation
   see also deregulation
   institutions 70–71
   petroleum 137
   requirements 221–222
   small-scale mining in Ghana 62
rehabilitation, small-scale mining areas 110

reporting 17
representation, ASM sector 113
research, agricultural lime 122–123
resources
   see also mineral resources
   curse 16, 25, 235, 236
   evaluation, aggregates 36–39
   maximisation of use 48
revenues 65, 221
risk 217–220
river aggregate mining 35–46
   impacts 40–42
   Jamaica 36–42
   key issues 44
   planning guidelines 42–43
river sand 36
river terrace deposits 40
Roan Selection Trust (RST) 73
Ross Mining NL 152, 155, 156
RST *see* Roan Selection Trust

sands
   concrete and mortar sand 227, 230
   glass 130
   manufactured 39–40
   marine 39
   mineral 131–132
   river mining 35–46
   Timor Leste 178–180
scenarios, mining wastes utilization 53–57
screening 64
scrubbing 64
seasonality, small-scale mining 96
secondary aggregates 40, 227
secondary materials 227–228
silt-rich deposits 170–172
simulations, mineral exploitation 218–219
slate waste 55–56
slavery 26
Slovenia case study 205–209
small-scale mining *see* artisanal and small-scale mining
smectite 169
SMRM *see* sustainable mineral resource management
social barriers 111
social impacts
   coal mining 144
   mining 'rushes' 101, 102
   responsibility 30–31
   river mining 42
   small-scale gold mining 65–66, 67
   Solomon Islands 156–157
   West Papua 239
social systems 155, 215
socio-economic conditions 122, 139
Solomon Islands, gold mining 149–160

stakeholders
   consultation 149–160
   holistic capacity building 189–191
   institutional relationships 186–188
   involvement 91, 206
structural 77, 78
sub-Saharan Africa, small-scale mining 95–120
subsistence farmers 154
surface mining, Ghana 63
sustainability
   minerals strategies, Zambia 83–85
   mining 233–242
   paradigm 6–7, 195, 196
   small-scale gold mining 70–72
sustainable development
   Bangladesh 127–134
   concepts 196–197
   environmental management 13–14
   goals 196
   indicators 204–209
   mineral resources 197–199
   Netherlands 227
sustainable mineral resource management (SMRM) 196, 201–204
   hierarchical model 201
   implementation process *203*
   Slovenia case study 205–209
synergy principal 72
system dynamics modelling 213–224
   complexity 214–215, *216*
   perceived risk 217–220
   policies and decisions 215–216
   variables and relationships 216–217

tailings disposal 236, 237
tantalum minerals waste 57
technical assistance 138–139
technical solutions 43
techno-centric perspective 197
technology 95, **97**, 110–112, 236–237
tenure rights 217–218
Timor Leste
   aggregates 177–180, 181
   artisanal and small-scale mining 181
   ceramic raw materials 165–173, 180
   development strategy 162–164
   mineral resources strategy 164–165
   ornamental stones 173–177, 180
titanium mining 92–93
TMCs *see* transnational mining companies
TNDP *see* Transition National Development Plan
trading 114–115
traditional authorities 71
training needs 110–112, 186–188, 190–193
Transition National Development Plan (TNDP) (Zambia) 83, 85

transnational mining companies (TMCs) 112–114
transparency 16–17, 114

underground mining 63
uneconomic resource 66
unsustainable growth 29

value cost ratios (VCRs) 124, **125**
Venilale clayey deposits 168–170

wastes *see* mining wastes
WBCSD *see* World Business Council for Sustainable Development
weak sustainability 197
wealth creation 66, 198, 209
West Papua, Grasberg mine 238–240
Western Mining Corporation (WMC) 238
women in small-scale mining 19, *100*, 105–106
World Bank 15–16, 185, 189

World Business Council for Sustainable Development (WBCSD) 12, 14–15
world population growth *28*
World Summit on Sustainable Development (WSSD) (2002) 15, 16, 17, 196, 198, 233

Zambia 73–86
  agricultural lime 121–122
  carbonate resources 122, 125
  exploration 82
  gemstones 76, 82, *83*, 84
  geological information 85
  large-scale mining 78–82, 85, 86
  licences **82**
  mineral resources 75–78
  national strategies 83–85
  small-scale mining 82, 84
  sustainable minerals development 77–78
Zambia Consolidated Copper Mines (ZCCM) 75, 78
zinc 217, 218